Size, Function, and Life History

Size, Function, and Life History

WILLIAM A. CALDER III

HARVARD UNIVERSITY PRESS
Cambridge, Massachusetts, and London, England   1984

Copyright © 1984 by the President and Fellows of
  Harvard College
All rights reserved
Printed in the United States of America
10 9 8 7 6 5 4 3 2 1

This book is printed on acid-free paper, and its binding
materials have been chosen for strength and durability.

LIBRARY OF CONGRESS CATALOGING IN PUBLICATION DATA

Calder, William A., 1934–
    Size, function, and life history.

    Bibliography: p.
    Includes index.
    1. Body size.   2. Animal ecology.   3. Physiology.
I. Title.
QL799.C29   1974   591.4   83-22815
ISBN 0-674-81070-8 (alk. paper)

To Lorene

I think I told you I was writing a book . . .

Contents

Preface

THERE CAN BE NO DOUBT that body size is a major factor in an animal's life history and requirements; ask any zookeeper to compare the responsibilities of care, such as feeding and cage cleaning, for elephants and chinchillas. This book is an attempt to collect quantitative representations of the consequences of size—a field more formally known as allometry.

I have several purposes in mind: (1) to understand the constraints on proportions and functions that apparently influenced the evolutionary design of larger and smaller animals; (2) to separate the background biology of size from that of other variables, so that these variables can be studied in their own right; and (3) to seek a new quantitative framework of basic principles that relate ecology and physiology, a framework from which specific adaptations have diverged and against which they can be measured. I seek to convince you that any biological study must first consider size as the most significant characteristic of an animal, and to give you a feeling for how allometry can be utilized. Success in our quest for explanation of the patterns of life will depend on how clearly we see those patterns.

This is an exciting time for the study of allometry, somewhat analogous to when the East and West crews building the first transcontinental railroad could dimly see each other in the distance; for we are beginning to see connections between allometric expressions, not just isolated regressions for statistical exercises. C. Richard Taylor, Ewald Weibel, and their colleagues have given us an integrated allometry of respiration and running locomotion. Hermann Rahn, Amos Ar, and Charles Paganelli have transformed what seemed like a random natural history of bird eggs into a neat allometric code of egg design and function. J. F. Eisenberg's 1981 treatise on mammalian evolution has a strong emphasis on size and its consequences. Reviews of scaling appear frequently, and every visit to the library's rack of current periodicals uncovers two or three more allometric analyses. A computer search for titles in allometry and scaling for the period 1969 through 1976 yielded 263 titles, an average of 33 per year. The period 1977 through 1981 yielded 392 titles, or 78 per year — an average annual increase of 14%. Some of these articles constituted significant steps in our understanding. Others having the potential to reveal exciting new patterns failed to do so. Still others contained serious errors that could have been easily avoided. In several instances, what had already been published was "rediscovered."

With this sort of proliferation, the linkage of ideas and functions becomes ever more likely. David Western, a resource ecologist with the New York Zoological Society in Kenya, stated (1979): "The concept of biological similarity is well established in biomechanics and physiology . . . I believe that there is now sufficient evidence to show that allometry is also fundamental in understanding life history strategies and ecological systems."

An allometric inventory is essentially a recycling of information. If a dozen biologists have studied a dozen species and published their data, we can assemble a dozen mean values for each structural, physiological, or ecological detail and correlate these with body mass. The generalized patterns of quantitative similarity between anatomical and physiological, or physiological and ecological, variables then provide a basis for speculation that may form bridges between artificial subdisciplines and produce a more coherent view of biology. For example, times for rapid physiological events such as heartbeat or breath duration, for reproduction, growth, life span, and population turnover all scale in similar fashion to body size. Are ecological dynamics and community structure therefore consequences of physiological pace at the tissue level, or is the physiological pace a product of ecological dynamics?

I must emphasize the speculative aspect here. This is a book of inquiry, not necessarily of truth. The techniques of a dozen biologists may not produce fully comparable data. The selection of animals may bias the results. The exponential functions may be crude. Correlation does not equal causation. However, the empirical descriptions do give us patterns, and an impressive degree of consistency encourages us to continue.

This book contains approximately 750 allometric equations. Most were gathered from the literature, where they appeared with varying amounts of statistical detail. Rather than go back and try to fill in all the specifics, I opted to devote my time to developing the broadest possible patterns that might transcend disciplinary lines. Such breadth comes at the sacrifice of depth. Nevertheless, the isolation of disciplines in biology has led to such under-emphasis and misunderstanding of the importance of body size in function and life history that I have been compelled to try. Even a crude framework is better than no bridge at all.

I am fortunate that my mentors have been willing to continue to teach me over the years since I completed my degree requirements. James R. King and Knut Schmidt-Nielsen have been of enormous assistance, providing ideas, explanations, and corrections long after their official responsibilities toward a graduate student were fulfilled. (They are, of course, not to be held responsible for my subsequent mistakes and far-fetched extrapolations.)

My colleagues, the faculty and graduate students in the Department of Ecology and Evolutionary Biology at the University of Arizona and at the Rocky Mountain Biological Laboratory in Colorado, have extended my ability to keep up with the literature by bringing many references to my attention. A seminar in Ecology 331 in the spring of 1972 was especially stimulating: as class projects, Steve Buskirk made a fundamental revision of the allometry of home ranges; Stan (Bud) Lindstedt initiated our ongoing collaboration in the allometry of physiological time; and Bill Reynolds specialized in skeletal allometry. After my first draft of this volume had been prepared, Sara Hiebert and the 1982 class in Ecology 586 helped immensely in the process of testing and editing; where I have been stubborn in resisting changes, they are not to blame. The final product would still have been rough without the editorial expertise of Vivian B. Wheeler of Harvard University Press.

My efforts have been facilitated and encouraged in many ways by my chairman, E. Lendell Cockrum, who ran a harmonious department in which we could all flourish without inhibitions. Whatever creativity emerged has come with a deep appreciation for academic freedom and diversity of views, and unusual oblivion to the chaos and contradiction of external politics. Sabbatical leave and a visiting professorship in the School of Zoology, University of New South Wales, have permitted me to complete the final revision.

In full awareness of my dreadful scrawl, I am indebted to Rebecca McKinley for getting the manuscript through the first draft and a portion of the final version. Because of staff cutbacks Lorene Calder later took over, typing and proofreading without pay or complaint. Lorene also contributed artwork, and much-needed encouragement throughout. Dawn Stanley was a patient and conscientious "gopher," retrieving information from the excellent library we are privileged to have at the University of Arizona.

I am grateful to Lynda Delph for the new figures prepared for this book, and for redrawing a number of previously published graphs. Authors and publishers have been generous in permitting reuse of their illustrations. I have taken material from several of my papers that previously appeared in journals, including some in collaboration with Eldon Braun and Bud Lindstedt.

Development of many of the ideas explored in this book has been greatly facilitated by research grants from the National Science Foundation and the National Geographic Society.

Size, Function, and Life History

1 The Biology of Body Size

Along with morphology or structure, let us add size . . . Structure is
quality, while size is quantity . . . Size may be an important difference
between two species in one genus and have consequences which permeate
into its ecology, its reproductive activities, its evolutionary progress, its
development, its physiological activities. In fact, size is as important as
morphology.

—J. T. BONNER, *SIZE AND CYCLE* (1965)

SIZE IS CERTAINLY one of the most significant characteristics of any
animal. This being the well-documented case, it should then follow that it is
one of the most tightly correlated and readily used factors for quantitative
analysis of patterns in the comparative physiology and life history of
animals. One might even anticipate that size would have a prominent role in
attempts to formulate biological theory. In actuality, size has until recently
been one of the most neglected aspects of biology.

Allometry: Why and So What?

If we choose to use it, allometry or the biology of scaling provides at the very
least a useful tool for comparative physiology. However, having a tool and
having an explanation are entirely different matters. As yet, we lack any
"special theory or general explanation for physiological adaptation going
beyond Darwin's general theory of natural selection" (Ross, 1981). Perhaps
this perceived deficiency of a theory of comparative physiology exists
because we have not completed our empirical homework, the homework

that reveals the basic patterns. C. R. Taylor (1977a) described the situation: "the million-fold range in body weight found in terrestrial vertebrates provides us with a powerful natural experiment for understanding the design constraints under which vertebrates have been built."

If there is a general theory or law for comparative physiology, it is the theory of similitude, or Newton's "law of similarity." Again, this is not something for which comparative physiology can take any credit, except in extending it to living organisms. In physics and engineering, the law of similarity states that models must be dimensionally and functionally similar to the structures they represent. In zoology, the so-called model is just another species of a size smaller or larger than the prototype (though in most cases we probably could not determine which of two sizes came first, which was the prototype and which the scaled-up or scaled-down model).

I must confess, however, that I have never been particularly conscious of the law of similarity as I have been describing the scaling of physiological processes to body size (allometry) and examining some consequences of size for animals at or near the size limits for their particular habits. Allometry has been, rather, a statistical or graphic tool for comparing animals quantitatively. Not only has it proved useful in comparative physiology, it seems to offer high potential for understanding other aspects of biology as well, potential that the leading students in these areas seem to be overlooking.

The quest for theory with respect to allometry could be pursued in at least two directions. One would be to seek a theory of allometry in its own right. Until now allometric descriptions have come piecemeal, with the limited purpose of describing single organs or functions. To formulate a theory from a few scattered regressions would put us in the situation of John G. Saxe's charming poem (1852) about the blind men and the elephant:

> It was six men of Indostan
> To learning much inclined
> Who went to see the Elephant
> (Though all of them were blind),
> That each by observation
> Might satisfy his mind.

So far we lack a unifying theory of allometry. Just as there are descriptions of the elephant's side, tusk, trunk, knee, ear, and tail, there are ideas of heat loss, percentages of active tissues, and problems of mechanical support to explain the observed scaling. We study animals by examining one or two properties or actions at a time, but we know that they do not function that simply in nature. A theory is premature until we get enough of the pieces to see how, or if, they fit together. In the process of trying to mesh them, gaps should become apparent. If there are consistent patterns (and indeed there are), then we become better oriented for future efforts.

An alternative direction in the quest for theory is to view allometry as a description in the formulation of broader theories. I shall use E. O. Wilson's monumental *Sociobiology* as an example to which allometry could provide a supporting role. Wilson (pp. 4–5) gives a scheme for the connections between phylogenetic studies, ecology, and sociobiology. In his figure 1-1, the "Theory of Sociobiology" is depicted as having implications for group size and age composition. At least in birds and mammals these characteristics, and much of what Wilson treats as the "relevant principles of population biology," are profoundly body-size dependent. Thus, before we get to a "theory of allometry," allometry is seen to be an empirical component of the "theory of sociobiology."

Let me balance my enthusiasm by quoting the last paragraph from a delightful article (Lin, 1982), which I recommend as an excellent first exposure to zoological scaling:

> A few caveats are in order. The material in this paper complements but does not substitute for careful, detailed analysis. In addition, a correct trend or number may emerge for entirely the wrong physical reasons; this is especially true in dealing with very complex systems. Finally, the agreement between theory and observation is admittedly fortuitous and guided largely by hindsight; on the other hand, to the irreverent, this is part of the pleasure of creative laziness.

We live in a world of change. Some changes are abrupt, like the displacement of an earthquake or a political coup d'etat. A new discovery or idea can revolutionize a science. The microscope, Darwin's theory of evolution, the cracking of the genetic code, plate tectonics, the computer—each has changed biology irreversibly, ushering in eras in which the questions of biology and the approaches to their answers have become more and more sophisticated. Each time a new pattern becomes obvious, the course of future data gathering can be redirected to expedite progress.

Dramatic changes can occur slowly as well, so slowly that we scarcely realize what is taking place. Mountain orogeny, sand dune drift, genetic drift, altered relationships between *Homo sapiens* and the ecosystems occur at small rates of slippage or compression, in small shifts by grains of sand, by slight statistical changes, or via modest incremental steps of progressive legislation—yet for every generation there is an invasion of new perceptions. Allometry has been creeping in.

With the launching of the Russian Sputnik in 1957, an exponential growth in science was stimulated. It has been said that more scientists were active in the 1960s than in the entire previous history of our species. Not all of us were revolutionizing our fields, however; most were gathering data on the more interesting species, filling tables with numbers, filling journals with tables, and filling libraries with journals. However, as Robert MacArthur (1972) put it, "To do science is to search for repeated patterns, not simply to accumulate facts."

We needed to compare our animals and data with other animals, if there was to be any answer to the question "So what?" If the data could be arranged into a pattern, then the inductive process could bring forth some principles. There are large birds and small birds, large eggs and small eggs; a hummingbird cannot lay anything like an ostrich can. An elephant requires more food overall than a mouse, but gram for gram the mouse needs more. If the elephant had a metabolic rate as intense as that of a mouse, it could not dissipate the heat produced fast enough to avoid being cooked. If a shrew weighing 3 g had a heart as big as that of a muskrat, the shrew would be "all heart." An LSD dosage rate per kilogram of body mass that only drugs a cat will kill an elephant (Schmidt-Nielsen, 1972). Comparison not only puts the facts into meaningful order, it gives us some predictive capacity.

Suppose we encounter a new beast that we wish to understand. What can it do, how fast, and at what cost? We could consult a biological compendium and seek information about its relative, if the relative has been studied. On the other hand, if we know only its "weight," we can predict (give or take 25% or so) a wide variety of its specifications and requirements: home range, heart and metabolic rates, life span—each from an empirical allometric equation based on body size.

While biology has been hampered from without by dogma, superstition, and ignorance, there are also phenomena within our ranks that retard our attainment of better understanding. One is a "sense of being engulfed by a flood tide of raw data" (Yates, 1979). Another is the uncritical acceptance and amazing persistence of qualitative statements which, when examined quantitatively, often turn out to be untrue. While we may find the anti-intellectualism of creationism, astrology, and biorhythm beyond penetration at times, allometric correlations can be very helpful in coping with raw data and in putting to rest some of the qualitative guesses, such as the recurring myth that bird skeletons are lighter than those of mammals.

The traditional biology curriculum orients us in taxonomy, morphology, physiology, embryology, genetics, and ecology. Body size is important in the consideration of all of these, whether as cause, consequence, or tool, but there is no study of allometry per se. This is reasonable enough; allometry is more descriptive than conceptual, yet it can be used to good advantage throughout biology. A single allometric equation may be doubted or disputed: the selection of animals may be biased or otherwise unrepresentative of the class, the methods used in gathering the data may incorporate errors or rest on dubious assumptions, or significant factors that should preclude grouping the data may have been ignored. Each line is thus less than precise and is in need of better alignment. However, when related life processes and characteristics all emerge in concordance, the general structure takes on a certain credibility.

Recently an entire issue of *Respiration Physiology* was devoted to a series of papers on "Design of the Mammalian Respiratory System" by Taylor, Weibel, and collaborators. Based on an allometric analysis that extended from the level of ventilatory gas exchange to cellular metabolism, their general hypothesis was that "animals are designed economically . . . animals don't build and maintain structures in excess of what they need." These studies were confined to support of the rate of maximum oxygen needs in locomotion, but if the principle holds for the support of oxygen requirements, it must also be true for removal of the wastes from this metabolism, supply of energy for it, and maintenance of territory from which the energy can be obtained. I am convinced that allometry will be a major bridge between ecology and physiology. Indeed, ecology can only benefit by considering its "strategies" in a broader functional context.

There is a practical side to allometry, too. Most of biomedical research depends on small animals that are easily bred and maintained. From guinea pig to man is a huge extrapolation. The correct slope or exponent for this extrapolation is as important to patient safety as is the correct scale-up from a wind-tunnel airplane model to an airliner defying gravity with 200 individuals on board.

Allometry, literally "of other or different measures," is used here, as by Gould (1966), to describe the "differences in proportions [in form and process] . . . correlated with changes in absolute magnitude of the total organism . . . the study of size and its consequences." These size relationships have been observed along time scales of phylogeny (evolution) and ontogeny (growth and development) and as static phenomena. The patterns of these three types of allometry, as seen in the quantitative characteristics of exponents and coefficients, are not necessarily the same (Cheverud, 1982). Conditions may have changed over the slow progress of evolution, and intermediate stages of ontogeny may be shielded from natural selection through parental care so that the neonate or juvenile need not scale like the adult of another species than never gets larger.

The significance of body size is pointed out periodically, often with the criticism that zoologists fail to give this preeminent factor sufficient attention (Haldane, 1928; Van Valen, 1973; Calder, 1981). While size relationships are now being examined in comparative studies, their importance is usually obscured by our professional specialization. At the risk of superficiality, we must attempt to put the pieces together.

The focus of the present survey is static allometry, the comparison of a diversity of animals at comparable times, usually as adults. Eggs and young are examined, but in an interspecific, static way. If we analyze comparable stages in growth, for example data when 50% of adult body mass has been attained, we can obtain generalities about the physiological time scale or pace of growth and life to which adaptive variations solve particular problems. Of course, distinctions may blur between evolution of lineages and diversity among contemporaries, but we must keep in mind that speculative license is being taken. Those who wish to explore the allometry of ontogeny or evolutionary change will find the following to be valuable: Gould, 1966, 1982; Alberch et al., 1979; Lande, 1979; Bonner and Horn, 1982; Cheverud, 1982.

Most of my discussion will focus on birds and mammals, because the emphases of biomedical, avocational, and agricultural interests have produced enough information on these two classes to permit a synthesis with fewer gaps than would be the case in some other area. I hope that this attempt will stimulate parallel syntheses for other taxa. The allometry of insects and reptiles shows many parallels with, and extensions from, the plots of mammals and birds, several of which will be noted here and are described elsewhere (Heinrich and Bartholomew, 1971; Bennett and Dawson, 1976; Bartholomew, 1981; Casey, 1981; and Bennett, 1982).

The Evolution of Body Size in Birds and Mammals

SHREW TO BALUCHITHERIUM, BEE BIRDS TO ELEPHANT BIRDS

Throughout our exposure to biology we have been shown family trees, phylogenetic diagrams, and cladograms which attempt to relate fossil and present-day forms. There appear to have been periods of linear succession, and then adaptive radiations when the "time had come" for a particular stock. A new environmental opportunity came into being, and a type of animal well suited for exploiting it entered a period of rapid evolution into a variety of closely related species, specialized so as to make best use of the resources while avoiding the inefficiency of excessive competition. This process of adaptive radiation is usually portrayed as a proliferation of bill shapes, plumage or pelage colors, and other qualitative characteristics. It is seldom cast in terms of adaptive radiation of body sizes, even though body size dictates with far greater priority than factors of color and shape what the requirements for food, water, and oxygen will be, and which opportunities can be profitably exploited. Evolution toward larger or smaller size in these ranges requires many changes in different (allo-) measures (metric), as can be seen in Figure 1-1.

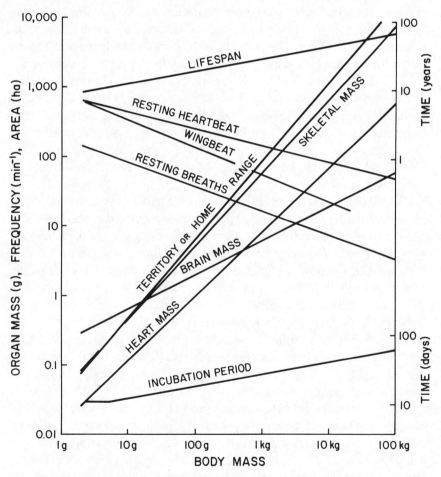

1-1 A sampler of the allometry of birds. None of the changes required for larger or smaller size is exactly proportional to the body-mass change, so the scaling involved in evolutionary size change is no simple matter.

Over time, evolution has produced birds that range in size from the bee hummingbird (*Mellisuga* [calypte] *helenae*), about 2 g body mass, to the elephant bird, which Amadon (1947) estimated to have had a mass of 457 kg, and mammals from the etruscan shrew (*Suncus etruscus*) of 2 g to the largest of land mammals, the extinct giant rhinoceros (*Baluchitherium*), at an estimated 20,000 to 30,000 kg (Schmidt-Nielsen, 1975; Economos, 1981a).

Thus the radiation in body size spans a factor of 2.86×10^5 for birds, and 1.50×10^7 for mammals. Within each class are many different forms, and of course each narrower taxon or more specialized ecological type tends to have less flexibility and therefore is suited for a smaller body-size range. For example, the larger birds exceed the maximum size for sustained powered flight, so that flying birds seem to have attained an upper size limit with the trumpeter swan (*Olor buccinator*), its 12.5-kg body mass being 6,250 times that of the smallest bird, the bee hummingbird. Diving birds have spanned a 2,000-fold size range. Within the hummingbird family (Trochillidae), the giant hummingbird (*Patagona gigas*), at 21 g, is only 11 times the mass of the bee hummingbird. The smallest anseriform is only 1.8% of the size of the trumpeter swan, and the Galliformes range from a 40-g painted quail (*Excalfactoria = Coturnix chinensis*) to the 12.7-kg turkey (*Meleagris gallopavo*), weighing 318 times as much. Most avian orders span 100-fold or smaller size ranges.

Mammalian divers (water shrew 11 g to blue whale over 100 metric tons) occur over a much wider size range than avian divers. On the other hand, the eutherian mammals exhibit a smaller range in body sizes of fliers than do flying birds. This full range occurs in only one order, wherein the 1,380-g *Pteropus edulis* weighs 373 times as much as a 3.7-g *Pipistrellis pipistrellus*. The largest rodent is the capybara, *Hydrochoerus hydrochaeris,* which at 50 kg is 5,000 times as heavy as *Micromys minutus.* The squirrels range 750-fold, from under 10 g (Bornean pygmy squirrel, *Exilisciurus*) to 7.5 kg (*Marmota*). The range of lagomorphs is only 67-fold, from the 105-g pika (*Ochotona princeps*) to a 7-kg hare (Lepus sp.) (Walker, 1975). The Artiodactyla start at a 2.5-kg mouse deer (*Tragulus javanicus*) and go up 1,800-fold to a *Hippopotamus amphibius* of 4.5 metric tons.

Even more significant than the existence of these extremes are the effects of size on the lives and supporting processes of each animal and the utility of size in making comparisons. Allometric equations in subsequent chapters will show, for example, that birds have larger hearts (that beat more slowly), reproduce quicker, travel faster (at lower cost), and live longer than mammals. Their basal metabolic rates and skeletal masses are similar, however.

INTRASPECIFIC VARIATION IN BODY MASS

A first rule in the search for body-size correlations and dependencies is that phylogenetic and ontogenetic data must not be mixed. The proportionality of neonatal mass to adult body mass varies across the size ranges in a quantitatively different way for birds and for mammals. The neonatal mass of eutherian mammals approaches proportionality to the adult, while in birds neonatal mass, like egg mass, comes closer to a metabolic scaling, the ratio of neonatal to adult mass decreasing with size (see Figure 10-1 later; Blueweiss et al., 1978). Another complexity is that growth rates are also size dependent. We must exclude variability in degree of growth and development from the analysis of evolution of body size, by selecting data from comparable stages in life history. Hence interspecific allometric treatments here are concerned with adults only (see Sweet, 1980).

Within a species body size can vary considerably. Geographic variation, for instance, appears to have adaptive significance, and often comes under the so-called Bergmann's Rule, to be considered later. For the present let me merely note that it is common to observe that specimens of birds from the higher latitudes of species distribution may weigh 7% to 16% more than specimens collected at the lower latitudes.

It is, of course, well known that feeding and activity levels produce a wide range of body mass, especially in domesticated species not currently subjected to entirely natural selection. The coefficient of variation (CV)—the standard deviation expressed as a fraction of mean value—is a useful parameter. The CV for body mass in *Homo sapiens* is 0.24, and for the domestic horse (*Equus caballus*) the CV of body mass is 0.33. The latter includes the range from a 150-kg shetland stallion to a 522-kg draft horse (Quiring, 1950). For eutherian mammals in general, Yablokov (1966) found a CV range of 0.12 to 0.15. The CV for brain mass is generally less than half that for body mass (Lindstedt and Calder, 1981), so brain mass could be used to predict other variables in form, function, and life history with greater precision; it is, however, easier to obtain data on body mass, and body size is more likely to have a causal relation to metabolic requirements than the size of one specialized organ.

ECOLOGICAL SEPARATION

Descendants of animals of a given body plan evolved allopatrically (in different locations) would, upon becoming sympatric (moving back to a common location), be expected to seek the same resources and thereby enter into competition. One will outcompete the other, sending the loser to extinction; or the two may specialize divergently, the subsequent ecological separation being known as character displacement. For example, since predators are larger than their prey, two predators could undergo a character displacement in body size so that they could utilize prey of different sizes. Hutchinson (1959) reviewed these principles and collected information on a number of congeneric and co-occurring pairs of bird and mammal species. He showed that the ratios of the bill (culmen) lengths of the bird pairs and of the skull lengths of the mammal pairs (quantitative measurement of the "trophic apparatus") varied from 1.1 to 1.4, with a mean of 1.28. This he used as "an indication of the kind of difference necessary to permit two species to co-occur in different niches but at the same level of a food web."

How would this difference be related to total body size? If all the linear dimensions of these animals scaled geometrically in such a way that each pair member had the same relative proportions, we would expect body mass (M) to be proportional to body volume (V), which would in turn be proportional to the cube of the linear dimensions (l^3) of length, breadth, and depth. Since the mean value for skull-length or bill-length ratio of the pairs was 1.28, we might expect

$$M_2/M_1 \propto (l_2/l_1)^3 = 1.28^3 = 2.10.$$

Thus the larger of the pair would be about twice the total mass of the smaller. Hutchinson used the weasels *Mustela erminea* and *M. nivalis,* both found in Great Britain, as one of his pairs. Extrapolation from skull length to body mass must be done with caution, as may be seen from an analysis of skeletal allometry in six species of mustelid carnivores (Sweet, 1980). His regressions relating postcranial measurements to femur length, jaw, and palatal measurements were linear (regression slopes 0.94 to 1.07, mean of 1.0); that is, they increased from species to species in the same proportions, so that a jaw measurement could be used to predict other measurements. However, the regression for skull length to femur length had a slope of only 0.79, and for braincase length versus femur length it was only 0.67. Brain size does not increase in linear proportion to body size, but has the same sort of slope as the skull-length versus femur-length regression obtained by Sweet (1980). Since skull length is related to both brain space and feeding dimensions, the intermediate slope of 0.79 seems appropriate, at least qualitatively. It is less clear why Hutchinson obtained size ratios for mammalian skull lengths (brain plus trophic aspects) similar to those he derived for bird bills (trophic only).

From the impressive ranges in body sizes of animals, Schmidt-Nielsen

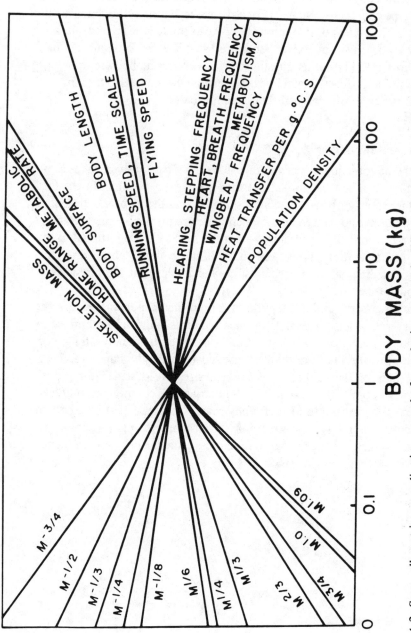

1-2 Some allometric generalizations, mostly for eutherian mammals. When plotted on log-log paper, the fractional exponential powers of body mass appear as the slopes of the lines. For example, body surfaces of animals, like surface areas of geometric figures, are proportional to volume (or mass, assuming constant density) raised to the ⅔ power. On the log-log plot the ⅔ becomes the slope, that is, an increase of 10² in surface when mass increases by a factor of 10³. (Adapted from Zar, 1968a; Calder, 1974; and Bartholomew, 1982.)

(1975), in a fascinating introduction to scaling in biology, identified two fundamental questions for biologists: "(1) What are the size-limits for a certain type of organism? and (2) For a given size, how must an organism be designed?" In a later consideration (1977) he expanded the second question to "For an organism of a given structure, what are the consequences of a change in size?" The earlier question suggests a focus on body morphology and physiology, while the later one could extend the analysis to the animal's niche in the environment, as I shall attempt to do.

SOME SCALING RULES

What are the consequences of a change in size? If all scaling were in the same proportion, it might be simple; but that is not even possible. In the first place, there are many inverse relationships, such as that between the length of a pendulum and its frequency. Then there are the exponential relationships we memorized in geometry—that the surface is proportional to the square of linear dimensions, and the volume to their cube. Allometry, literally "of different measurements," is inevitable from the physical dimensions of size change.

Allometrics seem to fall into patterns of repeated exponents. Figure 1-2 shows a series of size trends in mammals and birds that will be examined in subsequent chapters. Only a fraction of the data points actually fall right on the lines, but the lines are the best statistical fits for the data and give a graphic idea of how animals of different sizes are designed for proper function. Chapter 2 will explore the allometry of the components that take up space and contribute to total body mass. Chapter 3 will be a sort of workbook in allometry. The remaining eleven chapters will deal with major body, reproductive, and ecological functions, using the pervasive influence of size as the focus of analysis.

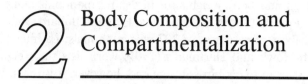

Body Composition and Compartmentalization

Animals are built reasonably . . . no more structure is formed and maintained than is required to satisfy functional needs.

—C. R. TAYLOR AND E. R. WEIBEL (1981)

HAVING CONSIDERED the range of adaptive radiation in body size that has been attained in different taxa and ecological life-styles among mammals and birds, we now examine the proportional structuring. Gould (1966) has defined allometry as "the study of size and its consequences," among which are "the differences in proportions correlated with changes in absolute magnitudes of the total organism." Organ weights are some of the most easily measured examples of these "differences in proportions," and it is to the allometries of organs that this section is devoted.

Some types of organs scale linearly, some by other than a direct proportion to body mass. The functional interrelationships among the various organs constrain organ allometry. Of course, the general allometric equations for organs, like all other allometric patterns, are obtained by regression of data obtained from as wide a size range as possible. This means that there can be considerable variation around predicted values, reflecting such factors as an animal's locomotory behavior, sexual specialization, state of nutrition, and health.

Much of the information on organ weights is contained in the literature. Quiring's *Functional Anatomy of the Vertebrates* contains an extensive list of organ and body weights for vertebrates; Brody's *Bioenergetics and Growth* lists organ and body weights for birds and mammals along with allometric equations that summarize the data; Pitts and Bullard (1968) give an extensive analysis of carcasses of 207 wild mammals; Latimer (1925) studied domestic fowl; and Hartman's (1961) work lists the weights of locomotory muscles in flying birds. In the end, however, it is the arduous task of dissecting animal carcasses and weighing the parts that provides the data for these lists. Such research is beginning to broaden our knowledge of how body space is allocated, which in turn helps us to understand both function and evolution.

There are problems associated with the standard gravimetric method of determining organ weights, however, and these should be kept in mind. Specific technical difficulties are discussed where appropriate, but in general gravimetric analysis suffers in precision because techniques are not standardized. Most studies have been concerned with one or a few component organs, so measurements of muscle mass and skeletal or internal organ mass have not come from the same animals. The determination of total blood volume is accomplished by dye dilution techniques using live animals. Total fat content is determined by extraction with ethyl ether from minced carcasses. Freshly dissected muscles and other organs still contain blood and fat that therefore is counted twice. During dissection there is a variable loss of water from the organs, resulting from evaporation, blotting, and handling. Skeletal masses are usually obtained from dried museum specimens, whose bones would be expected to weigh less than those of freshly dissected animals—not only because the latter contain more water, but also because fresh bones are difficult to clean completely.

Should the total body mass used in allometry include gut contents, a major source of variability (but representing mass that the animal must be designed to support and carry), or should gut contents be subtracted from live mass? Disallowing gut contents is not practical in studies wherein the animals are not, or should not be, sacrificed. The types of animals chosen for gravimetric studies are also important. For example, the use of organ and body weights from domestic animals bred for maximum meat production tends to skew the results, particularly if the overall sample size is small. Sick or injured animals often have inflamed organs or, as in the case of a chronic or long-term muscular injury, exhibit compensatory hypertrophy in other muscles.

With all these variables, it is understandable that the allometrically reconstituted mass may exceed 100% of body mass in some portions of the size range, or that in other portions we cannot quite account for all of the animal.

Data on body and organ weights are best documented for mammals. More extensive data on birds are just now beginning to be collected, and for this reason many of the avian allometries must be regarded as preliminary. Very few allometric studies have included the lower vertebrates.

INTEGUMENT

As animals increase in size, the relative surface area (area per unit mass) of the body decreases; consequently so does the relative area covered by the integument (which includes the skin and its derivatives—scales, feathers, horny coverings of the beak, claws, and hoofs). At the same time *absolute* surface area increases, and the thickness of the skin tends also to increase in order to maintain its mechanical strength. The drop in relative surface area is somewhat more than enough to offset the rise in unit mass of the skin, however, and the overall result is that the integument occupies a slightly smaller fraction of body mass in larger animals, for example, 19% of an 18-g mouse as opposed to 7% of a 460-kg cow (Pace et al., 1979; Lindstedt and Calder, 1981). Figures 2-1 and 2-2 illustrate this point for birds and mammals.

SKELETON

Since the time of Galileo it has been known that bone thickness must increase relatively more than bone length as body size increases in order to provide sufficient support for the added mass of the animal. Thus skeletal mass occupies a greater fraction of body mass in larger animals, as determined from cleaned and dried museum skeletal specimens. However, regression analysis of the data of Pitts and Bullard (1968) on unscraped fresh skeletons including marrow for Alaskan, Virginian, and Brazilian mammals ($n = 43$, body-mass range 7 g to 17 kg) shows no size trend away from about 10% of total body mass. This may be partially masked by a mass isometry of marrow tissue in parallel with the blood volume. In mammals the combined proportions of decreasing skin mass and increasing skeletal mass constitute approximately the same fraction of total body mass, regardless of body size (see Figure 2-1). It is worth noting that the metatarsal bone of a rabbit is 40% stronger when recovered with the fresh skin than when tested as naked bone

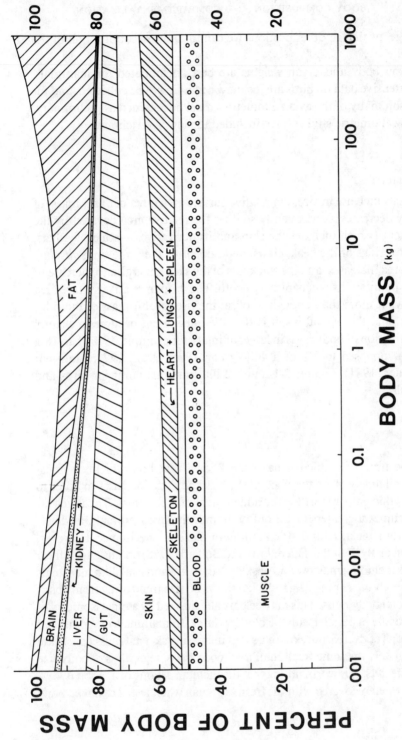

2-1 The body composition of eutherian mammals, reconstituted from available allometric generalizations (equations shown in Table 3-4). Skeletal muscle, blood, heart, lungs, and spleen account for about 52% of body mass, regardless of animal size. Skeleton and fat contribute proportionately more of the total mass in larger mammals, while the skin, brain, and many organs scale up in less than linear fashion. There is considerable variation in what the graph would predict for a particular size, though there are relatively few studies that give complete analyses for one species (see Table 3-6).

(Currey, 1968), so skin and bone can properly be considered together as supporting tissues.

One of the firmly entrenched notions of traditional biology is that birds have lighter, pneumatized skeletons as a weight-reducing "adaptation to flight." Prange and coworkers (1979) have shown, to the contrary, that skeletal mass in birds bears a relationship to body mass essentially indistinguishable from that in mammals (compare Figures 2-1 and 2-2). The birds have redistributed their skeletal mass without an overall reduction, devoting the mass lost from the pneumatized wing bones to the two, rather than four, more robust leg bones.

SKELETAL MUSCLE

Within a given class, skeletal muscle usually occupies a major fraction of total body mass, independent of body size (Figures 2-1 and 2-2). Furthermore, the masses of individual muscles (*M. semitendinosus* and *M. vastus medialis,* for instance) are essentially linearly proportional to total body mass (Mathieu et al., 1981). Another generalization based more on assumption than on data is that the proportion of body mass devoted to skeletal muscle decreases as one travels up the phylogenetic scale from fish to birds to mammals (Gold, 1973; Calder, 1974). The data in Table 2-1 suggest the contrary—that approximately the same proportion of body mass is also devoted to locomotion in each of these vertebrate classes. Mean values from limited tabulations on elasmobranchs, birds, and mammals are statistically indistinguishable.

There are advantages to being fast and powerful and exceptions to the general range of one-third to one-half muscle, but there are practical limits on how much muscle an animal can have. If a larger mass of muscle is to do proportionately more work, it must be supplied with more energy-yielding substrate (such as blood glucose), which in turn must be supplied by a larger digestive system. The kidney and liver would also require a greater capacity to process the metabolites and by-products of the processed nutrients. If the contractions of this larger mass of muscle were aerobic, the oxygen supply to the muscles would have to be increased, which means that a greater respiratory capacity, a bigger heart, and a greater blood volume would also be required. If the movement powered by these additional muscles occurred mostly in quick bursts, the muscles could operate anaerobically. The oxygen debt incurred during anaerobic bursts of activity could then be repaid slowly by a smaller cardiopulmonary system. Finally, contracting skeletal muscles bring about movement only by virtue of their attachment to rigid skeletal

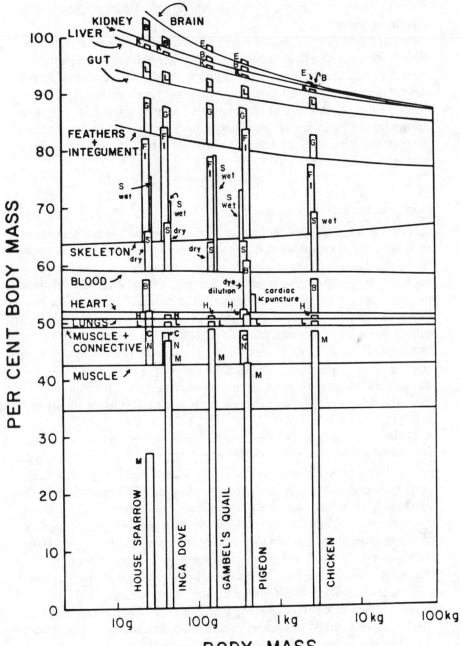

components; greater muscular forces would necessitate stronger bones. In the end, then, the enlargement of all of the various support organs and bones necessary for the operation of more muscles would increase total body mass, reducing the percentage of body mass occupied by the muscle.

The highest known proportions of body mass devoted to skeletal muscle are found in the scombrid *Katsuwonus pelamis* and the salmonid *Salmo irideus,* which are respectively 68% and 55–67% locomotor muscle (Bainbridge, quoted in Bone, 1978). Both the scombrid and salmonid fishes are noted for sustained swimming over long periods of time. While both have invested a higher proportion of their body mass in propulsive muscle than most birds and mammals, the goldfish (*Carassius auratus*), on the other hand, may have as little as 33% locomotor muscle (range 33–45%) and the blue shark (*Prionace glauca*) as little as 36%.

Mammals also vary considerably in muscle content; cats such as the lion (*Felis leo*), bobcat (*Lynx rufus*), and lynx (*L. canadensis*), which are 56–59% skeletal muscle (Pitts and Bullard, 1968; Munro, 1969), rival the salmon in the amount of body mass dedicated to locomotion. The record among mammals is held by the lion, 58.8% of whose body mass is occupied by skeletal muscle—1.31 times the average percentage of skeletal muscle for the members of its class (Davis, quoted in Munro, 1969). Thus distinctions in proportions of skeletal muscle must be made not at the level of phylogenetic class, but at a level that takes into account an animal's locomotory specializations.

2-2 The body composition of birds, reconstituted from an even more limited allometric base than that in Figure 2-1 for mammals. The first horizontal line (at 35%) represents the flight muscles. The vertical bars indicate deviations from the generalized patterns for specific birds. The house sparrow and pigeon were freshly killed, the Inca dove and Gambel's quail were relatively fresh road kills, and the chicken data are from the literature. M = muscle, CN = connective tissue, lower L = lungs, H = heart, B = blood (note discrepancy between dye dilution data and cardiac puncture data), S = skeleton (dry from museum specimens vs. "wet" for fresh bone mass), FI = feathers and integument, G = gut, upper L = liver, K = kidneys, B = brain, and E = eyeballs. Despite the limitations of the data, the pattern is generally similar to that of the mammals: over half the body mass is allocated, independently of size, to muscle and connective tissue, blood, lungs, and heart; the skeletal portion increases with size; and the integument and other major organs are proportionately smaller in larger birds. As size increases, there is a progressive increase in the "unaccounted" fraction. This is presumably fat, but the data in the literature are too variable to reduce to a reliable allometric curve (migratory fuel capacity at time of collection is far from constant).

Table 2-1 Proportion of body mass contributed by skeletal muscle.

Class	No. of species (n)	Body-mass range (g)	Skeletal muscle		References
			% of body mass	Range (%)	
Fish	8	—	49.3[a] ± 12.01 s.d.	33–68	Bone, 1978
Sharks	4	—	43.9[a] ± 9.94 s.d.	36–60	Bone, 1978
Teleosts	4	—	54.8[a] ± 12.61 s.d.	33–68	Bone, 1978
Snakes	4	2.3–703	46.0 ± 9.08 s.d.	33–57	R. Rusler and Calder, unpublished
Birds	5	25–2,600	42.6[b] ± 8.98 s.d.	27.2–48.4	S. Hiebert, unpublished; Latimer, 1925
		10–3,700	34.5[a]		Hartman, 1961
Mammals	14	4–4,987,000	44.4[c] ± 6.23 s.d.	31.3–58.8	Munro, 1969
	42	4–17,640	49.5[d] ± 4.20 s.d.	35.1–58.5	Pitts and Bullard, 1968

a. Locomotory muscle only.
b. Does not include neck muscles (49.3% with connective tissues and muscles of neck).
c. Includes connective tissues.
d. Percent corrected live weight (minus fur and gut contents).

Among birds the tinamou *Tinamus major* has the highest proportion of locomotor muscle—Hartman (1961), in an extensive tabulation of locomotor mechanisms, lists the tinamou as devoting 43.7% of its weight to flight muscles, 7.4% to thigh muscles, and 5.8% to leg muscles, for a total of 56.9%. The locomotor muscles of the roadrunner (*Geococcyx californianus*) constitute 35% of the body mass, a proportion almost identical to the mean for 42 species of flying birds (Hartman, 1961, table 3; Hiebert and Calder, unpublished). The roadrunner flies for short distances only, usually gliding from a running start, though occasionally making flapping flights of short duration. Its flying muscles contribute 13.5% of body mass and its leg muscles 21.7%, a reversal from the respective figures for Hartman's fliers, 24.8 ± 6.5% and 9.0 ± 4.5%. Among the mammals the lowest percentages of skeletal muscle occur in sedentary, domesticated animals. For example, a female domestic pig weighing 271 kg was only 31.3% muscle (Munro, 1969).

HEART AND BLOOD VOLUME

On the average, birds have hearts 41% larger than mammals of equal size (Lasiewski and Calder, 1971), but again there is considerable variation at the species level. It may not seem surprising that the lion has a heart 1.36 times heavier than the average mammal's along with its muscle mass 1.31 times the "normal," for the cardiopulmonary system supplies the muscles with necessary oxygen. On the other hand, species of *Lynx*, with similar high dedication to proportional muscle mass, have relatively smaller than average heart size (Pitts and Bullard, 1968; Munro, 1969). The tinamou, although it has the highest percentage of skeletal muscle of any bird listed by Hartman (1962), has one of the smallest hearts relative to body size. It is likely that its huge muscle mass is used anaerobically in short flights. Indeed, tinamous are characterized as "poor fliers" (Welty, 1975), without a definition of the criterion of "poor" as meaning slow, weak (why so much muscle?) or of short endurance (as suggested by the small heart).

Blood volumes for birds and mammals are statistically indistinguishable. However, accurate measurement is difficult, and most methods introduce significant error. Dye dilution techniques, such as the one employed by Bond and Gilbert (1958), are by far the most accurate. Unfortunately, these authors measured blood volumes from a limited number of aquatic and nonaquatic bird species, over a body-size range too limited (310 g–2,470 g) for definitive allometric analysis.

Blood volumes can be obtained from freshly killed animals, but because blood is distributed throughout the body, not only in the heart and major blood vessels but also in the capillary beds of all the somatic tissues, large amounts remain in the body even after bleeding. Cardiac puncture may remove only a third of the total blood mass. The problem is more acute in roadkills and other animals obtained some time after death has occurred, for the blood has usually already coagulated.

LUNG

Birds also tend to exceed mammals in their lung capacity (by 12%; Lasiewski and Calder, 1971). This correlation of larger lung with larger heart is not unexpected, since the circulatory system distributes the oxygen obtained by the lungs. Similarly, the lung of the lion, whose heart (relative to body mass) is the largest of any mammal, is 1.67 times its predicted mass.

DIGESTIVE TRACT

Since the digestive system processes the nutrients that provide the muscles with energy-yielding substrate, and since the muscles occupy a constant fraction of total body mass, the mass of the digestive system would also be expected to increase linearly in order to keep up with the added mass of muscle. Figure 2-1 illustrates that this is not the case for the mammals (the body mass contributed by the digestive system is 10% for a 7-g mammal, but only 4% for a 6.6-ton mammal). In birds, however, the digestive system shows a more isometric relation to body mass (Figure 2-2).

NERVOUS AND ENDOCRINE SYSTEMS

Control mechanisms such as the brain and endocrine glands decrease in relative size as total body mass increases, and a moment's reflection on the size of the controls on large and small radios or in large and small vehicles suggests why this may be so.

For almost every animal dissected and weighed, a certain fraction of body mass remains unaccounted for; this fraction climbs as total body mass increases. For example, we can account for 97% of the mass of a 7-g mammal, but only 81% of a 6.6-ton mammal. A large portion of this mass is probably body fat, which in turn indicates that bigger mammals are able to carry proportionately more ample reserves of energy in the form of fat.

Summary

The skin and skeleton together tend to occupy a constant fraction of body mass regardless of body size. This relationship is particularly noticeable in the mammals. In addition, the combined mass of skeletal muscle, skeleton, integument, lungs, heart, blood, and spleen constitutes approximately three-fourths of body mass. In the two homeothermic classes of vertebrates the proportions of muscle plus oxygen-supplying organs (lungs, heart, and blood) are isometric; that is, they increase linearly with increasing body mass and occupy a constant fraction of body mass. With an increase in body mass, the mass of the integument, brain, and other organs will not increase in the same proportion (mass exponent of less than one), whereas the skeleton increases proportionately more (mass exponent greater than one).

Those readers who are in a position to contribute field data and help fill gaps in our knowledge of body compartmentalization may want to consider the following. The mass of an excised organ includes whatever blood and fat it contains at the time of death. Furthermore, water is lost from body parts through evaporation, blotting, and handling during dissection. Skeletons freshly removed weigh considerably more than cleaned and dried bones from museum specimens. A thorough gravimetric analysis would include determination of blood volume by dye dilution; withdrawal of blood via cardiac puncture; and weighing of organs and carcass remaining after each removal; fresh skeleton and cleaned and dried skeleton. Progressive weighings of removed tissue give an indication of the rate of loss of mass due to evaporation.

Thorough gravimetric analysis requires considerable time and effort. Furthermore, it entails the sacrifice of animals. While I find acceptable the sacrifice of domestic and pest species (house sparrow, starling, pigeon, and chicken give a nice size series), I prefer to obtain data on other species from animals salvaged as roadkills and the like, even though determination of blood volume for individuals obtained in this way is not possible.

Symmorphosis

In an extensive analysis of the scaling of the mammalian respiratory system, Taylor and Weibel (1981) suggested the principle of symmorphosis, which they define as "a state of structural design commensurate to functional needs resulting from regulated morphogenesis, whereby the formation

. . . of structural elements is regulated to satisfy but not exceed the requirements of the functional system." Indeed, it is this principle that appears to underlie the allometric baselines by which body mass and volume are apportioned into the different organ systems, at least in adults.

Gans (1979) provided an alternative view of what he termed momentarily excessive construction: "Most aspects of phenotypes will, at any moment of an individual's life, be capable of fulfilling demands much greater than those routinely encountered. This capacity gives members of a population of organisms the capacity to modify their behavior." These two views may not be as divergent as it first appears, for Taylor and Weibel were describing a symmorphosis at peak aerobic demand, which does in fact represent "demands much greater than those routinely encountered."

Evolution may be qualitative or quantitative. It would seem that qualitative evolution, such as that resulting in a new kind of animal assignable to a different phylum or class, is much more complex than quantitative evolution, such as that resulting in a similar species of a different size. Yet quantitative evolution cannot be as uncomplicated as it seems. Reflect, for example, on the scaling of a "simple" increase in body size—the skeleton increases more than the heart, the heart more than the skin, and so forth. Ultimately these changes are orchestrated by the genome, but the nature of that genetic control is not understood. Do separate genes direct the size of each component, or does a single gene influence an entire morphogenetic field? How much of the change in an organ's size results from an increase in numbers of cells, and how much from an increase in size of cells? Eventually we hope to explain how adequacy and economy, or "momentarily excessive construction," are programmed into evolutionary size changes. At present we can only anticipate a more precise description of the patterns, by means of mathematical expressions and qualitative statements about the organ systems described by those mathematical expressions.

3 Physical and Quantitative Background

When we make generalizations about trends among animals and plants, such as changes in size, it is almost automatic to point out the exceptions and throw out the baby with the bath.

—J. T. BONNER, *SIZE AND CYCLE* (1965)

GRAPHIC PORTRAYAL of the relative allocation of body mass to the vital organs and structures has given a qualitative look at allometry. Beyond the fact that the shrew is more brain and skin and the elephant more bone and fat, we need a precise means of analyzing, comparing, and communicating these different scalings. This chapter contains such tools and the principles of their use, and concludes with a quantitative analysis of the body composition of mammals and birds that was treated qualitatively and graphically in Chapter 2.

Progress in science requires free and clear communication, so impediments to that communication are unacceptable. Any contribution to the further understanding of nature depends upon, and must be related to, the existing fund of knowledge. If our findings are obscured in dimensional errors, unfamiliar units, and nonassociable symbols, no one can use our results as stepping stones to something closer to the truth. It has been an impediment to preliminary attempts to relate independently derived allometric functions, for example, to confuse energy with power. Similarly, biologists often confuse pressure and force. These mistakes are eliminated by dimensional analysis, or the reduction of such physical quantities to the fundamentals of mass (m), linear dimensions (l), time (t), and temperature (T), as has been done in the listing of symbols and physical dimensions found in Appendix A. An excellent treatment of dimensions, symbols, and units was given by Riggs (1963), and the topic has been reviewed in an ecological context (Calder, 1982a).

The Allometric Equation

It is evident from Figures 1-1, 1-2, and 3-1 that there are many biological variables that can be correlated with the size of an animal, and that the quantitative nature of the correlation is not the same for all organs and functions. The basal metabolic rate of a cow is higher than that of a rabbit, but it does not differ by as large a factor as the absolute size difference. It turns out that the various scalings are described quite well by the allometric equation:

$$Y = aM^b, \tag{3-1}$$

where Y is any physiological, morphological, or ecological variable that appears to be correlated with size, in most cases body mass (M), in kg (or m in g). I will use M or m without a subscript to signify adult body mass, except where a distinction from mass of organs or other stages in the life history needs to be emphasized for clarity, cases in which the subscripts "body" or "adult" will be added.

The exponent b is called the scaling factor because it describes the effect of a change or difference in body sizes. If $b = 0$, size has no effect because $M^0 = 1$ and Y is then just a constant, the coefficient a. If $b = 1$, Y increases in linear proportion to M (for example heart mass, Figure 1-1). If $0 < b < 1$, Y does not increase as much as M (metabolic rate, Figure 1-2). If $b > 1$, Y is increasing out of proportion to M (skeletal mass, Figure 1-2). Finally, if $b < 0$ (negative exponent), size increase is accompanied by a rate decrease (heart, breath, and stepping frequencies, Figure 1-2).

The curvilinear relationship of Eq. (3-1) is difficult to manipulate or envision quantitatively. For ease in derivation and interpretation, as well as for normalization of the variance, (3-1) can be transformed into a logarithmic form:

$$\log Y = \log a + b \log M. \tag{3-1a}$$

Note that this has the form of a straight line:

$$Y' = \text{intercept} + \text{slope } x. \tag{3-1b}$$

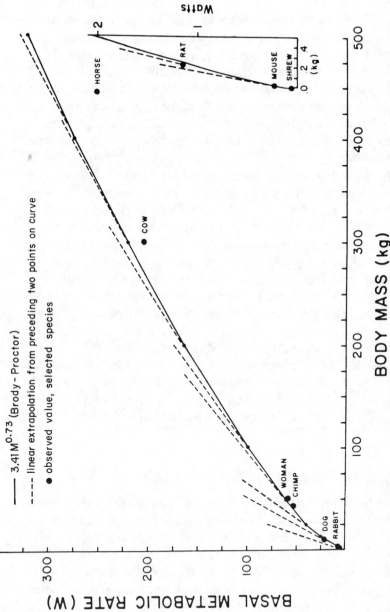

3-1 Basal metabolic rates of eutherian mammals plotted on linear scales to show the curvilinear consequences of $M^{3/4}$ scaling. The dashed lines show the consequences of a linear increase from the previous two points on the curve; note the progressive decrease in the slope (or progressive decrease in the increment of added cost of being larger). The two ungulates have even lower metabolic rates than the general equation predicts. Allometric equation by Brody and Proctor, cited in Brody (1945) and Kleiber (1961), units converted.

The coefficient a in Eq. (3-1), or the Y-intercept in Eq (3-1a), can be thought of as incorporating all the dimensions necessary for dimensional consistency in the equation — but since in most cases b is a decimal fraction, this would be a bit absurd. Equations based simply on observations (empirical equations) are exempted from the dimensional consistency requirement (Riggs, 1963), since they do not express *equivalence* but merely describe an observed correlation between two quantities. The allometric equations that abound on the pages that follow are empirical descriptions and no more, and therefore qualify for the exemption from dimensional consistency in that they relate some quantitative aspect of an animal's physiology, form, or natural history to its body mass, usually in the absence of theory or prior knowledge of causation.

Underlying the allometry is a similarity of biochemical pathways and morphological plans. Mammals are similar to one another more than they are different, whereas among the birds there is a common form and function. The differences among species are largely due to size differences. Because size has such a pervasive role in determining quantitative details of support against laws of gravity and thermodynamics, size correlates well with rates, capacities, and living space. The rates have the inverse of time as a dimensional component, so the time scale of the life history is also strongly size dependent. This makes Eq. (3-1) very useful in comparative zoology. Once the 75% or more of the variability due to size is accounted for, the general patterns help us to dissect the rest of the variability.

A comprehensive view of these general patterns, not in just one discipline, but clear across the life history, is needed before a "theory of allometry" can be formulated. But even without such a theory, we have a powerful tool at our disposal. The following examples give an indication of its uses.

Bird Eggs, Deer Antlers, and the Fire of Life

In 1927 Sir Julian Huxley published the results of an allometric (he called it heterogonic) study in which he examined the relation of egg size to body size in birds. Five years later this work was included in a broader treatment of "problems of relative growth," in a book by that title. Also in 1932 Max Kleiber measured the metabolic rates of animals of different sizes. He summarized his findings in a relationship (now known as Kleiber's Rule) which states that the basal or minimal resting metabolic rate of an animal is not a linear function of size, but is proportional to the three-fourths power of body mass ($M^{3/4}$). This means that an animal larger by a factor of 10^4 would have a metabolic rate larger by a factor of only 10^3. Six years later F. G.

Benedict added more animals, including nine species of birds, to the curve, which then became known as the "mouse to elephant curve." The data that had accumulated to that point yielded the following equation:

$$\text{basal metabolic rate} = 70.5\ M^{0.73}, \tag{3-2}$$

where M is body mass in kilograms and metabolic rate is in kilocalories per day.

These studies marked a significant point in the development of physiology, for Claude Bernard had stated in 1865: "I believe that the most useful path for physiology and medicine to follow now is to seek to discover new facts instead of trying to reduce to equations the facts which science already possesses. This does not mean that I condemn the application of mathematics to biological phenomena, because the science will later be established by this alone." We now have enough facts that we find it necessary to reduce biological phenomena to mathematical equations, to avoid what Yates (1979) called "a sense of being engulfed by a flood tide of raw data, and confused by complexity."

Until now allometry has been largely an empirical science, seeking simply to summarize the patterns we observe in equations, without necessarily explaining why the patterns exist. The urge to summarize data allometrically has been stimulated by the existence of all sorts of size-related patterns, and although allometry had its beginnings in growth, morphology, and physiology, it has now invaded community structure, ecological energetics, population dynamics, life history strategy, paleontology, and evolution as well.

Of course, the emergence of these patterns has inspired attempts to *explain* the empirically derived coefficients a and exponents b found in allometric equations. Just as the purely descriptive aspect of allometry has become an interdisciplinary undertaking, so the more analytical part will require the physiologist to look outside of physiology and the ecologist to venture beyond the boundaries of his own field. Unless this happens, the patterns that have been observed will remain isolated and uninteresting.

Let us now examine the variables that contributed to the beginnings of allometry — the size of bird eggs, deer antlers, and basal or standard metabolic rates. I use these three examples to illustrate how allometric equations are first formulated and later manipulated.

SIZE OF BIRD EGGS

In Table 3-1 is information suggesting that ostriches lay larger eggs than hummingbirds do, obvious enough without the benefit of a formal allome-

Table 3-1   Egg size and bird size. (Selected data from Brody, 1945.)

Species	Adult mass (g)	Egg mass (g)	Egg/adult ratio	Allometry
Ostrich	113,380	1,700	0.015	
Goose	4,536	165.4	.036	$r = 0.9979$;
Duck	3,629	94.5	.026	$0.198\ m^{0.77}$
Pheasant	1,020	34	.033	
Pigeon	283	14	.049	
Hummingbird	3.6	0.6	.167	

Comparisons: birds (Rahn et al., 1975): $n = 809$; $0.277\ m^{0.77}$
reptiles (Blueweiss et al., 1978): $0.41\ m^{0.42}$

tric equation. Less obvious, however, is the fact that egg size does not increase linearly with adult body size, for the egg of an ostrich weighs only 0.015 as much as the laying ostrich, while the egg of a hummingbird weighs 0.167 as much as the hummingbird, a proportion 11 times the ostrich proportion.

Mathematically, the relation of egg mass to adult body size for the six eggs listed in Table 3-1 can be summarized as

$$m_{egg} = 0.198\ m_{adult}^{0.77}, \tag{3-3}$$

where m = mass in grams. Now let us examine what the parts of this equation mean. Exponential curves like that generated by Eq. (3-3) do not lend themselves as easily to visual comparison as straight lines. We can, however, use a simple logarithmic transformation to convert the exponential equation (3-3) to

$$\log m_{egg} = \log 0.198 + 0.77 \log m_{adult}. \tag{3-3a}$$

Note that this equation is in the straight-line form of Eq. (3-1a,b).

In Eq. (3-3) the size of the exponent determines the shape of the curve, but what does it mean specifically about what happens to egg mass as adult body size increases? In this case the exponent, or the slope in (3-3a), is approximately ¾, which means that if two birds differ in body mass by a factor of four orders of magnitude (10,000 or 10^4), their eggs should differ in mass by approximately three orders of magnitude (1,000 or 10^3).

An exponent of one would tell us that egg mass and body mass are linearly proportional; that is, that an increase in adult body mass is accompanied by an increase in egg mass of equal proportions. The exponent in this case (0.77) is between zero and one, however, which means that if we were to plot egg-mass data on a graph with linear axes, the "best-fit" curve would be convex, seeking a horizontal asymptote as body mass increases (as metabolic rate does in Figure 3-1). Had the exponent been greater than one, the resulting curve would have been concave and would tend instead toward a vertical asymptote. The exponents of most allometric equations fall between -0.3 and $+1.1$.

From a more extensive data base (809 birds), Rahn and associates (1975) found:

$$m_{egg} = 0.277 \, m_{adult}^{0.77}. \tag{3-4}$$

Note that the exponent, which tells us what the effect of adult body size on egg mass will be, is the same; the coefficient, 0.277, is 40% higher. The small data base used to formulate Eq. (3-3) may have been biased toward species that produce relatively small eggs for their body size. Indeed, when the Rahn group separated the 809 birds by order, they discovered that the Columbiformes (pigeons) and Galliformes (fowl) lay smaller eggs for their size than many of the orders that provided data for Eq. (3-4). Furthermore, the egg of an ostrich is somewhat smaller than one would expect (by extrapolation) for a bird of its size, and the duck and goose eggs would have appeared small compared with body size, since they were obtained from domestic animals whose body mass probably exceeded that of their wild flying ancestors.

Recently it has been pointed out that Eq. (3-4) confused egg "weight" with egg "value," the latter calculated as length × breadth², thus having the dimensions of volume and exaggerating the intercept or coefficient a (Scott and Ankney, 1983; Grant, 1983).

HETEROGONY OF DEER ANTLERS

This section heading is borrowed from Huxley (1932), who provided one of the most fascinating and instructive case studies in allometry, that of relative size of deer antlers. It is a study that gives us some detailed statistics and at the same time is relevant to a wide range of disciplines: taxonomy based on body characteristics, nutrition, growth, and ecology.

Huxley analyzed antler and body size of red deer (*Cervus elaphus*), the European congener of the American wapiti or elk. Mean values for his data appear in Table 3-2. His data on antler mass (M_{ant}) relative to body mass (M_{body}) show a progressive heterogony or allometry, with relatively larger antlers (0.042 M_{body} vs. 0.022 M_{body}) in size classes 2 through 8. For these classes, Huxley calculated

$$M_{ant} = 0.0016 \, M_{body}^{\sim 1.6}. \tag{3-5}$$

Table 3-2 The heterogony of deer antlers. (From Huxley, 1932.)

Size class	Mean mass (kg)		Antler/body ratio (%)	Mean number of antler points
	Body	Antlers		
1	74.4	1.64	2.20	7.50
2	93.4	2.03	2.17	8.20
3	110.4	3.16	2.86	9.81
4	130.6	3.96	3.03	11.64
5	148.9	4.78	3.21	11.74
6	170.7	6.21	3.64	13.10
7	191.1	7.28	3.81	14.77
8	211.8	8.91	4.21	15.41
9	231.7	8.79	3.79	13.62
10	259.1	8.63	3.33	13.78

Using a modern calculator and log-log transforms of the listed mean values, we find

$$M_{ant} = 0.00089\, M_{body}^{1.72}, \qquad r^2 = 0.991. \tag{3-6}$$

Gould (1973, 1974) combined the first eight classes and found a scaling of $M^{1.65}$. Recalculated as an interregional plot using mean values from different populations, the scaling was a very similar $M^{1.67}$. If all ten classes are included (which probably mixes in additional variables of immaturity and approaching senescence), we note a slight decrease in the correlation coefficient and a rotation to a shallower slope if plotted on log-log scales:

$$M_{ant} = 0.0027\, M_{body}^{1.49}, \qquad r^2 = 0.972. \tag{3-6a}$$

The regression for the mean number of points on these antlers can be described as follows:

$$\text{Classes } 2-8: \quad n = 0.29\, M_{body}^{0.75}, \qquad r^2 = 0.976 \tag{3-7}$$

$$n = 6.08\, M_{ant}^{0.43}, \qquad r^2 = 0.982 \tag{3-8}$$

$$\text{Classes } 1-10: \quad n = 0.68\, M_{body}^{0.57}, \qquad r^2 = 0.887 \tag{3-9}$$

$$n = 6.33\, M_{ant}^{0.39}, \qquad r^2 = 0.957. \tag{3-10}$$

Now we see that the number of points is more closely correlated with antler mass than with body mass. This suggests that the growth of antler mass is a consequence of body mass, and that the branching pattern or distribution of antler mass toward the various terminal points may somehow be governed by the size of the antlers.

It is worth noting that while Eqs. (3-8) and (3-10) were obtained by running other allometric regressions directly, the relationships could be approximated by substituting the relationships of Eqs. (3-6) and (3-6a) into (3-7) and (3-9) respectively. (This, in fact, provides a good exercise in proper handling of compound exponents and coefficients.)

The stags of red deer of Scotland attain smaller body sizes than those of some parts of mainland Europe. They have relatively smaller antlers with fewer points (6–8) in districts where full size amounts to only 75–90 kg, compared to 20–25 or more points on the antlers of mainland stags of 200 kg or more. These characteristics might lead one to conclude that the Scottish red deer represents a distinct subspecies or race. In the context of Huxley's heterogony, however, this seems unlikely; for in number of points and in relative antler mass, the red deer seem to be reaching only size classes 2 or 3.

History and an unintended allometric experiment support Huxley's conclusion that this was differential growth, not genetic drift in an isolated population. First, there are subfossil skeletal remains in peat bogs of Scotland, only a few thousand years old, which show that deer had attained body and antler sizes as great as those recorded in mainland Europe. It appeared that the Scottish elk were only stunted animals, and that antler size had scaled down the same line upon which it had scaled up. Experimental proof came when Scottish red deer were introduced to New Zealand, an island continent that had lacked any terrestrial mammals until man came. The deer multiplied in this new, unbrowsed, lush paradise and regained not only the larger body masses (200 kg) of European red deer, but the relatively larger antlers (20 or more antler points). Furthermore, when the introduced populations had proliferated to the point of exceeding carrying capacity, leading to deterioration of the browsing conditions, there was a decline in body and antler mass.

Gould (1973, 1974), using linear dimensions, extended the allometry of cervid antler size to the case of the extinct "Irish elk" (*Megaloceros giganteus*). He found that this animal could be described on the same line as the other cervids:

$$l_{ant} = 0.0342 \text{ (shoulder height)}^{1.85} \qquad (3\text{-}11)$$

(units in inches), and concluded that *Megaloceros* was merely utilizing positive allometry in order to achieve large size and large antlers for display and competitive purposes. Equation (3-11) was derived as a reduced major axis. On the standard regression expressing length as the dependent variable and shoulder height as the independent variable, *Megaloceros* was slightly above the regression line of

$$l_{ant} = 0.0639 \text{ (shoulder height)}^{1.68}. \qquad (3\text{-}11a)$$

METABOLIC RATES OF MAMMALS

Metabolism is what Kleiber (1961) called "the Fire of Life." The metabolic rate is a measurement of the intensity of this fire, the rate at which energy obtained from food is expended to support the activity and the basic preservation of the living state.

Living systems are maintained in a dynamic steady state that is distinctly different from equilibrium. At equilibrium all potential energy has been dissipated, but in the steady state there is a significant potential energy resulting from the fact that concentrations, pressures, and temperatures in the animal's body are higher or lower than those of its surroundings. Because animals cannot be sealed off completely from the environment, they tend to "leak" materials and energy along these potential gradients. The leakage must be overcome by active, energy-requiring processes which maintain the steady state (homeostasis). The necessary energy is obtained from the environment in the form of food, and a substantial proportion of the daily intake is used just for homeostasis.

Table 3-3 Standard metabolic rates of mammals. (Data from Schmidt-Nielsen, 1979.)

Species	Adult mass (g)	ml O_2/(g · h)	ml O_2/h	Watts
Shrew	3.5	9.0	31.5	0.176
Mouse	25	1.58	39.5	0.221
Rat	226	0.872	197.1	1.100
Rabbit	2,200	0.466	1,025.2	5.72
Dog	11,700	0.318	3,720.6	20.77
Horse	700,000	0.106	74,200	414.3
Elephant	3,800,000	0.067	254,600	1,421.5
$n = 7$:		$6.59\ m^{-0.32}$	$6.59\ m^{0.68}$; $r = 0.988$	$0.037\ m^{0.68}$

Comparisons: Eutherian mammals, $n = 26$ spp. (Kleiber, 1961):	$0.018\ m^{0.756}$
Mammals, $n = 349$ points (Stahl, 1967), $r = 0.96$:	$0.020\ m^{0.76}$
Nonpasserine birds (Calder, 1974):	$0.029\ m^{0.73}$
Passerine birds (Calder, 1974):	$0.047\ m^{0.72}$
Small nonpasserine birds (Prizinger and Hanssler, 1980):	$0.038\ m^{0.72}$
Marsupials (Dawson and Hulbert, 1970):	$0.014\ m^{0.74}$
Reptiles (30° C) (Bennett and Dawson, 1976):	$0.0016\ m^{0.77}$

The energetic cost of this maintenance work can be determined in several ways. One method is to measure intake of food and output of feces and urine. Samples of food and waste are then analyzed in a bomb calorimeter to determine their energy content. Thus,

(g food/day)(joules/g food) − (g waste/day)(j/g waste)
= energy required/day.

Of course, corrections must be made for the calculated metabolic level if the animal is gaining or losing fat.

Greater precision over shorter time intervals can be attained by direct or indirect calorimetry. In direct calorimetry the amount of heat produced by the animal is measured in an insulated chamber fitted with sensors designed to measure heat flow. The indirect method involves the measurement of oxygen uptake or carbon dioxide production. Details of these methods were provided by Kleiber (1961) and Bartholomew (1982) and are familiar to most students of comparative physiology.

Obviously, metabolic rate can vary considerably. It will be at its lowest when the animal is sleeping comfortably in a warm nest and at its highest when the animal is shivering violently during extreme cold stress, or running at maximum speed. The minimum metabolic rate, that needed for basal homeostasis, is called the basal or standard metabolic rate.

Meaningful comparison of animals necessitates that the data be obtained in a uniform or standard manner. The criteria are that the subject be thermoneutral (that is, experiencing no heat or cold stress), resting, psychologically calm, and postabsorptive (fasted, not digesting and/or absorbing a meal).

Let us examine the standard or basal metabolic rates ($\dot{H}_{sm}$) of eutherian (placental) mammals.* Table 3-3 gives metabolic rates for seven selected species ranging in size from shrew to elephant. From this list we can derive:

$$\dot{H}_{sm} = 0.037 \, m^{0.68} = 4.06 \, M^{0.68}, \tag{3-12}$$

where $\dot{H}_{sm}$ is in watts and m is body mass in grams. From a larger sample of 26 species Kleiber (1961) reported (in kilocalories per day):

$$\dot{H}_{sm} = 67.6 \, M^{0.756}. \tag{3-13}$$

---

* Historically, the resting metabolic rate (whether measured directly or as oxygen consumption) results in 100% heat production, so H for heat energy is appropriate here, and also in discussing thermoregulation. However, about 25% of the metabolic rate of activity goes into work, not heat, so E for energy should be used there!

Of course, there can be no comparison until both equations utilize the same units. Since 1 kilocalorie is equivalent to 4184 joules, and 1 watt = 1 joule/second, 1 kcal day^{-1} is equal to 0.0484 watt. (The conversion from grams to kilograms, or kilograms to grams, of body mass in allometric equations where the exponent does not equal one is not so simple, however. To convert from m in grams to M in kilograms in Eq. (3-12), we must solve for $1,000^{0.68}$; conversely, if Eq. (3-13) is restated for use with m instead of M, we must solve for $0.001^{0.756}$.) Thus, for comparison with the summary for the seven species of mammals listed in Table 3-3, in watts, Kleiber's 26 metabolic rates are summarized as

$$\dot{H}_{sm} = 3.27 \ M^{0.76} = 0.018 \ m^{0.76}, \tag{3-13a}$$

or as the similar Brody-Proctor equation, shown in Figure 3-1:

$$\dot{H}_{sm} = 3.41 \ M^{0.73}. \tag{3-13b}$$

Note that because of the slight difference in exponents (0.76 vs. 0.68), the discrepancy between coefficients is not the same at the 1-g level (0.037/ 0.018 = 2.06, or 106% too high) as at the 1-kg level (4.06/3.27 = 1.24, or 24% too high). Keep this peculiarity in mind when comparing metabolic rates of reptiles, birds, and mammals, for example. Note also that if the difference between exponents is not statistically significant, the slight divergence means nothing. The magnitude of the effect of various exponents across a wide range in body mass can be seen in Appendix Table 1, Appendix B.

The Derivation of Allometric Equations

With these specific examples behind us, we can look more generally at both the method by which allometric equations are obtained from the available data, and at the limitations of this method.

The data themselves are sometimes collected in studies designed for the study of a particular allometric relationship; more often, however, the data are simply recycled from other studies for added insight into the natural world.

The allometric equations are, as we have already seen, exponential in form. In addition, because a straight line is easier than a curve to generate, describe, and handle, values for both body mass and the dependent variable are transformed to $\log_{10}$. Least-squares regression analysis (or derivation of the "estimating equation") is then performed on these points to determine the slope of the line of best fit. Whereas formerly it took two to three days to run a simple regression with a slide rule or mechanical calculator, with the appearance of programmable electronic calculators the chore lasts but a few minutes. In fact, the programmable calculator might be considered a major breakthrough for comparative physiology, in that it has greatly expedited the process of discovering patterns in the evolution of size.

Biological measurements incorporate a certain amount of variability that stems from technique, experimental conditions, and the state of the animal (psychological, nutritional, reproductive). In general, measurements falling within $\pm 20\%$ of the mean value are not considered extraordinary. Thus variability on an absolute scale is less for smaller animals than for larger animals. For example, metabolic rates for shrew and elephant with 20% variation are 0.176 ± 0.035 watt and $1,422 \pm 284$ watts, respectively; if, however, we consider the $\log_{10}$ of these values,

for the shrew:

$$\log (0.176 - 0.035) \text{ to } \log (0.176 + 0.035) = \bar{1}.1492 \text{ to } \bar{1}.3243;$$

for the elephant:

$$\log (1,422 - 284) \text{ to } \log (1,422 + 284) = 3.0561 \text{ to } 3.2320,$$

we find that the differences between the $+20\%$ and -20% values, 0.1751 and 0.1758, are about the same. This is important because for least-squares regressions to be legitimate, variability must be fairly evenly distributed at both extremes. As we have seen, data transformed to $\log_{10}$ satisfy this requirement (Lasiewski and Dawson, 1969).

Any technique that is so simple to use is bound to have its limitations (see Tanner, 1949; Kleiber, 1950; Zar, 1968; Lasiewski and Dawson, 1969; Yates, 1979; Smith, 1980; Heusner, 1982a,b). First, an impressive mathematical equation, although it summarizes trends in available data, is no substitute for an explanation of those trends. Further understanding can be gained only by delving into the physical dimensions involved (Yates, 1980); that is, exponents must be expressed as functions of the fundamental physical entities. The intricacies of such analyses are discussed in Chapter 5.

Second, logarithmic transformation may yield no better correlation than simple linear regression of untransformed data, and the smaller the body-size range represented by the data, the more likely this is to be true.

Third, log transformations of data can minimize variability in a deceptive way. When correlations are made over a size range of four or more orders of magnitude (10,000-fold difference), variation by a factor of two (real or caused by error) is extremely small by comparison. Even when such varia-tion occurs at one of the extremes of the size range, its effect on the slope of the log-log plot is relatively slight. The point is illustrated in Figure 3-2, which plots the metabolic rates of a hypothetical "hypo-ostrich" (with a metabolic rate half that of a normal ostrich) and a hypothetical "hyper-ostrich" (with a metabolic rate twice that of a normal ostrich). This argu-ment brings us to a related point, namely that the size range of the data has a much greater influence on the form of the regression equation than does the sample size. In other words, the fact that the data for metabolic rates range from quail to ostrich is much more important than the fact that sample size

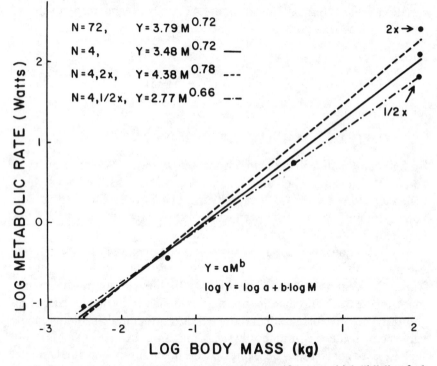

3-2 Basal metabolic rates of birds (except order Passeriformes, which "idle" at fuel consumption rates about 1.6 times those of "nonpasserines") are indicated by the solid line derived from only four species, but with a large size range (quail to ostrich). The line varies by only 8% from the standard equation for 72 species. Even if the ostrich datum were in error by a factor of two either way, the allometric line would be shifted only slightly. This can be taken to indicate either the robustness of the allometric regression or its limitations in predicting specific values. (Redrawn from Lasiewski and Calder, 1971, and converted to Système International—S.I.—units.)

ranges from 4 to 72. The correlation coefficient reflects not only the tightness of the biological relationship between body mass and some other characteristic, but also the range of values for the dependent and independent variables (Smith, 1980).

While allometric equations show the basic patterns, there are obviously differences superimposed by both the phylogenetic legacy of mechanisms and patterns to which a lineage has been committed and by adaptations to particular ecological niches. The allometry gives us the basic pattern and a reference base for studying the phylogenetic and ecological variations, which are examined in Chapter 12.

Logarithmic regressions should not be accepted uncritically, but the limitations of allometric methods need not prevent us from moving forward. At present, these techniques are justified somewhat circuitously by the very fact that they work. Even more convincing is the observation that the allometries of extremely different characteristics are compatible, and that derived regressions correspond with their direct empirical counterparts. Of course, the most powerful support for this method will come from sound theoretical explanations for the patterns expressed by allometric equations.

After running a linear regression of logarithmically transformed data, we want to know how well the resulting equation deals with the variance in the data. The tighter the fit, the more likely it is that the equation describes the underlying natural engineering, and the better it can predict values in advance of additional measurements.

This "goodness of fit" is expressed conveniently as the coefficient of determination (r^2), a fractional value between zero (no correlation) and one (perfect fit). The r^2 value represents the fraction of the variability in Y that is accounted for by X (in this book, M for mass) and is calculated as follows:

$$r^2 = \frac{[\Sigma(X - \overline{X})(Y - \overline{Y})]^2}{[\Sigma(X - \overline{X})^2][\Sigma(Y - \overline{Y})^2]} = \left[\frac{n\Sigma XY - \Sigma X \Sigma Y}{n(n-1)(s_X s_Y)}\right]^2, \tag{3-14}$$

where $\overline{X}$ and $\overline{Y}$ are the respective mean values and s_X and s_Y are standard errors of X and Y. The second form is a convenient way to obtain r^2 from a scientific pocket calculator if this is not built into the calculator and yielded automatically. Values of r^2 for allometric equations that span a logarithmic body-mass range of three or so often exceed 0.75, which means that body mass accounts for three-fourths or more of the range in log Y.

The least-squares regression of log Y on log M in Eq. (3-1a) is the simplest and most widely used estimate of best fit, but it is not the only model and it is not without criticism. This regression yields an unbiased line of best fit only if the independent variable X can be assumed to be free of errors in measurement and free of mutual dependency, with Y, on some other variable. The regression line minimizes the sum of the squares of vertical distances (Y-deviations).

The other models, or methods, are major-axis analysis and reduced-major-axis analysis, neither of which assumes an error-free abscissa. The major-axis model minimizes the sum of the squares of the distances perpendicular to the line, thereby compromising between X-deviations and Y-deviations. The reduced major axis minimizes the sum of the areas of the triangles bounded by the line of best fit and lines parallel to the X- and Y-axes extending from the data points to the best-fit line. When the correlation is strong (high r^2), the results are not appreciably different (Alexander et al., 1979), but there is a consistent bias when r^2 is small.

Each method has its disadvantages, and a good review is given in Harvey and Mace (1982). At present, there are no solutions to some of these problems; hard and fast rules are lacking. One example that Harvey and Mace selected to determine the line of best fit is the relationship between home-range size and body size, a relationship that has been customarily treated by the least-squares regression method. The slopes or allometric exponents obtained thereby tend to be smaller than those yielded by reduced-major-axis analysis. The authors state that they "are not arguing that reduced major axis is the correct model for analysis here. Clearly, since home range size probably depends on body size (via metabolic needs), a regression model *that incorporates measurement error on the abscissa* would be more appropriate" (emphasis in original). They also point out that there is a possibility of producing, via extrapolation, artificial differences in elevation (intercept coefficients, a) when comparing taxa, and they caution against uncritical acceptance of apparent differences in exponents when the coefficients of correlation are low.

Additional Considerations

PRELIMINARY REGRESSIONS FROM LIMITED DATA

There are many situations in which we need an approximate relationship to guide our thinking. How long should we wait for adequate data? It takes only two points to describe a straight line, but if one or both come from an atypical animal, the line will not be representative; the more data points and the wider the size range, the better.

Seeking a rough perspective on the effects of body size upon avian respiration, Lasiewski and Calder (1971) published a tentative relationship between tidal volume (V_t, the volume of air inhaled in one breath) and body mass M, based on values from only six species, some obtained indirectly:

$$V_t = 13.2 \, M^{1.08}. \tag{3-15}$$

This they compared with tidal volume of mammals (Stahl, 1967; $n = 688$):

$$V_t = 7.7 \, M^{1.04}. \tag{3-16}$$

Although tidal volume scaled similarly for birds and mammals, the birds seemed to have greater tidal volumes (1.7 times those of mammals of the same body size). Equation (3-15) was frequently cited in papers by other authors, for whom it provided some framework for comparative purposes.

In 1979 Bech and colleagues added published values from nine additional species of birds, more than doubling the data base. They found:

$$V_t = 16.9 \, M^{1.05}. \tag{3-17}$$

The relationship of Lasiewski and Calder (1971) was thus 22% low; birds are actually pumping not 1.7 but 2.2 times as much air as mammals per resting breath. Qualitatively the picture was the same: birds take deeper breaths, but are affected proportionately in the same way as mammals in meeting the physiological demands of changes in body size. Thus, even while we lacked the quantitatively correct proportionality, we had an approximate scaling factor on which to base comparisons of tidal volume and dead-space volume, lung morphometrics, and so on.

In conclusion, while "the danger in developing allometric relations based on a small sample of partly indirect and estimated values" (Bech et al., 1979) has been overstated, a large number of data points considerably improves the accuracy of comparisons between the two classes. It seems worthwhile at this point to pursue our exploration of the consequences of body size, regarding these allometric equations from limited data points, transformed to log-log coordinates, as the best we can do now, suitable for a preliminary exploration that will doubtlessly be refined again and again in the future.

Unfortunately, a study that states the body mass of its animal subjects is more the exception to the rule than the rule. There is a wealth of Y-data in the literature that lacks the X-values for allometric regression analysis. It has become standard practice to "borrow" body-mass figures for the species being considered from other sources, so that the new physiological, morphological, or ecological data need not languish for lack of analysis. Because we are usually dealing with body-mass ranges of several orders of magnitude, a 25% "error" displacement from true body mass will have only negligible effect on the overall relationship. Hence, in this volume I frequently show an equation as "calculated from Doe, 1945," which means that Doe provided Y and other authors provided X.

The potential for a conceptual linkage between comparative physiology and comparative ecology seems too high to delay an attempt at synthesis until the statistical methods are perfected. Hence I will proceed with regression, but at the same time I recommend that the reader consider other treatments by Kermack and Haldane (1950), Clarke (1980), and Harvey and Mace (1982).

WHY NOT USE MULTIPLE REGRESSION?

When we run a linear regression of logarithmically transformed data, we want to know how well the data fit the resulting equation. The tighter the fit, the more likely it is that the equation (and the underlying engineering through natural selection) can be used to predict values for animals that have not actually been measured.

Fit can be improved by incorporating other factors into multiple regressions (Tanner, 1949; Kleiber, 1950; von Schelling, 1954; Sacher, 1959; Sacher and Staffeldt, 1974). However, once allometric relationships are thus complicated, there is no longer a universal basis for comparison and manipulation of equations, such as by allometric cancellation. For example, we have seen that metabolic rate ($\dot{H}_m$) is proportional to $M^{3/4}$, and we will see in a later chapter that life span (t_{ls}) is almost proportional to $M^{1/4}$. The allometric cancellation, as follows:

$$\frac{(\dot{H}_m)(t_{ls})}{M} \propto \frac{(M^{3/4})(M^{-1/4})}{M^{1.0}} \propto M^{-0} \tag{3-18}$$

suggests that in a lifetime the total energy metabolism per kilogram of body mass is size independent (M^0). Harris and Benedict (1919; cited in Kleiber, 1961), however, found that metabolic rates (in kilocalories per day) of men and women can be predicted as follows from M, height in centimeters, and age in years:

men: $\dot{H}_m = 66.4730 + 13.7516\,M + 5.0033\,(\text{height})$
$$- 6.7750\,(\text{age}); \tag{3-19}$$

women: $\dot{H}_m = 65.0955 + 9.5634\,M + 1.8496\,(\text{height})$
$$- 4.6756\,(\text{age}). \tag{3-20}$$

These equations incorporate refinements to handle variability due to aging, obesity, or leanness. For wild animals additional terms for reproductive state and season could be included. In addition, Sacher (1976) found an excellent correlation for mammalian life span:

$\log t_{ls} = 0.62$ (log brain weight) $- 0.41$ (log adult body weight)
$$- 0.52 \text{ (log specific metabolic rate)}$$
$$+ 0.026 \text{ (body temperature)} + 0.90. \qquad (3\text{-}21)$$

What basic generalizations about longevity or lifetime energetics can be abstracted from the complexity of these relationships?

$$\frac{(\dot{H}_m)(t_{ls})}{M} = \frac{1}{M} \times [66.47 + 13.75\, M + 5.00 \text{ (height)} - 6.78 \text{ (age)}]$$

$$[0.62\, M_{brain} - 0.41 \log M - 0.52\, (\dot{H}_m/M)$$
$$+ 0.03 \text{ (body temperature)} + 0.90]. \qquad (3\text{-}22)$$

In general, it is body mass (how much animal tissue must be sustained and regulated), rather than mass of its constituent parts, topographical layout, or history of use, that determines basic support costs, opportunities, and homeostatic needs. Thus, body mass is not only an expedient measure of size, it is a biologically appropriate one. Because the provision for size, and the consequences of size, are the focus of this study, body mass, logarithmic transformations, and least-squares regressions are the descriptive tools necessary for understanding the evolution of size. Greater precision might have been obtained by considering other factors as well, but in the interest of finding allometric patterns I have accepted cruder approximations based on body size only.

This does not mean that there is anything wrong with multiple regressions, for indeed the remaining scatter does beg statistical explanation. With this as a starting point, Cabana and colleagues (1982) proceeded to show that the clutch masses of birds are described better when the adult masses of both the males and the females are considered.

Stearns (1983 and in press) has made effective wider use of multiple variables in principal-component analysis of life-history traits, to show the constraints of both size and phylogeny on life-history traits of mammals and reptiles. This type of analysis is dependent on the availability of data sets that are complete, for all the traits being considered, in each of several species. Since there are few species that have been fully measured, multiple-variable regression is not yet suitable for the goal of this book: to integrate morphology, physiology, reproduction, growth, and ecology by means of body-size analysis, using existing resources in the form $Y = aM^b$.

POTENTIAL HAZARDS

We seek patterns in nature; for it is pattern, not bulk of data, that leads to understanding. With such a premium on pattern and its identification, we can become biased and prone to forcing the data, to confusing correlation with cause, and to attaching unreal significance to coincidence.

The chances of committing these errors can be reduced by rigorous application of dimensional analysis. If there is an allometric relationship between metabolic rate and body mass, one might anticipate that a similar power function or exponent would describe the relationships (1) between body size and capacity of the animal to capture or harvest the food that provides the energy to support the metabolism; (2) between body size and processing capacity of the teeth or the digestive system; and (3) between body size and area of the home range from which the animal obtains its food. When a regression yields an exponent approaching the "magic three-fourths" of Kleiber's Rule, the temptation to seize it as evidence of the logically expected interrelationship is great; likewise, an exponent significantly different from three-fourths can lure us into attaching special significance to the body size at which two different regression lines cross. If the two lines have different dimensions, however, where they cross is merely a consequence of how they are plotted. Work incorporating obvious errors of this sort has been published. Empirical power functions can scarcely be analyzed amid such dimensional confusion.

Practical Applications

Examples of allometry selected so far have been simple descriptions of isolated variables. If these descriptions are reasonably accurate, they can be substituted in known relationships such as the Fick Principle to derive allometric statements when data for direct analysis are not available. Allometric equations also are useful for predictions or first approximations of animal requirements.

In order to grasp the basics of nature, we must initially study one function, organ, or level at a time; still, it is obvious that in the complex living state, all functions are interrelated. The metabolic fire requires both fuel and oxygen, so ultimately we must relate it to cellular biochemistry, to ecological feeding strategies, and to the logistics of respiration and circulation.

The need for considering these interrelationships is more widely appreciated by biologists looking at the metabolic needs of an individual or a species. What is not universally realized is that the interrelationships can be analyzed by means of allometric expressions for groups as large as a phylogenetic class or subclass. The possibilities for allometric correlations were elegantly set forth by Edward Adolph over three decades ago, but his observation is unfamiliar to many recent recruits to allometry. In 1949 he stated:

> Many physiological properties go hand in hand with one another; their "determination" is reciprocal and not unique. It seems likely that an organism is an integrated system by virtue of the fact that none of its properties is entirely uncorrelated, but that most are demonstrably interlinked; and not just by single chains, but by a great number of criss-crossed linkages.

By means of allometric equations and an alignment chart, Adolph demonstrated this fact quantitatively for the eutherian mammals. Later on, Stahl (1962, 1963, 1967) extended the allometric correlation of physiological and morphological variables in mammals with multiplication and division (allometric cancellation of two or more different expressions).

Before we explore allometric cancellations, we should perhaps review the rules for manipulation of exponential expressions:

$$(M^{b_1})(M^{b_2}) = M^{b_1+b_2},$$
$$M^{b_1}/M^{b_2} = M^{b_1-b_2} = 1/M^{b_2-b_1},$$
$$(M^{b_1})^{b_2} = M^{b_1 b_2}.$$

While individual studies may go only as far as generating an equation for the quantitative dependence of a single physiological variable on body size, we can also relate one process to another quantitatively by combining two or more allometric equations. For example, the rate of clearance from the blood of the substance para-amino hippurate (PAH) by the kidney can be used as a measure of plasma flow through the kidney. Adolph (1949) related PAH clearance to body mass as

$$\text{ml plasma/hr} = 5.4 \ m^{0.80} = 1,356 \ M^{0.80}. \tag{3-23}$$

Thus it is possible to derive scaling relationships by substituting into equations involving multiplication or division allometric expressions that differ in mass exponents. For example, the scaling of average metabolic turnover times is estimated by dividing a capacity ($\propto M^{1.0}$, such as total body water or gut volume) by a flow rate ($\propto M^{3/4}$, such as rate of urine production or food intake). However, quantities cannot be added or subtracted unless they share the same dimensions (Stahl, 1962; Riggs, 1963). When we make comparisons or hypothesize compromises between theoretical scalings that differ significantly in exponents (for example, $M^{1/2}$ for heat loss, $M^{2/3}$ for surface area, and $M^{1.0}$ for gravitational force), we are effectively adding or subtracting. The Gompertz equation used to describe growth or mortality (Eqs. 10-36 and 11-50) involves subtraction.

This confronts us with the distinction between "ratio numbers" suitable for allometric equations, as discussed by Stahl (1962), and the dimensional homogeneity required for meaningful addition or subtraction. In these instances one can sidestep the problem by using an approximation method, which we shall do in Chapters 10 and 11. Each of the individual allometric relationships is solved for a homoscedastic series of hypothetical sizes (10 g, 100 g, 1,000 g, and on up). These values are substituted into the complex equation, which is then solved for each of the hypothetical sizes. The final figures are logarithmically transformed and submitted to least-squares regression to yield a single aM^b expression which approximates the scaling that might result from complex consideration.

Stahl (1967) subsequently related cardiac output (rate of blood pumping by the heart) to body mass as

$$\text{ml blood/min} = 187\, M^{0.81}. \tag{3-24}$$

If we assume that the plasma volume is 55% of the blood volume (hematocrit = 45%), blood flow to the kidney is

$$\text{ml blood/min} = (1{,}356 \div 0.55 \div 60 \text{ min/hr})\, M^{0.80} = 41.1\, M^{0.80}.$$

Thus the proportion of blood flow that goes to the kidney in eutherian mammals is

$$41.1\, M^{0.80}/187\, M^{0.81} = (22\%)(M^{-0.01}). \tag{3-25}$$

When the confidence intervals of the original equations are considered, the residual mass exponent (r.m.e., or difference between exponents in an allometric cancellation) of $M^{-0.01}$ is insignificant, and the proportion of blood flow to the kidney can be regarded as essentially the same for mammals ranging in size from shrew to elephant. The effects of various r.m.e. values across such a range in body size may be appreciated from Appendix Table 1.

Similarly, Stahl (1967) showed that the heart generally beats 4.5 times per breath in resting mammals, regardless of their size (and remarkably close to the five heart compressions to one mouth-to-mouth breath in emergency cardiopulmonary resuscitation of humans):

$$241 \ M^{-0.25}/53.5 \ M^{-0.26} = 4.5 \ M^{0.01}. \tag{3-26}$$

Relating breathing frequency to life span, we see that the average lifetime of terrestrial mammals of all sizes is long enough for 200 million resting breaths. These derived relationships are seen to be "design criteria" for evolutionary changes in size among the Eutheria; we could proceed to compare these criteria with those for other groups of vertebrates.

Such insight into the constraints on evolution is not limited to the study of comparative physiology, but is beginning to be appreciated in the ecological context as well. Western (1979) pointed out: "Those life history variables which are scaled to size will also be correlated with each other and will therefore not be independent variables but rather will be mutually entrained by size."

The interrelationship of animal characteristics thus gives us considerable ability to predict, via derived relationships, values for functions that have not even been measured yet. With this mechanical background on the basics of allometry, we are ready to return to the scaling of body components. The five problems in Appendix C will serve to check your assimilation of the preceding material before you go on.

Body Compartments — An Allometric Summary

We return now to the topic of the previous chapter and summarize our scaling in proper allometric form (Table 3-4). To repeat, some of the bird equations cannot be taken as more than a first approximation for comparing allocations within the bodies of birds and mammals. The exponents are perhaps of greater interest. If the allometry for both birds and mammals has the same exponent, then size must be exerting the same proportional effect. If the exponents are not significantly different, then we can take a ratio of the coefficients to see how absolute organ sizes compare on an equal-body-size basis (see Appendix B). For example, note that the liver and kidneys have essentially the same allometric exponents. The coefficients in bird and mammal expressions are also the same, so we can generalize widely that birds and mammals have the same liver sizes. Bird kidneys tend to be 20% larger than those of mammals of the same size. Both of these organs can be said to scale hypoallometrically, that is, contribute a decreasing proportion

Table 3-4 Allometric comparison of the internal proportions of body mass in mammals and birds. (From Figures 2-1 and 2-2.)

Component	Body mass in —		Mammal/bird	n_M, n_B	r_M^2, r_B^2	References
	Mammals	Birds				
Fat	0.075 $M^{1.19}$	—	—	54,—	0.969,—	Calculated from Pitts and Bullard, 1968 and unpublished
Eyes	—	0.0069 $M^{0.70}$	—	—,50	—,0.80	Calculated from Quiring, 1950
Brain	0.011 $M^{0.76}$	0.007 $M^{0.58}$	1.56 $M^{0.18}$	309,180	0.96,0.93	Brody, 1945; Martin, 1981
Thyroid	0.0001 $M^{0.92}$	0.0001 $M^{0.86}$	1.11 $M^{0.06}$	>100,—	0.947,0.889	Brody, 1945
Gonads, male	—	0.0077 $M^{0.76}$	—	—,8	—,0.80	Calculated from Ricklefs, 1974
Gonads, female	—	0.059 $M^{0.92}$	—	—,7	—,0.952	Calculated from Ricklefs, 1974 (ovary plus oviduct)
Mammaries	0.045 $M^{0.82}$	—	—	14,—	—,—	Hanwell and Peaker, 1977
Adrenal	0.0003 $M^{0.80}$	0.0002 $M^{0.89}$	1.54 $M^{-0.09}$	>100,—	0.935,0.922	Brody, 1945
Kidneys	0.007 $M^{0.85}$	0.009 $M^{0.91}$	0.83 $M^{-0.06}$	138,334	0.980,—	Brody, 1945; Turcek, 1966
Liver[a]	0.033 $M^{0.87}$	0.033 $M^{0.88}$	1.00 $M^{-0.01}$	175,—	0.984,0.955	Brody, 1945
Gut	0.075 $M^{0.94}$	0.090 $M^{0.99}$	0.83 $M^{-0.05}$	41,—	0.925,0.939	Brody, 1945
Integument[b]	0.134 $M^{0.92}$	0.142 $M^{0.92}$	0.94 $M^{0.00}$	8,5	—,—	Pace et al., 1979; Lindstedt and Calder, 1981; Hiebert and Calder, unpublished
Integument[c]	0.106 $M^{0.94}$	0.142 $M^{0.92}$	0.75 $M^{0.02}$	42,—	0.992,—	Calculated from Pitts and Bullard, 1968 and unpublished

						Reference
Fur, feathers	—	0.032 $M^{0.98}$	0.064 $M^{0.95}$	0.50 $M^{0.03}$	40,—	0.885,—	Calculated from Pitts and Bullard, 1968 and unpublished; Turcek 1966
Skeleton	0.061 $M^{1.09}$	0.065 $M^{1.07}$	0.93 $M^{0.02}$	49,311	0.992,0.993	Prange et al., 1979
Spleen	0.003 $M^{1.02}$	0.001 $M^{0.92}$	1.72 $M^{0.10}$	55, 5	—,0.904	Brody, 1945; Calder, 1976
Lungs	0.011 $M^{0.99}$	0.013 $M^{0.95}$	0.89 $M^{0.04}$	>100, 51	0.922,0.85	Stahl, 1967; Lasiewski and Calder, 1971
Heart[d]	0.006 $M^{0.98}$	0.009 $M^{0.94}$	0.68 $M^{0.04}$	568,—	0.980,—	Stahl, 1967; Kilgore, personal communication
Blood	0.069 $M^{1.02}$	0.090 $M^{0.99}$	0.77 $M^{0.03}$	840, 11	0.990,0.902	Calculated from Bond and Gilbert, 1958 (very small sample); Stahl, 1967; Cohen, 1967; Nirmalen and Robinson, 1972
Connective tissue	—	0.06 $M^{0.98}$	—	—, 2	—[e]	—
Muscle	0.450 $M^{1.0}$	0.48 $M^{1.08}$	1.02 $M^{-0.08}$	14, 5	0.974[e]	Munro, 1969; Hiebert and Calder, unpublished
Muscle	0.468 $M^{0.99}$	0.48 $M^{1.08}$	0.98 $M^{-0.09}$	42, 5	0.992[e]	Hiebert and Calder, unpublished; calculated from Pitts and Bullard, 1968 and unpublished
Flight muscles	—	0.35 $M^{1.04}$	—	—, 41	—,0.980	Hartman, 1961

a. Prothero (1982) summarized liver mass as 0.035 $M^{0.89}$ for 114 species of mammals.

b. Includes fur or feathers.

c. Body mass minus fur and gut contents, so regression exaggerates fraction of field mass.

d. Heart of marsupials = 0.0077 $M^{0.98}$ (Dawson and Needham, 1982).

e. Very small sample; not significant, but no other information.

of body mass in progressively larger animals. This has been referred to in the past as negative allometry (Huxley and Tessier, 1936), a term I prefer to avoid because of possible confusion with a negative exponent, which signifies an inverse relationship (absolute decrease), as contrasted to only a proportional decrease. In contrast, skeletons of both classes scale hyperallometrically, the total skeletal requirements being indistinguishable in both classes, as noted previously, but amounting to about 5% of a shrew's or a hummingbird's mass, 8% of a swan's, and 27% of an elephant's.

The mammal/bird skeletal ratio, $0.93\ M^{-0.02} \approx 0.93$ suggests that the mammal skeleton is actually 7% lighter than is the bird skeleton. This apparent difference was shown by means of analysis of covariance to be statistically insignificant at the 0.01 level (Prange et al., 1979), a finding that will be examined in more detail in the next chapter.

The scaling of the integument ($M^{0.92}$ to $M^{0.95}$) is in good agreement with theoretical expectation. The mass of the integument should be proportional to the product of surface area ($M^{0.67}$) and thickness. If the thickness is a characteristic linear dimension which conforms to the elastic similarity model, it would be scaled as $M^{0.25}$; if geometrically similar, it would scale as $M^{0.33}$:

$$\text{(area)(thickness)} \propto (M^{0.67})(M^{0.25}) = M^{0.92}$$
$$(M^{0.33}) = M^{1.00}. \tag{3-27}$$

The allometry of gonad size has not been analyzed for mammals on a class-wide basis, but the tabulation for primate testicular mass by Harcourt and colleagues (1981) shows:

$$M_{testes} = 0.0037\ M_b^{0.67}; \quad r^2 = 0.604; \quad n = 33\ \text{spp.} \tag{3-28}$$

while a limited sample of bird data tabulated by Ricklefs (1974) yields:

$$M_{testes} = 0.0077\ M_b^{0.76}; \quad r^2 = 0.800; \quad n = 8\ \text{spp.} \tag{3-29}$$

$$M_{ovary+oviducts} = 0.059\ M_b^{0.92}; \quad r^2 = 0.952; \quad n = 7\ \text{spp.} \tag{3-30}$$

This limited sample is similar in scaling to, or not distinguishable from, that of other glandular organs and the brain.

The most variable component of the total body mass is, of course, body fat. I have gathered the allometric equations in Table 3-4 from a number of sources, not with the goal of accounting for the composition of the whole body, but rather of studying one or a few organs. If we calculate the mass of each component for a series of animal sizes (as I did to construct Figures 2-1 and 2-2) and add up all components except the variable amount of fat, we account for 97.5% of a 7-g mammal, but for only 81% of a 6,600-kg mammal. Is the balance all fat? If so, this means that the larger the animal, the greater its proportional stored reserve of energy.

Allometrically, this unaccounted-for amount of body mass, scaled for a series of eight body sizes from shrews to elephants, is

$$M_? = 0.09 \, M^{1.13}. \tag{3-31}$$

Because this expression was derived as a composite of several previous "best-fits," an expression of variance here would be small but meaningless. It does provide a basis for comparison with the allometry of total body fat, as extracted with ethyl ether from carcasses of wild mammals collected in Alaska, Virginia, and Brazil, and taken from published reports on four domestic species, including men (Pitts and Bullard, 1968). I have excluded cetaceans and marsupials from this regression, for consistency with other equations for terrestrial eutherian mammals:

$$M_{fat} = 0.075 \, M^{1.19}; \qquad n = 54; \qquad r^2 = 0.969. \tag{3-32}$$

The standard error of the exponent, s_b, = 0.03. At the 95% level of confidence, Eq. (3-32) does not differ from Eq. (3-31). Furthermore, Eq. (3-31) predicts within 8% of the observed fat composition of mouse, cat, normal man, and horse — although it predicts only 21% of fat amounts observed in shrews. Thus, in general it seems safe to assume that the portion of body mass not accounted for by the sums of the contributions to mass by the various organs represents a normal amount of body fat. This hyperallometry ($\propto M^{1.13}$ or $\propto M^{1.19}$) of fat reserves, which will be used at a hypoallometric metabolic rate ($\propto M^{0.75}$ or less) has interesting consequences for the advantage of larger body size, as expressed in Bergmann's Rule (see Chapter 8).

Brain Size

How much brain tissue is needed to ensure the successful integration of physiology and behavior? With the evolution of greater body mass, there was more animal for the brain to govern, but the brain did not expand in linear proportion to body mass. For example, a 1-g hamster's brain controls 107 g of its body, whereas a gram of an elephant's brain rules, on the average, 1,160 g of the elephant; and a hummingbird's brain controls 24 times its mass, whereas an ostrich's brain supervises a mass 2,900 times its own (from tabular data of Quiring, 1950). What factors affect the scaling of the brain? On the one hand, we could reason that since a mouse and an elephant perform the same number of functions, a brain of a constant size would be sufficient to control either of them. In other words, the size of the brain could be mass independent ($\propto M^0$) in the same way that an ignition switch need be no bigger for a large bus than for a subcompact car. A study of brain weights, however, indicates that this is not the case. Alternatively, we could speculate that because the elephant lives longer and accumulates more information during its lifetime (since, of course, elephants don't forget), it could use a brain at least proportional to that lifetime, which in turn is proportional to $M^{1/4}$ (see Chapter 6). However, observed brain weights are not consistent with this idea either.

In fact, the scaling of the brain among the vertebrates differs with class. The mass of the reptilian brain is approximately proportional to $M^{0.5}$. The body-mass exponent for avian brains is not statistically distinct from that for reptile brains, although at any body size the avian brain is larger (Jerison, 1968; Martin, 1981; see also Table 3-5). Birds and reptiles seem to establish a functional minimum scaling for brain vs. body size.

Birds weighing less than 75 g tend to have larger brains than their mammalian counterparts, but the exponent for mammalian brains is significantly greater. Most previous studies of mammal brain size have yielded exponents close to those for body surface area ($M^{0.65}$ to $M^{0.70}$).

Although the components of the brain await separate allometric analysis, we know that in mammals the cerebral hemispheres dominate in mass as well as in function. The area of the cerebral cortex (the surface layer composed of sensory, motor, and association neurons) has been increased by the folding of the cortex into sulci and gyri. A smooth cerebral cortex would have a surface area proportional to its hemispherical volume ($V_{1/2}$):

$$A_{1/2} \propto V^{0.67}. \tag{3-33}$$

The normal foldings of the brain, however, increase the surface area out of proportion to the normal surface/volume ratio. In a study of marsupials, carnivores, cetaceans, and man, Elias and Schwartz (1969) derived the following relationship:

$$A_{brain} = 3.2 \, V_{brain}^{0.93}. \tag{3-34}$$

Combined, Eqs. (3-33) and (3-34) suggest that the degree of folding has the following relationship to brain volume:

$$A_{brain}/A_{1/2} \propto V_{brain}^{0.26}. \tag{3-35}$$

However, a more recent analysis of a larger sample (305 species of mammals) by Martin (1981) yields a scaling for brain mass similar to that

Table 3-5 Allometry of the brain.

Taxon	Brain size	n	r	S_b	S_r	References
Mammalia	g brain mass = $9.96\,M^{0.697}$ ($0.081\,m^{0.697}$)	>200 indiv.	0.973	—	+64%, −39%	Brody, 1945
	g brain mass = $11.3\,M^{0.664}$ ($0.115\,m^{0.664}$)	16 spp.	—	±0.012	—	Jerison, 1961
	g brain mass = $0.129\,m^{0.666}$	63 spp.	—	±0.025	$S_{x.y}^2 = 0.0668$	Sacher, 1959
	g brain mass = $14.3\,M^{0.645}$ ($0.166\,m^{0.645}$)	202 spp.	0.93	—	—	Economos, 1980
	g brain mass[a] = $0.059\,m^{0.76}$	309 spp.	0.96	±0.03	—	Martin, 1981
	cm² brain surface[b] = $3.2\,V_{br}^{0.93}$	15 spp.	—	—	—	Elias and Schwartz, 1969
Chiroptera	cm³ cranial capacity = $0.043\,m^{0.80}$	225 spp.	—	—	—	Eisenberg and Wilson, 1978
By families	cm³ cranial capacity[c] $\propto m^{0.57\text{-}0.74}$	—	—	—	—	Eisenberg and Wilson, 1978
Frugivores	cm³ cranial capacity = $0.089\,m^{0.74}$	20 spp.	—	—	—	Eisenberg and Wilson, 1978
Insectivores	cm³ cranial capacity = $0.049\,m^{0.73}$	45 spp.	—	—	—	Eisenberg and Wilson, 1978
Felidae	g brain mass[d] = $0.227\,m^{0.611}$	6 spp.	—	—	—	Davis, 1962
Aves	g brain mass = $6.60\,M^{0.498}$ ($0.212\,m^{0.498}$)	—	0.906	—	+59%, −37%	Brody, 1945
	g brain mass = $0.198\,m^{0.51}$	53 spp.	—	—	—	Calder, 1976
	g brain mass[a] = $0.129\,m^{0.58}$	180 spp.	0.93	±0.04	—	Martin, 1981
Reptiles	g brain mass[a] = $0.017\,m^{0.54}$	59 spp.	0.96	±0.04	—	Martin, 1981

a. Martin used major-axis regressions.
b. Four marsupials, three carnivores, one primate, seven cetaceans.
c. Mean exponent, eight families: 0.65 ± 0.06.
d. Derived as "log brain weight" vs. log (body weight − heart weight).

for metabolic rates:

$$m_{brain} = 0.059 \, m^{0.76}. \tag{3-36}$$

On the basis of this, Martin suggests that the brain may be scaled to maternal metabolic rates, although he reminds us that correlation does not necessarily indicate a causal connection. Attempts by others to ascribe a life span – limiting function to relative brain size or "encephalization index" have not been so cautious (see Calder, 1976; Lindstedt and Calder, 1981). Brain-mass scaling seems to attract many speculations, particularly those guided by correlation coefficients.

Checks on Allometric Predictions for Body Components

The allometric equations given in Table 3-4 come from such a diversity of sources and studies of just a few organs at a time that it is appropriate to compare their predictions with what has been dissected out in the few relatively complete studies available. The ratios of actual measurement to predicted value are given in Table 3-6. Of the 56 predictions, 21 are within 10%, that is, if Economos' (1980) equation for brain size is used. The greatest discrepancy is the prediction for spleen size for the horse, apparently because of hypertrophy in the animals dissected (14 of 20 having been diseased or injured). While the heart size of the goat confirms the prediction well enough, the dog, deer, and horse had larger hearts than anticipated, which may reflect the needs of running animals that can sustain high speeds for extended periods. The skeletal muscle masses all exceeded the predictions except in the wapiti, seemingly appropriate for a predator and a horse, but not intuitively obvious for the case of the goat (Toggenburg domestic variety). The relative hyperdevelopment of heart and muscle is not paralleled in the lung size of the domestic dog, goat, and horse.

The skeleton of the horse was considerably heavier than predicted, but the prediction was based on dry bones (museum specimens), which we will see in Chapter 4 are 55% to 60% of fresh mass, in general, as in the case of the horse. The dog and goat bones had been autoclaved to aid in removal of adhering soft tissue, which may have also taken out some of the moisture; nevertheless, it appears that the goat is very lightly supported by its skeleton. However, being a ruminant, the goat has more of its mass invested in the gut, compared to the small gut of the dog, which feeds on highly concentrated food and needs less intake of food mass to get an equivalent amount of energy. The salt and protein content of the meat diet could explain the

Table 3-6 Ratios of actual body compositions of eutherian mammals to predictions for body mass. Average body mass of animal is given in parentheses.

Body component	Dog (15.89 kg)	Goat (33.01 kg)	Deer (63 kg)	Wapiti (233 kg)	Horse (258.31 kg)
Fat	—	—	—	1.26[b]	0.32
Eyes	—	—	—	—	—
Brain[a]	0.94	0.84	0.92	0.73	1.05
Thyroid	—	—	1.25	0.94	0.92
Adrenal	—	—	0.76	0.39	0.67
Kidneys	1.29	1.00	0.69	0.61	1.32
Liver	1.04	1.09	0.99	0.68	0.99
Gut	0.68	1.27	—	1.00	1.17
Skin	0.88	1.11	—	0.71	1.06
Skeleton	1.15	0.83	1.13	1.10	1.81
Spleen	0.89	0.90	—	—	3.96
Lungs	0.82	1.00	1.71	1.34	1.03
Heart	1.39	1.06	1.54	1.05	1.43
Blood	—	—	—	—	1.06
Muscle	1.24	1.16	1.00	0.98	1.14

Data source: Davis et al., 1975; Meadows and Hakonson, 1982; Webb and Weaver, 1979.

a. Because the equation of Martin (1981) tended to overestimate, the equation of Economos (1980) was used.

b. Includes fat, cartilage, and connective tissue.

large kidneys of the dog, but the disproportionately large kidneys of the horse have no ready explanation.

This comparison has overlooked the fact that the actual values may have fallen within the confidence intervals of the allometric equations in many cases. Even though those intervals have not been stated for most of the equations, we can get a crude idea of what to expect in the way of the predictions from allometric equations.

Form and Structural Support

4

THE CLASSIC APPLICATIONS of exponential or log-log expressions to egg sizes and metabolic rates, which we examined in Chapter 3, are a half-century old. Allometric thinking goes back considerably farther in science, at least three and one-half centuries, to Galileo Galilei.

Galileo, astronomer and son of Vincenzo Galilei, composer of music for the lute, realized that increases in bone diameter had to exceed increases in bone length. In his famous bone-scaling problem he used a proportionality of

$$d \propto l^2; \quad l \propto d^{1/2}. \tag{4-1}$$

This would provide a margin of safety, in a size-independent resistance to static stress. Otherwise an animal's bones could be crushed under its own body mass. A scaling that would preserve a similarity in resistance to this static stress regardless of animal size would fit a static-stress model.

In 1726 Jonathan Swift, under the pseudonym of Lemuel Gulliver, published *Travels into Several Remote Nations of the World*. The exercise in scaling by which Gulliver's needs for food and drink were calculated by the Lilliputian mathematicians is a favorite classic error (Thompson, 1917,

1961; Moog, 1948; Kleiber, 1967; Schmidt-Nielsen, 1972): "Having taken the height of my body by the help of a quadrant, and finding it to exceed theirs in the proportion of twelve to one, they concluded from the similarity of their bodies, that mine must contain at least 1728 of theirs, and consequently would require as much food as was necessary to support that number of Lilliputians." This calculation was based upon a simple geometric scaling, with all linear dimensions increased by the same proportions,

$$d \propto l. \tag{4-2}$$

Since body mass is proportional to volume,

$$M \propto V \propto l\pi(d/2)^2 \propto ld^2. \tag{4-3}$$

Substituting (4-2) in (4-3):

$$M \propto l^3. \tag{4-4}$$

Hence the Lilliputian calculation of $12^3 = 1,728\ x$, compared to what Galileo would have calculated to be 248,832 as

$$M \propto ld^2 \propto l(l^2)^2 \propto l^5. \tag{4-5}$$

Both Galileo and Swift predated Kleiber's metabolic scaling of $M^{3/4}$, so Gulliver's food requirements according to Galilean static-stress scaling and Kleiber would be $248,832^{3/4} = 11,141\ x$.

This suggests that Gulliver received only 16% of his food requirement. However, the application of engineering principles to the scaling of support has contributed the elastic similarity model (McMahon, 1973, 1975a,b). In order to have sufficient elastic recoil to resist buckling under self-loading, the required proportionality for torso and limbs is

$$d \propto l^{3/2} \tag{4-6}$$

and

$$M \propto l(l^{3/2})^2 \propto l^4. \tag{4-7}$$

The basis for this model is shown in Figure 4-1.

Had the Lilliputians been versed in elastic similarity and Kleiber's metabolic allometry, they would have calculated:

$$M \propto 12^4 = 20,736\ x, \qquad \dot{E}_m \propto 20,736^{3/4} = 1,728\ x.$$

Thus Swift and/or the Lilliputians were correct, but only because their error in assuming a geometric scaling was exactly offset by their ignorance of $M^{3/4}$ metabolic scaling (Harris, 1973; McMahon, 1980)!

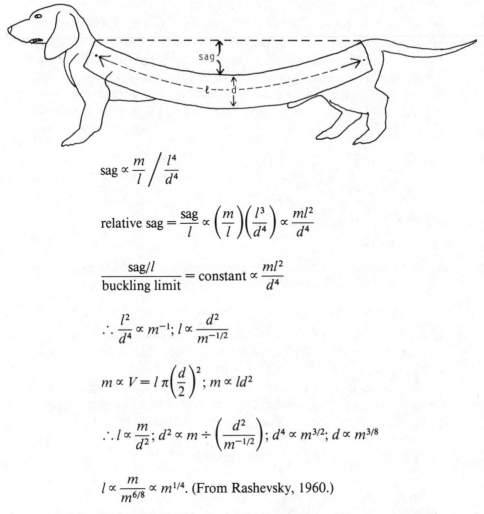

$$\text{sag} \propto \frac{m}{l} \Big/ \frac{l^4}{d^4}$$

$$\text{relative sag} = \frac{\text{sag}}{l} \propto \left(\frac{m}{l}\right)\left(\frac{l^3}{d^4}\right) \propto \frac{ml^2}{d^4}$$

$$\frac{\text{sag}/l}{\text{buckling limit}} = \text{constant} \propto \frac{ml^2}{d^4}$$

$$\therefore \frac{l^2}{d^4} \propto m^{-1}; \; l \propto \frac{d^2}{m^{-1/2}}$$

$$m \propto V = l\,\pi\left(\frac{d}{2}\right)^2; \; m \propto ld^2$$

$$\therefore l \propto \frac{m}{d^2}; \; d^2 \propto m \div \left(\frac{d^2}{m^{-1/2}}\right); \; d^4 \propto m^{3/2}; \; d \propto m^{3/8}$$

$$l \propto \frac{m}{m^{6/8}} \propto m^{1/4}. \text{ (From Rashevsky, 1960.)}$$

4-1 "It is well recognized that certain breeds of dogs are more prone to disc protrusions than others. Dachshunds are particularly commonly affected and have been shown to develop degeneration of their disc nuclei at a very early age" (Christoph, 1975). The problem is somewhat exaggerated in this sketch, which illustrates the consequences of increasing linear dimension without adequately increasing diameter to prevent sagging or buckling. The elastic similarity proposed by Rashevsky (1960) and McMahon (1973, 1975a,b) would have made life better for dachshunds.

The elastic similarity model predicts many allometric exponents much better than does either a geometric or a static-stress similarity model (Table 4-1). Nevertheless, some exceptions to elastic similarity have been noted, and the case is not unequivocal (Table 4-2). McMahon's elastic similarity model also follows Rashevsky's (1960) dimensional derivation for the relationship between columnar lengths and body mass (Figure 4-1),

$$l \propto d^{2/3} \propto (m^{3/8})^{2/3} \propto m^{1/4}, \tag{4-7a}$$

as examined empirically with measurements of limb bones.

The above discussion has considered only the static support of mammals and has ignored the greater dynamic stresses on moving animals. Hill (1950) observed that "similar animals should carry out similar movements not in the same time but in times proportional to their linear dimensions." In terms of the elastic similarity model,

$$t \propto l \propto m^{1/4}. \tag{4-8}$$

The frequency of cyclically repeated movements should then be

$$f \propto 1/t \propto m^{-1/4} \tag{4-9}$$

and, indeed, many do scale this way. However, there are other similarity models to consider with regard to movement which may predict scaling as in (4-8) and (4-9) (Alexander, 1982; Chapter 7).

Endoskeletons

The two basic functions of the vertebrate skeleton are to serve as support against gravity and as the rigid framework with which the contracting muscles can accomplish articulated movement. When we relate skeletal mass to body mass, as in Table 4-3, and then to structural strength, we assume that the composition and strength per gram of bone is the same throughout the size range from shrew to elephant. Also, when we reconstitute the body allometrically, or account for total body mass as in Table 3-4 and Figures 2-1 and 2-2, we are using available equations from a variety of techniques and sources.

Note that the scaling equations for skeletal mass have been obtained from cleaned, dried museum specimens. How much of the live body mass was water and other components of bone and marrow tissues? Considerations of either body composition or structural support might profitably be preceded by comparison of museum and fresh bone, made possible by the extensive

Table 4-1 Three alternative similarity models. (After McMahon, 1975b.)

Characteristic	General formula	Similarity formulas			Observed	Observer
		Geometric	Elastic	Static stress		
Proportions	$l \propto d^x$	$x = 1$	$x = \tfrac{2}{3}$	$x = \tfrac{1}{2}$	$x = 0.66$, Artiodactyla	McMahon, 1975a
Lengths	$l \propto M^{x/(x+2)}$	$l \propto M^{1/3}$	$l \propto M^{1/4}$	$l \propto M^{1/5}$	height $\propto m^{0.24}$, cattle; $l \propto m^{0.35}$, mammal bones	Brody, 1945; Alexander et al., 1979
Diameters	$d \propto m^{1/(x+2)}$	$d \propto m^{1/3}$	$d \propto m^{3/8}$	$d \propto m^{2/5}$	girth $\propto m^{0.36}$, cattle; $d \propto m^{0.36}$, mammal bones	Brody, 1945; Alexander et al., 1979
Total body surface	$A_b \propto dl$	$A_b \propto m^{2/3}$	$A_b \propto m^{5/8}$	$A_b \propto m^{3/5}$	$A_b \propto m^{0.65}$, mammals	Stahl, 1967
Cross-sectional area	$A_d \propto d^2$	$A_d \propto m^{2/3}$	$A_d \propto m^{3/4}$	$A_d \propto m^{4/5}$	$A_{aorta} \propto m^{0.72}$, mammals	McMahon, 1975a

Period for first geometry	$t \propto m/A_d$	$t \propto m^{1/3}$	$t \propto m^{1/4}$	$t \propto m^{1/5}$	$t \propto m^{0.25}$, mammal heart	Stahl, 1967
Frequency for first geometry	$f \propto A_d/m$	$f \propto m^{-1/3}$	$f \propto m^{-1/4}$	$f \propto m^{-1/5}$	$f \propto m^{-0.25}$, mammal heart	Stahl, 1967
Muscle work per stroke	$W \propto \tau_{muscle} A \cdot \Delta l$	$W \propto M$	$W \propto M$	$W \propto M$	skeletal muscle, $W \propto M$	Hill, 1950
Basal metabolic power	$\dot{E}_m \propto A_d$	$\propto M^{2/3}$	$\propto M^{3/4}$	$\propto M^{4/5}$	$\propto M^{0.75}$, mammals	Kleiber, 1961
Maximum aerobic power	$\dot{E}_{max} \propto A_d$	$\propto M^{2/3}$	$\propto M^{3/4}$	$\propto M^{4/5}$	$\propto M^{0.79}$, mammals	Taylor et al., 1978
Literature example		Swift, 1726	Rashevsky, 1960; McMahon, 1973	Galileo, 1637		

Table 4-2 Dimensions of body and body support members compared to similarity models.

Subject	Length	Width, diameter	Length	Diameter	Expected geometric exponent	References
Geometric similarity	$l \propto d^{1.00}$	$d \propto l^{1.0}$	$\propto m^{0.33}$	$\propto m^{0.33}$	g	McMahon, 1975b
Elastic similarity	$l \propto d^{0.67}$	$d \propto l^{1.5}$	$\propto m^{0.25}$	$\propto m^{0.375}$	e	McMahon, 1975b
Static-stress similarity	$l \propto d^{0.50}$	$d \propto l^{2}$	$\propto m^{0.20}$	$\propto m^{0.40}$	s	McMahon, 1975b
Spider, *Lycosa lenta*: Cephalothorax	$\propto d^{0.98}$	$0.696\, l^{1.03}$			g	Prange, 1977
(m = 2.6 to 1,218.3 mg) Leg segment (mm)	$\propto d^{1.06}$	$0.174\, l^{0.94}$	$5.62\, m^{0.35}$		g	Prange, 1977
Roach, *Periplaneta americana*: Body	$\propto d^{1.04}$	$0.355\, l^{0.95}$			g	Prange, 1977
(m = 3.1 to 1,075.0 mg) Leg segment (mm)	$\propto d^{1.00}$	$0.186\, l^{1.00}$	$2.77\, m^{0.34}$		g	Prange, 1977
Snake, *Crotalus adamanteus*: Body			$\propto m^{0.32}$		g	Prange and Christman, 1976
Vertebral width, body	$l \propto d^{0.81}$	$6.650\, l^{1.23}$			e/g	Prange and Christman, 1976
Centrum ball diameter, centrum	$l \propto d^{0.77}$	$0.385\, l^{1.30}$			e/g	Prange and Christman, 1976
Birds, limb bones (mm): Humerus		$0.169\, l^{0.804}$	$118\, M^{0.48}$		$< g$	Prange et al., 1979
Ulna		$0.111\, l^{0.812}$			$< g$	Prange et al., 1979
Femur		$0.057\, l^{1.132}$	$61.6\, M^{0.36}$		$\sim g$	Prange et al., 1979
Tibiotarsus		$\propto l^{1.13}$			$\sim g$	Prange et al., 1979
Running birds: Horizontal bones			$\propto m^{0.28}$	$\propto m^{0.40}$	$e?$	Maloiy et al., 1979
Vertical bones			$\propto m^{0.43}$			Maloiy et al., 1979
Ungulates, limb bones (mm)	$\propto d^{0.60}$ to 0.72 mean: $\propto d^{0.65}$				e	Maloiy et al., 1979
Artiodactyla	$\propto d^{0.65}$ to 0.75 mean: $\propto d^{0.67}$				e	Maloiy et al., 1979
Perissodactyla	$\propto d^{0.20}$ to 0.87 mean: $\propto d^{0.55\pm0.26}$				?	Maloiy et al., 1979

			e/g?	
Bovidae (antelopes): Legs	$\propto m^{0.26}$	$\propto m^{0.34}$ ($\propto m^{0.31 \text{ to } 0.38}$)	e/g?	McMahon, 1975a
Body length (m)	$0.315\,M^{0.28}$		e	McMahon, 1975a
Shoulder height (m)	$0.280\,M^{0.27}$		e	McMahon, 1975a
Hind height (m)	$0.329\,M^{0.24}$		e	McMahon, 1975a
Tibial mean diameter (cm)		$0.17\,M^{0.35}$	g/e	McMahon, 1975a
Mammals: Humerus	$\propto m^{0.36}$ ($\propto m^{0.27 \text{ to } 0.38}$)		g/e?	Alexander, 1977
Ulna	$\propto m^{0.36}$ ($\propto m^{0.29 \text{ to } 0.40}$)		g	Alexander, 1977
Metacarpal[a]	$\propto m^{0.37}$ ($\propto m^{0.19 \text{ to } 0.40}$)			Alexander, 1977
Femur	$\propto m^{0.36}$ ($\propto m^{0.27 \text{ to } 0.39}$)		g	Alexander, 1977
Tibia	$\propto m^{0.32}$ $\propto m^{0.22 \text{ to } 0.32}$		g e?	Alexander, 1977
Metatarsal[a]	$\propto m^{0.30}$ ($\propto m^{0.20 \text{ to } 0.32}$)		e/g?	Alexander, 1977
Body length (m)	$0.341\,M^{0.31}$		g	Economos, 1982, 1983
Body length (m); body mass ≤20 kg	$0.329\,M^{0.34}$		g	Economos, 1982, 1983
Body length (m); body mass >20 kg	$0.441\,M^{0.27}$		e	Economos, 1982, 1983
Insectivora, mean of five bones	$\propto m^{0.38}$		g	Alexander et al., 1979
Primata, mean of five bones	$\propto m^{0.34}$		g	Alexander et al., 1979
Rodentia, mean of five bones	$\propto m^{0.33}$		g	Alexander et al., 1979
Fissipedia, mean of five bones	$\propto m^{0.36}$		g	Alexander et al., 1979
Bovidae	$\propto m^{0.26}$		e	Alexander, 1977

a. Mean values exclude Artiodactyla.

Table 4-3 The scaling of skeletal or supportive structural mass in or around animals.

Organism	n	Body-mass range	Equation	r	s_b	$\dot{X}$	References
Exoskeletons							
Spider	61	0.025 to 1.2	$0.078\,m^{1.14}$	0.970	0.131	1.35	Anderson et al., 1979
Mollusk, fresh water	106	0.00085 to 55	$0.231\,m^{1.10}$	0.995	0.116	1.31	Anderson et al., 1979
Bird egg	368	0.5 to 8,500	$0.048\,m^{1.13}$	0.995	0.085	1.22	Anderson et al., 1979
Endoskeletons—Aquatic							
Petromyzon marinus (lamprey)		1 to 1,000	$0.0095\,m^{0.97\pm0.018}$*	—	—	—	Reynolds, 1977
Teleost fish		3 to 1,200	$0.033\,m^{1.03\pm0.069}$*	0.996	0.156	—	Reynolds, 1977
Whale	170	8,000 kg to 130,000 kg	$0.024\,m^{1.11}$	0.975	—	—	Anderson et al., 1979
Endoskeletons—Terrestrial							
Crotalus adamanteus (rattlesnake)	12	48.5 to 3,500	$0.020\,m^{1.17}$	0.993	0.080	1.20	Anderson et al., 1979
Bird	311	3.1 to 81,000	$0.040\,m^{1.07}$	0.993	0.119	1.32	Anderson et al., 1979; Prange et al., 1979
Mammal	49	6 g to 6,600 kg	$0.033\,m^{1.09}$	0.992	0.144	1.39	Anderson et al., 1979; Prange et al., 1979

Note: Body-mass range in grams, unless otherwise noted; r = correlation coefficient; s_b = standard error of exponent; $\dot{X}$ = upper limit of 95% confidence interval;
* = 95% confidence interval.

data of Pitts and Bullard (1968) for mammals. Prange and associates (1979) related dried skeletal mass $(M_{skel,d})$ to body mass (6 g to 6,600 kg) ($r = 0.992$; M in kg; S_{yx} (standard error of intercept) $= 0.154$; $n = 49$):

$$M_{skel,d} = 0.061 \, M^{1.090}. \tag{4-10}$$

Fresh, unscraped bone mass $(M_{skel,f})$ of smaller eutherian mammals (mean, 1,835 g; range, 7 g to 17.6 kg) includes normal moisture of live bone but also possible error for adhering bits of connective tissue, and scales as follows:

$$M_{skel,f} = 0.110 \, M^{0.96} \tag{4-11}$$

($r = 0.996$; $S_{yx} = 0.2518$; $S_b = 0.0977$). Compared at 1 kg, the museum skeletal mass is $0.061/0.11 = 55\%$ of fresh mass; at the mean body mass for Eq. (4-11), the ratio of museum to fresh bone mass is 60%. Thus the allometrically reconstituted body composition of Figure 2-1 has not taken into account the loss of moisture and organic material when skeletal material is prepared for museum storage.

Pitts and Bullard (1968) also determined the proportion of bone ash in body mass from their sample of 42 species of eutherians. The ratio of bone ash to fresh bone, expressed as a function of live body mass (minus fur and gut contents), scales as $M^{-0.13}$; the correlation is not significant, accounting for only 3% of the variability, so the mineral (ash) portion of fresh bone can only be expressed as a mean value of $35 \pm 6.1\%$ with no significant size trend. The matrix of mammalian bone is generally thought to be about 80% inorganic calcium phosphate and 20% organic material (Young and Hobbs, 1975). The 80% inorganic material is then 80% of the 55% dry portion or 44% of the fresh bone mass, compared to the mean ash proportion of 35% calculated from the data of Pitts and Bullard. The difference between 44% and 35% must therefore reflect the nonbone contribution of the bone marrow to fresh whole bone mass and/or scatter in the data. While I have not myself come across an allometry of bone marrow, Ganong (1981) states that the bone marrow approximates the liver in proportion of body mass.

With this perspective on what we are examining or overlooking in preserved skeletal material, we can return to the allometry of the bony endoskeleton, which has been analyzed in terms of skeletal mass, bone length, and bone width or diameter. We have models for scaling bone dimensions to preserve geometric, elastic, and static-stress similarity, and we might hope to find agreement between prediction from the correct model and empirical dimensions of length and diameter on the one hand, and between empirical linear dimensions and empirical skeletal mass on the other hand.

The elastic similarity model is supported by measurements of limb bones in the Artiodactyla. To preserve elastic similarity, the somatic exponent for bone and limb lengths would be ¼, and for their diameters it would be ⅜, predictions borne out by the bone-length measurements (mean of 5 bones, $l \propto m^{0.25}$) of McMahon (1975) and Alexander (1977; $m^{0.26}$). The mass of a bone would be proportional to its volume:

$$m \propto V \propto ld^2 \propto (m^{0.25})(m^{0.375})^2 \propto m^{1.0}. \tag{4-12}$$

Collectively, then, the skeleton should be linearly proportional to body mass, not the observed hyperallometric $m^{1.09}$ for eutherian mammals. In other words, the elastic similarity model predicts the slenderness or thickness of the bones required to protect against buckling—that is, the proportions, not the total mass. An elephant would therefore not be expected to devote 27% of its body mass to skeleton when a shrew can get by with carrying only 4.8% as skeleton.

When other orders of eutherian mammals are considered, the bone lengths scale as if geometric similarity ($l \propto m^{0.33}$) were adequate, with a mean somatic exponent of 0.345 (extreme range, 0.19 to 0.40). The bone diameters (mean exponent 0.395) relate to total body mass as they would if obeying elastic similarity ($d \propto m^{0.37}$); recall, however, that it is not the relationship between diameter and body mass, but the slenderness exponent, $l \propto d^{2/3}$, that is important for elastic similarity. Using these scalings of bone lengths and diameters for eutherians in general, the bone volume and mass should be in these proportions:

$$m \propto V \propto ld^2 \propto (m^{0.345})(m^{0.395})^2 \propto m^{1.14}. \tag{4-13}$$

When the Bovidae were included, except for the allometry of metapodial bones which scale atypically, the mean exponents obtained by Alexander and colleagues (1979) can be substituted to predict bone mass:

$$m \propto V \propto ld^2 \propto (m^{0.35})(m^{0.36})^2 \propto m^{1.07}. \tag{4-13a}$$

Note that the scalings of Eqs. (4-12) and (4-13) bracket the empirical derivation of Eq. 4-10. Thus the slenderness exponent of $l \propto d^{-1}$ for mammals other than the Bovidae is that of geometric similarity, which would give greater relative length and stride, but at the expense of more dedication of body mass to skeleton—as mass that must be carried for strength in articulation but is useless as far as other functions are concerned. Galileo built a good theory from a limited sample size, but in other than the Artiodactyla it is not supported statistically.

Birds

The pneumatization of avian bones, especially those of the wing, and the reduction in total number of bones by fusion gave rise to the wrong but long-held assumption that "the avian skeleton is specialized for strength and for lightness" (van Tyne and Berger, 1976). The allometry does not support this, as Prange and coworkers (1979) showed ($n = 311$; $r = 0.993$; $S_{yx} = 0.102$):

$$M_{skel,d} = 0.065 \, M^{1.07}. \tag{4-14}$$

Obviously this does not differ significantly from the mammalian equation (4-10) either in exponent or in coefficient (at $p = 0.01$; see Figure 4-2). Considering that for elastic similarity the diameters should be proportional to length raised to the 1.5 power, compared to 1.0 for geometric similarity, it is especially surprising that the diameters of wing bones are scaled in proportion to (humerus length)$^{0.804}$ and to (ulnar length)$^{0.812}$ and thus are even more slender than a mere geometric scaling. That the greater lengths of wing bones per unit mass would reduce the oscillating mass of the flapping wing is discounted by the Prange group as relatively insignificant.

There is, however, an apparent geometric scaling of leg bones. The widths of femur and tibiotarsi of birds scale as (length)$^{1.13}$, and femur length is proportional to the 0.36 power of body mass. Both of these exponents are closer to expectations based upon geometric similarity than upon elastic similarity. As Prange and coworkers point out, the legs of a bipedal bird support twice as much mass as the legs of a quadrupedal mammal. The avian femur is thicker than the artiodactyl femur (Figure 4-3). While the greater exponent for the width of the latter decreases the differences between the classes, within the range of femur lengths common to both the avian femur is always thicker in comparison.

When the study of avian leg bones is confined to running birds, ranging in size from a 75-g quail to a 41.5-kg ostrich, an interesting difference exists in the scaling of bone lengths, depending upon whether they oscillate from a horizontal position (femur and toe) or from the vertical (tibiotarsus and tarsometatarsus). The femur and toe lengths are proportional, on the average, to $m^{0.28}$, whereas the vertical tibiotarsal and tarsometatarsal lengths scale as $m^{0.43}$ and $m^{0.49}$. These length scalings are as if the horizontal bones were obeying elastic similarity, while the vertical bone lengths increase even more with body mass than a geometric similarity would predict. However, the diameters of both horizontal and vertical bones are scaled to body mass approximately as would be expected for elastic similarity (Maloiy et al.,

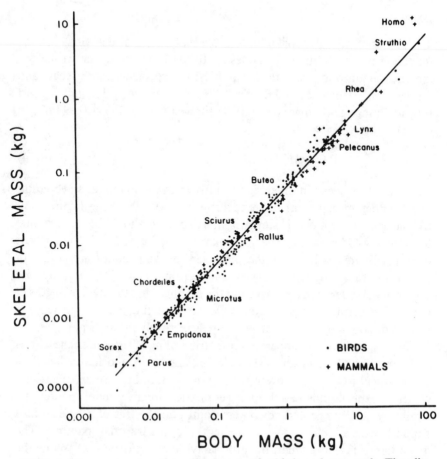

4-2 Skeletal mass as a function of body mass for birds and mammals. The allometric line plotted is actually for mammals, but the line for bird skeletons is
indistinguishable; see Eqs. (4-10) and (4-14). (Redrawn from Prange et al., 1979;
© 1979 by the University of Chicago Press.)

1979). The authors were thus not satisfied with the elastic similarity model
when applied to the legs of running birds, feeling that inertial forces in
running were stronger and more important than gravitational forces, a
matter which will be considered with the allometry of locomotion in
Chapter 7.

Exoskeletons

For samples of exoskeletal allometry we can consider the dimensions of
body and leg segments of wolf spiders and cockroaches. As seen in Table 4-2,

the intraspecific allometries of both species approximate a geometric similarity. The widths of spider cephalothorax and metatarsus of the third leg, as well as cockroach body and tarsus of the second leg, are more or less linearly proportional to their respective lengths, and the lengths are approximately proportional to the cube root of body mass.

The shells of bird eggs are the most simple exoskeletons for geometric allometry, since they are symmetrical in shape, regular in macroscopic surface detail, and lack the complications of articulation. Furthermore, they have been accurately measured as objects of old-time natural history and

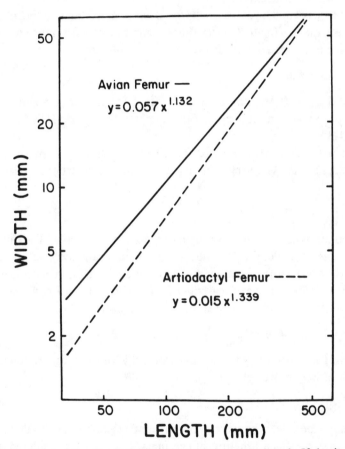

4-3 A comparison of scalings of femurs from birds and mammals. If elastic similarity were maintained, the width (diameter) would scale as length$^{1.5}$; if geometric similarity prevailed, it would be $d \propto l^{1.10}$. The actual femoral scalings both appear to be in between, with birds closer to the geometric model. (From Prange et al., 1979, with permission; © 1979 by the University of Chicago Press.)

later analyzed, in elegant functional perspective, in the respiratory studies of Rahn, Ar, and Paganelli.

Though it lies in repose without having to move, the eggshell represents the outcome of opposing factors in natural selection. While breakage by an inbound pebble, twig, or foot is a threat, breakage of the shell by the outbound chick is a necessity. There must be a compromise between being thick enough to resist breakage during incubation and turning, yet thin enough so that the length of the pores does not constitute an unreasonable impediment to respiratory diffusion and/or to the escape of the hatchling (though it takes less force to break the shell from within the concavity than from outside its convexity).

The masses of 368 birds' eggshells (m_{sh}) have been related to the full range in fresh egg mass of 0.5 to 8,500 g, the latter an estimate from the volumes of eggs of the extinct elephant bird *Aepyornis* as follows:

$$m_{sh} = 0.048 \ m_{egg}^{1.132} \tag{4-15}$$

$(r = 0.995; S_{yx} = 0.085;$ antilog $S_{yx} = 1.22)$. The exponent is slightly higher than most, but within the range of other exponents for skeletal vs. body mass (Table 4-3).

From a tabulation of the eggs of 29 species (house wren to ostrich; Paganelli et al., 1974), the egg length (l_{egg}, in cm) is proportional to fresh egg mass as

$$l_{egg} = 1.51 \ m_{egg}^{0.34} \tag{4-16}$$

$(r = 1.00)$, thus appearing to conform to geometric similarity. The thickness of the shells (x_{sh}), like the thickness of an invertebrate body wall, is analogous to the bone width or diameter which confers strength in endoskeletons. From an amazing sample of eggshells of 3,434 species (a 40% coverage of the world's avifauna!), Ar and colleagues (1979) report:

$$x_{sh} = 54.06 \ m^{0.448}. \tag{4-17}$$

Combining Eqs. (4-16) and (4-17) to express thickness (in micrometers) as a function of length, we get:

$$x_{sh} \propto l_{egg}^{1.32}, \tag{4-18}$$

which falls indecisively between geometric and elastic similarity. The scaling of shell thickness as a function of egg mass, Eq. (4-17), considered alone, appears to exceed even that required for static-stress similarity.

The product of shell thickness and surface area, $(A_{sh}$, calculated by Paganelli et al., 1974) gives the volume of shell material as a function of egg mass:

$$V = (x_{sh})(A_{sh}) = (0.005406 \, m_{egg}^{0.448})(4.835 \, m_{egg}^{0.662})$$
$$= 0.026 \, m_{egg}^{1.110}. \tag{4-19}$$

Equations (4-15) and (4-19) can be used to calculate eggshell density (ρ_{sh}):

$$\rho_{sh} = 0.048 \, m_{egg}^{1.132}/0.026 \, m_{egg}^{1.110} = 1.846 \, m_{egg}^{0.022}. \tag{4-20}$$

This is similar to the density of dried bone, which is about 2 (Alexander, 1968).

The protective value of the exoskeleton has been measured directly in the case of the avian eggshell. Ar and associates (1979) measured lengthwise the breaking strength of fresh eggs from 47 species. The force (F, in newtons, N) required to break the shell was related to egg mass as

$$F = 0.499 \, m_{egg}^{0.915}. \tag{4-21}$$

Shell thickness, expressed in centimeters, when squared is

$$x_{sh}^2 = 3.21 \times 10^{-5} \, m^{0.916}. \tag{4-22}$$

Equations (4-21) and (4-22) show that the square of shell thickness and the breaking force have very similar mass exponents:

$$F/x_{sh}^2 = 0.499 \, m^{0.915}/3.21 \times 10^{-5} \, m^{0.916}$$
$$= N/cm^2 = 1.55 \times 10^4 \, m^{0.001} \tag{4-23}$$

or

$$F = 1.55 \times 10^4 \, x_{sh}^2. \tag{4-24}$$

An empirical correlation closely confirmed this relationship:

$$F = 1.68 \times 10^4 \, x_{sh}^2. \tag{4-25}$$

Let us return now to the potential hazard of the weight of the hen, which rests in part upon the egg. As we saw in Eq. (3-4), egg mass scales less than linearly with adult body mass. Ar (1979) combined Eqs. (3-4) and (4-22) to calculate a dimensionless safety factor (n_s), which represents the extent to which the strength of the eggshell exceeds the weight of the adult that settles on the eggs in the nest. This safety factor is size dependent, in such a way that with the evolution of greater size of body and egg, there is less of a safety factor. The size dependence becomes apparent if the strength in excess of adult weight is expressed as a ratio to that adult weight:

$$n_s = \frac{F - (m_{adult})(g)}{(m_{adult})(g)} = \frac{F}{(m_{adult})(g)} - 1, \tag{4-26}$$

where g = gravitational acceleration, 9.8×10^{-3} N/g. In Eq. (4-21), F is correlated with egg mass. By Eq. (3-4) F can be related to adult mass:

$$F = 0.499 \, (0.277 \, m_{adult}^{0.77})^{0.915} = 0.154 \, m_{adult}^{0.71}. \tag{4-27}$$

This value is then substituted in Eq. (4-26):

$$n_s = (0.154 \, m_{adult}^{0.71}/0.0098 \, m_{adult}) - 1 = 15.7 \, m_{adult}^{-0.29} - 1. \tag{4-28}$$

This is only a crude approximation, because Eq. 3-4 incorporates a wide variability in the coefficients of egg vs. body-mass relationships of various orders of birds. With this limitation in mind, we can predict n_s for birds at opposite extremes of body size. A 2-g hummingbird would have a safety factor of 11.8, while any hen of over 13.3 kg body mass could not put her full weight on a single egg without crushing it. Bypassing the variability that is glossed over by Eq. (4-28), we can go directly to the data (in g-force) of Ar and colleagues (1979) to find the ostrich's safety factor:

$$\frac{F_{egg}}{F_{adult}} - 1 = \frac{75,750}{84,000} - 1 = 0.90 - 1 = -0.1.$$

In other words, the ostrich has no safety factor. However, ostriches have clutches of ten or more eggs, sometimes contributed by two or more hens, so the weight of the incubator is distributed over a number of eggs.

The idea of a safety factor does suggest that there is an upper size limit for oviparity in birds. The largest bird known, the extinct elephant bird *(Aepyornis maximus),* had an estimated body mass of 457 kg (Amadon, 1947), for less than half what the legs of the largest quadruped mammal *(Baluchitherium)* supported. *Aepyornis'* egg is estimated to have weighed 9,130 g (Paganelli et al., 1974). The breaking force of such an egg can be predicted from Eq. (4-21) to be 2,099 gf (20.6 N), compared to the adult weight 457,000 gf (4,481 N). Thus the breaking strength is less than a half-percent of adult mass; even if the hen's weight were applied over 200 eggs, there would be no extra safety factor.

There appears to be an upper limit of under 20% for the maximum proportionality of skeletal to body mass, except in the aquatic/marine mollusks (Anderson et al., 1979). *Aepyornis* is at this maximum. To increase shell strength further would necessitate a shell thicker than 3.2 mm. The shell thickness is, of course, the shortest diffusional path for gas exchange. Such a dimension greatly retards the time required for oxygen to diffuse from the environment to the inside of the eggshell, where the blood vessels of the chorioallantoic membrane must pick it up. Relative to embryonic mass, there is also relatively less gas-exchange surface area, further slowing the fulfillment of oxygen requirements.

If food supply or other niche characteristics did not set the upper limit to body size in birds, eggshell geometry may have done so.

Tensile "Skeletons"

The possessors of skeleton-like structural elements that resist only tensile forces are not warm-blooded mammals and birds, and are therefore beyond the scope of this book. However, I cannot resist digressing for a moment. The rigid skeletons discussed in this chapter showed good correlation for both length and diameter as functions constrained by body mass. In contrast, some animals and plants have supporting stalks which have skeletal functions that need only resist tensile forces. Examples of these given by Peterson and colleagues (1982) are the pendant fruitstalk of the sausage tree *(Kigelia pinnata);* the stipe that runs from the holdfast on a rock to the buoyant float and blades of the elk kelp *(Pelagophycus porra),* which are pulled by the current; and the stalks of setae on the adhesive toe pads by which *Anolis* lizards can hang from smooth vertical and overhead surfaces. For each of these diverse stalk types there is at best a weak correlation between length and diameter. The length, like that of a rope or cable, is independent of the load. However, the diameters show high correlation (0.890, 0.811, and 0.938, respectively). This highlights the need to know what sort of forces are involved in structural support if the scaling is to be explained.

5 Homeostasis

The interrelationships found imply quantitative orderliness among characteristics so diverse as urinary flow, renal clearance, duration of heartbeat, and oxygen consumption. Organisms may be pictured as systems of precise multiple interrelations.

—E. F. ADOLPH (1949)

FOR A SPECIES to survive, individual members must live long enough to reproduce successfully. This prerequisite of self-preservation is attained by behavioral and physiological mechanisms. The individual must avoid not only predation, disease, and natural disaster, but equilibrium with the thermochemical surroundings. As indicated previously, homeostasis is maintenance of the nonequilibrium state necessary for life.

Metabolism

This chapter is concerned with the allometry of the homeostatic or life-support systems that provide energy and oxygen, regulate internal physico-chemical conditions, and eliminate metabolic waste products. Included are digestion, respiration, circulation, osmotic balance and excretion, and temperature regulation. These are the processes that make the distinction between the internal and external environments, thereby achieving a dynamic balance between environments separated at the interfacing (and leaky) body surfaces.

The losses and gains that constitute dynamic balance occur across the body surfaces. The cost of regulating and/or offsetting these exchanges appears as the basal or standard metabolic rate. Quite logically, then, one would try to relate metabolic cost to body-surface area. In fact, 1983 was the centennial of Rubner's "surface law" or "surface rule," which attempted to relate metabolic rates of dogs to body-surface area rather than to weight. It is true that the correlation of metabolism with surface is better than that with mass, that is, if the surface of an animal scales as the surfaces of geometric bodies do, proportional to $M^{2/3}$. Nevertheless, the metabolic rate per unit of surface area is not a size-independent constant, even for Rubner's dogs (see Schmidt-Nielsen, 1972, for a review). This does not eliminate body surfaces as the location of energy and material exchanges, so it is necessary to have a quantitative appreciation of both the external and internal surface areas involved. First, consider the external surface area.

BODY-SURFACE AREA

The surface area, A, and the volume, V, of geometrically similar objects are proportional to linear dimensions:

$$A \propto l^2; \quad l \propto A^{1/2},$$
$$V \propto l^3; \quad l \propto V^{1/3}.$$

If we express linear dimensions as a function of volume and assume that density (ρ) of tissues is a size-independent constant within a class, then mass and volume are directly proportional and can be used interchangeably with the same proportions. Thus area is proportional to body mass as

$$A = km^{2/3},$$

or $\quad A \propto l^2 \propto (V^{1/3})^2 \propto V^{2/3} = km^{2/3},$

(5-1)

where k depends upon the shape and density of the animal. A value of 10 is generally assumed for birds and mammals when A is in square centimeters, m in grams. This is the so-called Meeh equation used to estimate body surfaces of animals for calculating surface-specific exchange rates, as in thermal conductance or gas diffusion.

Surface area has been variously measured to check on such estimations for real animal shapes. Drent and Stonehouse (1971) verified the reliability of $A = 10 \ m^{0.67}$ for surface area of the skin in birds. Walsberg and King (1978) compared external surface area (the outside of the "envelope" of feathers) with that of the plucked carcass. The surprising revelation of their

Table 5-1 The relationships between body surface area and body mass of birds and mammals.

Subject	Range of body mass	Equation	r^2	Slope (95% confidence interval)	Method	References
INTERSPECIFIC (phylogenic):						
Birds, skin area (cm^2)	19–649 g	$10.0\ m^{0.67} = 1,000\ M^{0.67}$	—	—	Measurement	Drent and Stonehouse, 1971; Calder and King, 1974; Walsberg and King, 1978
Birds, external plumage area (cm^2)	3.8–649 g	$8.11\ m^{0.67} = 813\ M^{0.67}$	0.996	—	Polyethylene film over frozen carcass	Walsberg and King, 1978
Marsupial mammals (cm^2)	0.016–28.7 kg	$10.0\ m^{2/3} = 1,002\ M^{2/3}$	—	—	Pinned fresh skins	Dawson and Hulbert, 1970
		$17.74\ m^{0.61} = 1,215\ M^{0.61}$	0.970	—	Same, log-log regression	Dawson and Hulbert, 1970
Eutherian mammals (cm^2)	—	$12.3\ m^{0.65} = 1,110\ M^{0.65}$	>0.903	0.04	Various; from literature	Stahl, 1967
INTRASPECIFIC (mostly ontogenetic):						
Domestic fowl	30–4,242 g	$5.29\ m^{0.74} = 678\ M^{0.74}$	0.996	—	Spread fresh skins (plucked)	Leighton et al., 1966

Animal	Size range	Equation			Method	Reference
White rat (cm²)	~13–380 g	$13.2\,m^{0.64} = 1{,}098\,M^{0.64}$	—	—	—	Brody, 1945
		$12.54\,m^{0.60} = 791\,M^{0.60}$	—	—	Same method, unstated; different investigators	Brody, 1945
		$7.47\,m^{0.65} = 666\,M^{0.65}$	—	—	Same method, unstated; different investigators	Brody, 1945
Squirrel monkey (cm²)	630–1,181 g	$10.8\,m^{2/3} = 1{,}082\,M^{2/3}$	—	—	Spread fresh skins; equation from eye fit, assuming ⅔ slope	Stitt et al., 1971
		$13.0\,m^{0.64} = 1{,}081\,M^{0.64}$	0.920	—	Log-log regression of data	Stitt et al., 1971
Rhesus monkey (cm²)	0.82–6.6 kg	$12.48\,m^{0.66} = 1{,}192\,M^{0.66}$	0.99	—	Lacquered skin, cut in sections	Lee and Fox, 1933
Dog (cm²)	3.3–52 kg	$8.18\,m^{0.70} = 1{,}030\,M^{0.70}$	—	—	—	Brody et al., 1928
Swine (cm²)	2.5–300 kg	$12.2\,m^{0.63} = 970\,M^{0.63}$	—	—	—	Brody et al., 1928
Man (cm²)	1.2–99 kg	$8.81\,m^{0.69} = 1{,}000\,M^{0.69}$	—	—	—	Brody, 1945
Dairy cattle (cm²)	14–600 kg	$31.3\,m^{0.56} = 1{,}500\,M^{0.56}$	—	—	Integrating roller	Brody, 1945
Stretched cylinder		$\propto m^{0.63}$	—	—	Calculated; $A = 3$ (sphere at 8 g); $A = 2$ (sphere at 70 kg)	McMahon, 1973

study is that the external area is less than the skin area, both scaling as $M^{0.67}$ (Table 5-1). The plumage serves as a fairing which covers the infoldings of appendages and skin and reduces the surface area involved in heat exchanges to $8.11/10 = 81\%$ of the area of the skin. This means that the plumage conserves heat, not only as a resistance to heat flow, but as a reducer of exposed area. The surface area of mammals has a similar relationship to body mass, the 95% confidence interval of the slope including the theoretical 0.667 exponent, as well as McMahon's 0.63 for a stretched cylinder.

When regressions are made on data for single species, the exponents are often considerably different but not in any consistent fashion. The exponents for domestic fowl, and in one study of the white rat, are greater than those for interspecific allometries of surface area for birds and for mammals, while man and the white rat in two other studies (cited by Brody, 1945) are similar to $m^{2/3}$, and for dairy cattle there is a smaller exponent (Table 5-1). Such single-species regressions could include variables of growth (see Cheverud, 1982), obesity, artificial selection, and/or insufficient body-size range. However, with these exceptions, surface areas of birds and mammals appear to conform to the expectation that area is proportional to $m^{2/3}$.

The intraspecific allometric relationships for body-surface area of mammals (Table 5-1) may be used to predict representative values for typical body sizes of adults (Table 5-2). The regression of these values, given in the note to Table 5-2, is statistically indistinguishable from Stahl's (1967) equation for "100" data points (which included multiple points for some species, his only weighing being against "excessive bias towards human values").

STANDARD METABOLIC RATE

The basic cost of homeostasis in a thermoneutral environment is reflected in the basal or standard metabolic rate (metabolic power, ml^2t^{-2}). The allometry of standard metabolic rates of eutherian mammals was examined in Chapter 3 as an allometric sample. We have seen that the standard or basal metabolic rate scales approximately as $M^{3/4}$ within a phylogenetic group of vertebrates such as the avian order Passeriformes, other avian orders pooled, the marsupial mammals (infraclass Metatheria), and the placental mammals (Eutheria). This appears to be true whether the regression is for ectothermic or homeothermic animals (Bennett and Dawson, 1976; Bartholomew, 1982). Since these metabolism/body-mass regressions have essentially the same exponent, they may be compared simply as ratios of

Table 5-2 Surface areas of mammals, using intraspecific predictions for typical mature body mass of each species.

Species	Adult body mass (kg)	Surface area (cm^2), predicted from Table 5-1	Exponent
White rat	0.30	508.08 ⎤	0.64 ⎤
		384.21 ⎬ 399.14	0.60 ⎬ 0.63
		304.40 ⎦	0.65 ⎦
Squirrel monkey	0.933	1,033.62	0.64
		(1,033.56)	
Rhesus monkey	2.68	2,282.4	0.66
Dog	20	8,044.83	0.70
Swine	50	11,540.37	0.63
Man	70	18,360.82	0.69
Dairy cattle	550	51,368.20	0.56
			mean: 0.64 ± 0.045 s.d.

Note: $n = 7$ spp.; $A = 1,045\ M^{0.64}$; $r^2 = 0.99$; 95% confidence interval of exponent $= 0.58–0.70$.

constants, a, to the 3.27 in the Kleiber equation (3-13a) for eutherian mammals. When data from extant forms do not encompass a size range sufficient for regression analysis, we can make comparisons directly with predictions for specific body masses from the Kleiber equation. The minimal metabolic power consumption of various groupings compares with eutherians as follows: passerine birds, 163% (Aschoff and Pohl, 1970a,b) to 183% (Lasiewski and Dawson, 1967); nonpasserine carinate or flying birds, 112% (Calder and Dawson, 1978); ratite birds (except kiwis), 83%; marsupials, 66% to 68% (MacMillen and Nelson, 1969; Dawson and Hulbert, 1970); kiwis, 64% (Calder and Dawson, 1978); monotremes, 25% to 65% (Schmidt-Nielsen et al., 1966; Grant and Dawson, 1978; Dawson et al., 1979; see also Fig. 5-1). We might expect that natural selection would have minimized the energy cost of homeostasis within resting eutherian mammals and nonpasserine birds. How then can marsupial and monotreme mammals and ratite birds (hitherto erroneously labeled "primitive") maintain homeostasis with considerably less expensive standard metabolic levels than appear necessary for their "advanced" relatives, without sacrificing alertness and rapid evasion? The metabolic factoral scope seems to be relatively constant among homeotherms, so Dawson (1973) suggested that perhaps a higher basal expense could preserve a higher peak capacity. However, in a later study Dawson and Dawson (1982) reported that the

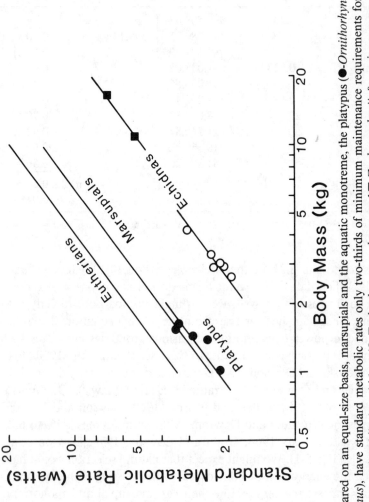

5-1 Compared on an equal-size basis, marsupials and the aquatic monotreme, the platypus (●–*Ornithorhyn-chus anatinus*), have standard metabolic rates only two-thirds of minimum maintenance requirements for eutherian mammals, while the echidnas (○–*Tachyglossus aculeatus* and ■–*Zaglossus bruijni*) require energy at rates less than one-third of what the eutherians need for basal maintenance. Within these taxa, however, size has essentially the same effect, the $M^{3/4}$ scaling first described by Kleiber. (After Dawson et al., 1979; with permission.)

metabolic scope of marsupials is eight to nine times the standard metabolic rate, compared to three to five times for eutherians during cold-stress exposure.

The often-demonstrated proportionality to $M^{3/4}$ has led biologists into many efforts to explain "Why ¾?" and sometimes even into unfortunate attempts to force other allometries to scale in proportion to standard metabolic rates. The metabolic rate could not be scaled up linearly, because the heat produced would be excessive for loss across a body surface that only increased as $M^{2/3}$, so perhaps $M^{3/4}$ was a compromise between surface and volume proportionalities (reviewed by Kleiber, 1961). On the other hand, poikilothermic or ectothermic animals also appeared to follow the same sort of metabolic scaling of $M^{-3/4}$ (reviewed by Bennett and Dawson, 1976).

More satisfying than a compromise between two scaling proportionalities would be a model that predicts the ¾ exponent directly. The elastic similarity model does this (McMahon, 1973, 1975); it provides resistance to buckling under self-loading by maintaining the proportionality between length and the ⅔ power of diameter of a limb or the torso. For simplicity, consider a limb as a column, the length of which is proportional to the ⅔ power of the diameter (Eq. 4-6). The weight of a column is proportional to its volume. This volume is the length multiplied by the cross-sectional area, or $l\pi(d/2)^2 \propto ld^2$. Therefore

$$mg = (\rho d^{2/3}\, \pi\, d^2/4) \propto d^{8/3}, \tag{5-2}$$

where g = gravitational acceleration and ρ = density. From this we can solve for the diameter:

$$d \propto m^{3/8}. \tag{5-3}$$

Recall now the principle of symmorphosis (Taylor and Weibel, 1981), by which the formation of structural elements "is regulated to satisfy but not exceed the requirements of the functional system." These requirements include locomotion, in which the muscles contract, exerting a force that is proportional to the cross-sectional area of the muscle. The power output of a muscle $(\dot{E})$ is the product of that force and the distance per unit time, or the velocity of shortening:

$$\dot{E} = \pi(d/2)^2(\Delta l/\Delta t). \tag{5-4}$$

According to Hill (1950), the velocity of shortening of a particular muscle is a size-independent constant, from species to species (m^0 within a class). If we substitute this m^0 constant for $\Delta l/\Delta t$ and the proportionality in Eq. (5-3) into (5-4), we see that the power output of the particular muscle is

$$\dot{E} \propto \frac{\pi}{4}(m^{3/8})^2(m^0) \propto m^{3/4}. \tag{5-5}$$

This reasoning would explain the scaling of metabolism during heavy exertion or activity, and it is indeed in agreement with the empirical relationships of running metabolism and size. It would also explain the scaling plan for the logistical variables involved in maintaining the flow of energy to the working muscles, the requirements satisfied according to the principle of symmorphosis. We can then return to the resting or standard metabolic level via the empirical observation of a metabolic factoral scope, or ratio of maximal aerobic activity to resting metabolism, that is approximately independent of size.

McMahon's (1973, 1975) explanation of the $M^{3/4}$ metabolic scaling for mammals has not been directly examined in birds, but the following avian allometric relationships are compatible with it: locomotory muscle mass, and basal and flight metabolic rates. As seen in Chapter 4, the wing bones of birds do not scale according to the elastic similarity model; but dimensionally more relevant to the consideration of metabolic activity would be the relationship between muscle diameter or cross-sectional area, muscle length, and shortening velocity. Wing-beat frequency does not scale at all like the stride frequency of mammals; since the latter conforms to the elastic similarity model, something different must happen in avian flight.

THE GEOMETRY OF FOUR DIMENSIONS

Beyond our everyday perception (well beyond mine!) is the concept of "hyperspace," in which there are more than the three dimensions of length, width, and height, the product of which encloses a volume.

Equation (5-1) expresses surface area in terms of our familiar three dimensions. In a more general way, if we signify the number of dimensions by n:

$$A \propto V^{(n-1)/n} \propto V^{(3-1)/3} \propto V^{2/3} \propto m^{2/3}. \tag{5-6}$$

Suppose there actually are four dimensions. Blum (1977) has suggested this possibility. If so, we would expect

$$A \propto V^{(4-1)/4} \propto V^{3/4} \propto m^{3/4}. \tag{5-7}$$

This would mean that the standard metabolic rate is directly proportional to surface area—a very comfortable finding intellectually, because the ¾ exponent could then be explained as a consequence of maintaining homeostasis against leakages that are proportional to the area exposed.

What is the fourth dimension? While it might be expected to be a spatial dimension, Blum (1977) acknowledges that it could be otherwise. Physiological times scale approximately to $m^{1/4}$ (Lindstedt and Calder, 1981). Hainsworth (1981) points out that if time were the fourth dimension, the total metabolism in a lifetime would be directly proportional to mass:

$$\text{lifetime metabolism} \propto (m^{3/4})(m^{1/4}) = m^{1.0}.$$

This would mean that the specific metabolism per kilogram in a lifetime would be size independent:

$$\text{specific metabolism} \propto (m^{3/4})(m^{1/4})/m^{1.0} \propto m^0. \tag{5-8}$$

Such does appear to be the case (Lindstedt and Calder, 1981), but the fact remains that body-surface area is not proportional to $m^{3/4}$, but to $m^{\sim 2/3}$ (Table 5-1). Thus the metabolic rate per unit of body area appears to increase slightly with size:

$$\text{metabolic rate per unit} \propto (m^{0.73} \text{ to } m^{0.75})/m^{0.67} \propto m^{0.06} \text{ to } m^{0.08}. \tag{5-9}$$

The space that most of us are coming from is three-dimensional. However, there is some evidence for an intraspecific metabolic scaling of $M^{2/3}$ (for instance, figure 1 of Wilkie, 1977). This could be in part the result of elevated metabolic rates, the sum of maintenance and growth in the smaller (younger) animals, but we must be open to the possibility that the ¾ exponent is not representative of smaller size and phylogenic ranges, as will be discussed in Chapter 14.

DIMENSIONAL ANALYSIS

We started Chapter 3 with the reminder that anything physically measurable in biology is a composite of the basic dimensions of mass (m), linear dimensions (l), time (t), and temperature (T). With these any physical entity can be defined:

$$y = m^\alpha l^\beta t^\gamma T^\theta. \tag{5-10}$$

where α = the exponent of the mass ratios of two different species, m_1/m_2; β = the exponent of the length ratio l_1/l_2; γ = the exponent of the ratio of the times they require to perform the same function, such as one heartbeat cycle or one pace, t_1/t_2; and θ = the exponent of the ratio of temperatures. It would be satisfying indeed if the body-mass exponent for standard metabolic rates could be accounted for in this manner (Günther, 1975; Yates, 1979). Since temperature is not dependent on size, the last term in Eq. (5-10) can be dropped:

$$y = aM^b = m^\alpha l^\beta t^\gamma. \tag{5-10a}$$

Equation (5-10a) can be simplified as follows. We can assume that all animals built on the same model or plan, such as all eutherian mammals, have the same basic chemical composition and tissue densities. That being the case, mass is proportional to volume, which has the dimensions l^3, and we can replace m^α with $l^{3\alpha}$. In simple geometries time periods are proportional to lengths: a long pendulum has a longer period than a short pendulum; a long string on a harp has a shorter vibration cycle, so that it can have more cycles per second (higher pitch). Thus l^γ can be substituted for t^γ. These can all be combined to yield a mass proportionality of $l \propto m^{1/3}$:

$$y = aM^b \propto l^{(3\alpha+\beta+\gamma)} \propto m^{(3\alpha+\beta+\gamma)/3}. \tag{5-11}$$

The metabolic rate has the dimensions of rate of energy conversion per unit time or power, ml^2t^{-3}, that is, $\alpha = 1$, $\beta = 2$, and $\gamma = -3$. If we substitute these in the above equation, (5-11), we obtain

$$y = aM^b \propto m^{(3+2-3)/3} \propto m^{2/3} \neq m^{3/4}. \tag{5-12}$$

In the same fashion the dimensions of frequencies such as heart rate should be $\propto m^{-1/3}$, not the observed $m^{-1/4}$. Günther (1972, 1975) and Günther and Martinoya (1968) have attempted the most extensive application yet of dimensional analysis to the similarity of mammals. They have resolved the discrepancies between dimensional predictions and empirical exponents by adding a correction factor, Γ, such that:

$$y = aM^b \propto m^{(3\alpha+\beta+\gamma+\Gamma)/3}, \tag{5-13}$$

where Γ is the difference between ¾ and ⅔, or 0.065. While this makes everything "come out right," there is still no more of a theoretical explanation for the 0.065 than there was for the 0.75.

It is interesting, however, that things *do* come out right if the proportionality between m and l of McMahon's elastic similarity model is used. For if $m \propto l^4$, Eq. (5-11) is modified to read:

$$y = aM^b \propto l^{(4\alpha+\beta+\gamma)} \propto m^{(4\alpha+\beta+\gamma)/4}. \tag{5-14}$$

So for metabolic rate dimensions of ml^2t^{-3}, we obtain, instead of Eq. (5-13), the following:

$$y = aM^b \propto m^{(4+2-3)/4} \propto m^{3/4}. \tag{5-15}$$

It is disappointing that body support for animals other than the hoofed mammals does not conform to the elastic similarity model (see Table 4-2).

COMPROMISING WITH GRAVITY

One of the products of the jet and space age is a new specialization in gravitational physiology, with applied interests in the problems of hypergravity and hypogravity. Hypergravity is in effect the multiplication of the normal force of gravity at the earth's surface by acceleration away from the ground, as when an aircraft pulls up suddenly and dissipates the downward velocity developed in a dive. Hypogravity is experienced as a gradient between normal gravitational loading on the earth's surface and the weightlessness of outer space. Since both situations have their meaning relative to normal conditions, the effects of gravity on structural design and basal maintenance are important and gravity has been considered as a possible factor in $M^{3/4}$ basal metabolic scaling.

Tolerance to hypergravity is proportional to $M^{-0.14}$ of mice, rats, and dogs, so a reciprocal scaling of $M^{0.14}$ has been assumed to describe the incremental cost of maintaining the body against static gravitational loading (Economos, 1979). On this basis Economos derived a binomial equation for basal metabolism with a surface-area term ($\propto M^{0.67}$) and a gravity term ($\propto M^{0.75+0.14}$), which together in the $5:1$ proportion of their relative influences fit mammalian $\dot{E}$ slightly better than a simple $M^{0.75}$.

It has also been suggested that the $M^{0.75}$ scaling is a simple compromise between $M^{1.0}$ for the force of gravity (M $\cdot$ g) and $M^{0.5}$ for heat loss (Pace and Smith, 1981; Chapter 8). Because $M^{0.5}$ and $M^{0.75}$ are not linear functions, a simple linear averaging of $M^{1.0}$ and $M^{0.5}$ does not yield $M^{0.75}$. Instead, there must be a weighting of $3M^{0.5}$ to $1M^{1.0}$ to approximate $M^{0.75}$ scaling.

Considering the range of environmental extremes on the earth, from severe cold stress to almost continuous warm temperatures on low-latitude islands, for example, one would not expect natural selection to be so inflexible. Two variables cannot be simultaneously optimized, so the influence of gravity should be inversely proportional to the degree of cold stress imposed by the environment for which an animal has evolved. It is difficult, therefore, to accept these numerical compromises as the explanation of $M^{3/4}$ scaling.

Oxygen and Carbon Dioxide

At the cellular level the oxygen vital for aerobic metabolism moves by diffusion. There is no molecular pump that can speed the movement of oxygen across the cell membranes, as there is for glucose, sodium, or potassium. The removal of the waste product, carbon dioxide, is similarly dependent upon diffusion.

The rate of diffusion of oxygen ($\dot{V}_{O_2}$) is described quantitatively by Fick's law of diffusion:

$$\dot{V}_{O_2} = -DA \frac{\Delta p_{O_2}}{x},$$
(5-16)

where D is the diffusion coefficient, a proportionality constant for the combination of the molecule size of a particular gas and the medium such as protoplasm and membrane through which the diffusion is occurring; A is the surface area in cross-section available for diffusion exchange; Δp_{O_2} is the difference in the partial pressures of oxygen between two points separated by distance x, such as between the immediate environment outside the cell and the center of the cell. The difference divided by the distance, $\Delta p_{O_2}/x$, is the partial-pressure *gradient,* which serves as the potential, "pushing" the diffusion. (The product of Δp_{O_2} and the volume involved is the potential energy.) The gradient can be raised either by increasing the pressure difference or by decreasing the distance over which it occurs.

If there were an evolution of a geometrically similar but larger cell, say by doubling the diameter, the distance x would be twice as great, but the area A would increase by linear dimensions squared, $2^2 = 4$. The combined effect of A/x would be $4/2 = 2$. In other words, the rate of diffusion would be doubled, but meanwhile the volume of the cell would have increased as the cube of linear dimension, 2^3, or eightfold. If D and Δp_{O_2} were unchanged, the metabolic rate per microgram of protoplasm would be reduced to one-fourth of the previous level because of the limits of diffusion.

If we view this as a reciprocal to the rate of diffusion—as the time required for, say, a step change in Δp_{O_2} to reach 95% of equilibrium within the cell—the relationship is similar to that for nitrogen, which is

$$t_{0.95} = 6.31 \, x^2,$$
(5-17)

where $t_{0.95}$ is time in seconds and x is distance in millimeters (Riggs, 1963). In a very revealing table Riggs shows that $t_{0.95}$ for nitrogen in water at 20° C is 0.063 s if x is 1 μ, 6.3 s if x is 1 mm, 18 *hr* if x is 0.1 cm, and 73 *days* if x is 1 cm. The nitrogen molecule is only slightly smaller than the oxygen molecule; oxygen should diffuse 94% as rapidly. Thus it was not practical to remain unicellular in the evolution of larger body size. The "physiological time" for diffusion as a function of cell mass would have been proportional to $m^{0.67}$ or $m^{0.50}$, vs. the $m^{-0.25}$ common in vertebrates.

This is confirmed, for example, in sizes of cells (mean cell cross-section) for liver, thyroid, and renal epithelium, and pancreatic acinar cells of mammals from mouse to ox, which (to quote Munro) show that "in glandular tissues the larger mammals have more cells, not larger cells." Muscle mean cell mass (wet weight per milligram of DNA) increases by a factor of two from the smallest to the largest mammals (Munro, 1969). A mere doubling over the mammalian size range is approximately equivalent to an $M^{0.05}$ scaling (see Appendix Table 1).

Thus it comes as no surprise that the mean dimensions of red blood cells (rbc's), in microns, from the *Biology Data Book* (Altman and Dittmer, 1972) show no significant body-size trend (slope not significant, $p > 0.05$):

$$l_{rbc} = 6.29 \ M^{-0.02}. \tag{5-18}$$

If there were a positive or negative allometry to erythrocyte size, the animal with the larger erythrocyte would simply suffer a slower equilibration time for the loading and unloading of oxygen or carbon dioxide, clearly a poor design feature.

The basic function of both the respiratory and circulatory systems, as seen in Fick's law of diffusion (Eq. 5-16) is to maintain as high a Δp_{O_2} as possible, to drive the diffusion as rapidly as possible. It is in these systems that we will see the scaling necessary to meet the needs of the total mass of the animal body, the concept of "more cells, not larger cells" described by Munro (1969).

Metabolic rate is now usually determined by measuring oxygen consumption, assuming that the metabolism is all aerobic. If so, the animal must have attained a steady state such that rate of consumption is equal to rate of oxygen delivery by the respiratory and circulatory systems in series, along the "respiratory cascade." We should therefore be able to account for the animal's oxygen consumption rate by the conservation of volume (Stahl, 1962):

$$(\text{pumping frequency}) \times \begin{pmatrix} \text{stroke or} \\ \text{tidal volume} \end{pmatrix} \times \begin{pmatrix} \text{fractional removal} \\ \text{of cycled air} \\ \text{or blood} \end{pmatrix} = \begin{pmatrix} \text{oxygen} \\ \text{consumption} \\ \text{rate} \end{pmatrix}$$

$$\underbrace{(f) \qquad\qquad \times \qquad (V) \qquad\qquad \times \qquad (F_{O_2})}_{\substack{\text{minute volume or} \\ \text{cardiac output} \\ \text{(ml/min)}}} \quad = \quad (\dot{V}_{O_2})$$

$$\text{number/min} \ \times \ \text{volume (ml)} \ \times \ \begin{array}{c} 0.2095 - 0.1595 \\ (\text{ml } O_2/\text{ml air}) \end{array} \ = \ \text{ml } O_2/\text{min}$$

$$\tag{5-19}$$

Such a determination should be obtainable, not only with values for an individual animal, but with allometric expressions for each term, using the allometric cancellation technique (Stahl, 1967; Calder, 1968). Furthermore, if such a relationship is true, it follows that if we have allometric expressions for all but one variable, we can derive the missing allometric variable by solving for the unknown (Stahl, 1962). We shall use these applications shortly for both respiratory and circulatory systems. From this will emerge a rough pattern, an apparent set of evolutionary guidelines or design principles:

1. *Considerations of body composition.* From the graphs and equations of Chapters 2 and 3 we can see that the sum of muscles, skeleton, skin, gut, liver, and brain claims, at a minimum, 70% of body mass. Within the remaining 30% all organs must be scaled up and down, with limits on hyperallometric increase so that one organ cannot usurp entirely the space or mass devoted to another function. The oxygen delivery systems of lungs, heart, and blood serve the aerobic needs of the muscles during heavy exertion. The muscle mass appears to claim an essentially constant fraction of body mass, $0.45\ M^{1.0}/M$. The delivery systems scale up linearly, $\propto M^{-1.0}$, to provide the oxygen needed by a linearly scaled muscle mass. Furthermore cyclic output, volume per breath or heartbeat, is an approximately constant fraction of total capacity used, so volumes tend to scale approximately linearly.

2. *Conservation of volume.* Stahl (1962) pointed out that this principle, (represented by Eq. 5-19) is one of two key principles in biological analysis.

3. *Synchronism of times.* This is Stahl's (1962) second key principle. Cyclic periods are proportional to linear dimensions, that is, to $M^{1/3}$ to $M^{1/4}$ (see also McMahon, 1975). This is the governor of the physiological time scale (Hill, 1950), a subject to which I shall devote the next chapter.

4. *Size independence.* The range of pressures and expansions that the animal can attain and/or withstand, the costs of high velocities and of dead space, and the amount of fractional removal or extraction that is practical, all confine certain variables within such narrow limits that the very small or large mammals are treated equally. Thus pressures, flow velocities, ratios such as dead space to tidal volume, oxygen uptake to tidal volume, and diastolic to systolic ventricular (end) volumes tend not to scale up or down but are usually independent of size ($\propto M^0$).

How should respiratory and circulatory systems be designed (scaled) to meet the needs of a particular body size, and to meet them most efficiently? The needs could vary tenfold to twentyfold; what is adequate for the resting animal would not suffice during sustained exertion (as in running or migratory flight). In order to pump the oxygen-carrying air and blood to the tissues requiring it, pressure is required. Pressure ($ml^{-1}t^{-2}$) times volume (l^3) pumped equals work (ml^2t^{-2}), and to perform the work requires oxygen for the respiratory and heart muscles, a sort of overhead cost that a good manager would require be kept at a minimum. Unless we can think of a reason why a very small animal or a very large one could afford to invest more than 1.2% (at rest) to 3% (in heavy exercise) of its metabolic oxygen consumption into the pumping of that oxygen, we might expect the work rates of respiration and circulation to be a constant function of metabolic rate, that is $\propto M^{0.75}$, and for the mass of pumps, plumbing, and circulating fluid to be held to a minimum fraction of body mass, or $\propto M^{1.00}$.

The lung volume is a fixed proportion of body mass (0.6%), and the heart mass in mammals is a fixed proportion of body mass (1.1%). Then we might expect the linear dimensions of these organs to be directly proportional to the linear dimensions of the body. The oxygen must ultimately travel by diffusion from the lungs to the blood and from the blood into the metabolizing cells. Since diffusion rates are inversely proportional to the distance that must be crossed, it would not make sense for these distances to be scaled as are other linear dimensions of the body. Thus we might expect that the ultimate dimensions of alveoli, membrane thicknesses, red blood cells, and capillary diameters would be a size-independent M^0, for mice and elephants alike. Going from lungs and hearts ($\propto M^{1.0}$) to alveoli (M^0), where do the exponents begin to decrease — in the tracheae and aortae, in the bronchi and arteries, or in the bronchioles and arterioles?

Another approach to the anticipation of oxygen-supply-system scaling is to consider the nature of the transport process and how it changes throughout the system. The process is initiated by organs that are allowed only a fixed fraction of the body mass, perhaps according to the principle of symmorphosis. These are the $M^{1.0}$ lung + thorax pump and the $M^{1.0}$ heart pump. The pumping activity of each is oscillatory or cyclical. The minimum energy cost for operating an oscillator is at its natural frequency, which is inversely proportional to its characteristic linear dimension (l):

$$f \propto \frac{1}{l} \sqrt{\frac{(\sigma/\epsilon)}{\rho}}, \tag{5-20}$$

where σ/ϵ is Young's modulus, or the ratio of stress (F/A) to strain ($\Delta l/l_o$), and ρ is mass density. Both σ/ϵ and ρ are constants for a particular material, such as muscle (Alexander, 1968; McMahon, 1975).

If the scaling is geometrical ($l^3 \propto m$), frequency would then be proportional to l^{-1} or to $m^{-1/3}$; if the scaling conforms to elastic similarity, with

$l^4 \propto m$, the frequency would be a function of $m^{-1/4}$ (McMahon, 1975). It is the product of frequency and volume displacement per pump cycle that gives us the metabolically proportioned minute volume, or cardiac output (Eq. 5-19):

$$(m^{-1/4})(m^{1.0}) \propto m^{3/4}. \tag{5-21}$$

When the flow of air reaches the alveolar or capillary level, it is no longer pulsatile and the transport becomes a matter of diffusion in which

$$\dot{V}_{O_2} = -DA\frac{\Delta p_{O_2}}{x}, \tag{5-16}$$

where D is the diffusion coefficient, A is the surface area available for diffusion, Δp_{O_2} is the pressure difference across the site or surface of diffusion, and x is the thickness or path length. If, as we might initially expect, Δp_{O_2} and x are size independent, M^0, then we might expect also that the area available for diffusion, A, is directly proportional to the rate of oxygen uptake at maximum aerobic level:

$$A \propto \dot{V}_{O_2} \propto M^{3/4}. \tag{5-22}$$

However, we shall soon learn otherwise.

The blood capacity of the oxygen is a function of the hemoglobin concentration and the hematocrit (proportion of blood volume taken up by red cells). Thus if we double the hematocrit, the blood would only have to travel half as fast; or we could get by with half as much, but going from a hematocrit of 45% to 90% would approximately triple the blood viscosity, and the cost of overcoming the friction would be prohibitive. So we might anticipate a size-independent M^0 for hematocrit as well.

In order to understand respiratory and circulatory design and check the validity of the above expectations, it is necessary to consider the allometry of these systems — and the allometry is not possible without comparative data over a range of animal sizes. Thus, even if one's interests extend no further than the workings of our own species, it is obvious that the other species are needed — further justification for allocating resources to the preservation and study of species that may not be of immediate economic value.

RESPIRATION

Mechanics. Respiratory physiology is one of the fields that has been most extensively analyzed with respect to the requirements imposed by body size, and this allometry has had a significant effect on the quantitative matura-tion of the field. Most of the necessary measurements in the past have been obtained from resting animals. Recently the allometry has been extended to maximal oxygen uptake during sustained exertion. We start here with the morphological allometry, then proceed to resting respiration, about which more is known than respiration during activity.

The functioning of the respiratory systems of birds and mammals is perhaps appreciated more readily if we first compare the proportions of the structures involved. From Table 5-3 we see that the lung of a bird is only slightly heavier than that of a mammal of the same body mass, but the bird lung is more dense, occupying about half as much volume as the mammal's lung. This reflects the fact that the mammalian lung is an inflated "tree" of bronchial and bronchiolar airways, terminating in the expandable alveolar airspaces. The lung tissues, including the blood vessels, constitute about 20% of lung volume during breathing at rest, but when the full inspiratory reserve is inhaled, these same tissues constitute only about 10% of lung volume. The bird's lung inflates only slightly if at all, the secondary bronchi and parabronchi receiving unidirectional through-ventilation by means of the bellows action of the air sacs (see Schmidt-Nielsen, 1972). The avian lung is about 30% tissue (Dubach, 1981). The air sacs contribute about four-fifths of the volume of the respiratory system in birds and are the major reason why the total respiratory system volume in birds is nearly three times that of mammals equivalent in size.

There is remarkable agreement between the allometry of resting oxygen consumption rates and the separate allometric expressions for the factors which should, according to the principles of conservation of mass and dimensional consistency, account for that consumption (Table 5-3). Our theoretical expectations to this point are also satisfactorily met.

There are undeniable exponential similarities between primary ($\dot{V}_{O_2}$) and secondary derivations and between the two homeothermic classes. Note, however, that if we express the oxygen consumption rate (ml O_2/min) as a fraction of the total air volume pumped per minute (ml air/min) we get the milliliters of oxygen extracted per milliliter of air breathed: 0.03 $M^{-0.02}$ for eutherians. This figure appears to be low (normal for resting man is 0.05). The low fractional extraction may result from differences in animal re-sponse during measurement of air required and during measurement of oxygen consumption. The latter requires less restraint, has no extra cost of ventilation in a face mask, and is probably less upsetting to the animal. Assuming interclass similarity in this bias, birds seem to remove more

Table 5-3 Allometric comparison of the respiratory systems and oxygen transport of mammals and birds.

RESPIRATORY SYSTEM VOLUMES[a]

	Lung mass (g)	Lung volume (ml)	Tracheal volume (ml)	Total system volume (ml)
Nonpasserine birds	$12.6\,M^{0.95}$	$27.8\,M^{0.97}$	$3.70\,M^{1.09}$	$155\,M^{0.92}$
Eutherian mammals	$11.3\,M^{0.99}$	$53.5\,M^{1.06}$	$0.82\,M^{1.18}$	$54\,M^{1.06}$
Birds/mammals	$1.12\,M^{-0.04}$	$0.55\,M^{-0.12}$	$4.51\,M^{-0.09}$	$2.85\,M^{-0.14}$

OXYGEN TRANSPORT IN RESTING BIRDS AND MAMMALS[b]

	Frequency (cycles/min)	× Tidal or stroke volume (ml)	× Oxygen extracted[c] (ml O_2/ml air or blood)	= Oxygen consumption rate (ml O_2/min)
Nonpasserine birds				
Respiration	$17.2\,M^{-0.31}$	$16.9\,M^{1.05}$	$0.04\,M^{-0.05}$	$\left.\begin{array}{l}\\ \\ \end{array}\right\}11.3\,M^{0.72}$
Circulation	$156\,M^{-0.23}$	—	—	
Circulation/respiration	$9.1\,M^{0.08}$	—	—	
Marsupial mammals				
Respiration	$22.1\,M^{-0.26}$	$12.1\,M^{0.98}$	$0.03\,M^{0.00}$	$\left.\begin{array}{l}\\ \\ \end{array}\right\}7.5\,M^{0.72}$
Circulation	$116\,M^{-0.24}$	$1.22\,M^{1.07}$	$0.05\,M^{-0.11}$	
Circulation/respiration	$5.2\,M^{0.02}$	$0.10\,M^{0.09}$	$0.53\,M^{-0.11}$	
Eutherian mammals				
Respiration	$53.5\,M^{-0.26}$	$7.69\,M^{1.04}$	$0.03\,M^{-0.02}$	$\left.\begin{array}{l}\\ \\ \end{array}\right\}11.6\,M^{0.76}$
Circulation	$241\,M^{-0.25}$	$0.78\,M^{1.06}$	$0.06\,M^{-0.05}$	
Circulation/respiration	$4.5\,M^{0.01}$	$0.10\,M^{0.02}$	$0.46\,M^{-0.03}$	

a. From Hinds and Calder, 1971, and modified from Lasiewski and Calder, 1971, with lung and respiratory volumes revised by incorporation of data from Dubach, 1981.

b. Equations from Stahl, 1967; Lasiewski and Calder, 1971; Bech et al., 1979; and Dawson and Needham, 1981.

c. Calculated as $\dot{V}_{O_2}/\dot{V}_{air\ or\ blood}$.

oxygen from the respired air than do mammals, even at rest. This is attributed to their superior respiratory design, a unidirectional flow as compared to the mammals' cul-de-sac dilution process.

While the respiratory volume per minute of birds appears to be 55% to 75% that of eutherian mammals, owing to birds' higher resting extraction, an even greater difference between the avian and mammalian systems is the relative contributions of respiratory frequency and tidal volume to the minute volumes. Compared at the 1-kg body-mass level (because of the small differences in exponents for these variables), a 1-kg nonpasserine bird would have a respiratory frequency only 17.2/53.5 = 0.32 that of a mammal of such a size. Conversely, the 1-kg bird's tidal volume is greater, 16.9/7.69 = 2.2 times that of the mammal (Bech et al., 1979). This points out a difference in respiratory mechanics. The avian system has voluminous air sacs that are highly compliant, and when expanded occupy a considerable portion of the visceral space, while the bird's lung itself is smaller and more dense. The compliance is the ratio of volume change to pressure change exerted to obtain it. Total compliance includes the rib cage and musculature as well as the respiratory system. A more compliant structure inflates to a larger volume with less pressure differential applied to it. It is thus the reciprocal of the "stiffness" expressed by Young's modulus in Eq. (5-20), which could be rewritten in terms of compliance, κ:

$$f \propto \frac{1}{l}\sqrt{\frac{1}{\rho\kappa}}. \qquad (5\text{-}23)$$

Thus the lower respiratory frequency of birds could result in part from this reciprocal relationship to $\kappa^{1/2}$.

Is the total compliance of the avian lung/air sac respiratory system greater than that of mammals? The late Bob Lasiewski and I enjoyed a collaboration in which we compared, insofar as possible, the scaling in avian and mammalian respiratory systems. At the start of the 1970s there was only one published value for respiratory compliance, in the domestic chicken. We were finding so many parallels between avian and mammalian allometry that we included an "equation" based on that *one point*, borrowing the slope of $M^{1.04}$ from the more-studied mammalian system:

mammals: $\kappa = 1.56\ M^{1.04}$;   (5-24)

birds: $\kappa = 5.83\ M^{1.04}$.   (5-25)

Recent reexamination of respiratory compliance in mammals (25.9-g mouse to 26-kg dog) has confirmed the scaling of Eq. (5-24):

total κ: inflation $= 1.34\ M^{1.10}$;   (5-26)

deflation at 20 cm $H_2O = 1.80\ M^{1.11}$;   (5-27)

deflation at 8 cm $H_2O = 1.56\ M^{1.06}$.   (5-28)

lung κ: inflation $= 1.94\ M^{1.08}$;   (5-29)

deflation at 20 cm $H_2O = 3.09\ M^{1.15}$;   (5-30)

deflation at 8 cm $H_2O = 2.20\ M^{1.09}$.   (5-31)

thoracic κ: inflation $= 5.59\ M^{1.20}$;   (5-32)

deflation at 20 cm $H_2O = 4.61\ M^{1.07}$;   (5-33)

deflation at 8 cm $H_2O = 4.90\ M^{1.16}$.   (5-34)

All of these relationships had high correlation coefficients, $r = 0.959$ or greater (Bennett and Tenney, 1982). Note that compliance is the reciprocal of stiffness, and that the total system stiffness is the sum of lung and thoracic stiffnesses. Hence the relationship of κ_{total} to κ_{lung} and $\kappa_{thoracic}$ is the sum of the reciprocals:

$$1/\kappa_{total} = 1/\kappa_{lung} + 1/\kappa_{thoracic}. \tag{5-35}$$

This explains the range of coefficients in Eqs. (5-26) to (5-34). The mean exponent is 1.11, standard deviation (s.d.) 0.047. While they all appear to scale hyperallometrically, the exponents are not significantly different from 1.0.

The tentative bird equation was subsequently confirmed (and even given the dignity of a correlation coefficient, $r = 1.0$) when Crawford and Kampe (1971) published their value for κ_{total} of the pigeon, which I combined with the chicken value as

birds: $\kappa_{total} = 5.86\ M^{1.03}$. (5-25a)

It would be reassuring if this relationship were derived from a larger data base, but until such time we have a provisional index for comparing avian and mammalian total respiratory compliance:

$$5.86\ M^{1.03}/1.56\ M^{1.04} = 3.76\ M^{-0.01}. \tag{5-36}$$

From (5-23) we would expect

$$f_{bird}/f_{mammal} \propto 1/(3.76)^{1/2} \propto 0.52. \tag{5-37}$$

This could go a long way toward explaining the observed difference in respiratory frequencies, that of birds being 0.32 of mammals $(1/0.52 \div$

1/0.32 equals 62%, which may be the proportion due to a difference in compliance).

The trachea serves to conduct respiratory air exchanges from the environment in front of the nose or on the bill toward the lungs within the thorax. While this "windpipe" is obviously necessary, it interposes two disadvantages. There is frictional resistance to the airflow along the tracheal wall, and the volume of air contained in the trachea is dead air—the remainder of the last breath outbound that did not get dumped, and the tail end of the next inspiration that will never reach the lungs before the flow is reversed again. Birds have longer necks and therefore longer tracheae, which means that they must have a potential requirement for greater effort to draw enough fresh air into the system, and more dead-space air that must be drawn for nothing and compensated for.

Most of the flow of air in the trachea should be nonturbulent and laminar, described by the Poiseuille equation, a form of Ohm's Law which states that flow ($\dot{V}$) is proportional to the pressure difference (Δp) and inversely proportional to the resistance ($8\eta l/\pi r^4$):

$$\dot{V} = \frac{\Delta p \pi r^4}{8\eta l}.$$ (5-38)

We see that the greater the length l of the tube or the higher the viscosity η, the greater the resistance; but the larger the radius r, the less the resistance. Because it appears as r^4 (the combined effect of bigger cross-sectional area, πr^2, and the velocity gradient made possible by a larger radius), a small increase substantially decreases the resistance, enhancing the flow rate.

Excluding those birds whose tracheae are convoluted for vocal resonance, the essentially straight tracheae of birds on average are 2.7 times as long as those of mammals of the same body masses. Bird tracheae also have greater radii, 1.29 times those of mammals (see Table 5-6 later). If we consider the mammalian trachea as the reference standard, the combined effects on airflow are almost exactly compensatory (Hinds and Calder, 1971):

$$\dot{V} \propto \frac{(1.29)^4}{2.7} = \frac{2.77}{2.7} = 1.03.$$ (5-39)

The dead-space volumes (V_{ds}) of bird tracheae are considerably larger than those of mammals. Figured as a simple cylinder of length times cross-sectional area, the larger dimensions of bird tracheae increase the volume to

$$V_{ds,bird}/V_{ds,mammal} \propto (2.7)(1.29^2) = 4.5.$$ (5-40)

In mammals dead-space volume is scaled as a relatively constant proportion of the resting tidal volume (V_t), to minimize the wasted (nonexchange) volume but allow for warming, humidification, and buffering of the composition of air going to the alveoli (equations from Stahl, 1967):

$$\frac{V_{ds}}{V_t} = \frac{2.76\ M^{0.96}}{7.67\ M^{1.04}} = 0.36\ M^{-0.08}. \tag{5-41}$$

Simultaneously, Tenney and Bartlett (1967 and personal communication) examined the allometry of mammalian tracheae, which contribute to the total dead-space volume:

$$V_{tr} = 0.82\ M^{1.18}. \tag{5-42}$$

The proportion of this dead space to the total volume ventilated depends upon whether the ventilation is deep and infrequent, or shallow and of a high frequency. One way to overcome the effect of the greater V_{ds} for birds ($4.5 \times$ mammalian V_{ds}) would be to increase the tidal volume by the same factor and breathe at a slower frequency, to get the proper total volume of fresh air per minute. We have noted that the avian resting tidal volume is larger than that of mammals, but by a factor of two, less than half what we would have expected, to compensate for the V_{ds} increase. The amount of oxygen extracted (0.04 to 0.05 $M^{-0.05}$) is greater than that calculated for mammals (0.03 $M^{-0.02}$, Table 5-3) and could make up most of the remaining compensation that is apparently necessary for offsetting the large tracheal dead-space volume. This scheme for the evolution of respiratory allometry in birds appears in Figure 5-2.

The mean velocity of air during inspiration through the trachea (u_{in}) is equal to the volume flow rate (minute volume) divided by the cross-sectional area of the trachea. This is

$$\text{mammals:}\quad u_{in} \propto M^{0.80}/(M^{0.39})^2 \propto M^{0.02}, \tag{5-43}$$

$$\text{birds:}\quad u_{in} \propto M^{0.74}/(M^{0.35})^2 \propto M^{0.04}. \tag{5-44}$$

Therefore it appears that air velocity in the trachea is essentially independent of body size in both classes. The actual values depend on the proportions of the respiratory cycle devoted to inspiration and expiration. The velocity of the tracheal mucus movement (u_{tm}, in millimeters per minute) that serves to clear the tract of foreign material is, however, strongly size dependent ($r^2 = 0.90$) (Felicetti et al., 1981) in mammals:

$$u_{tm} = 3.02\ M^{0.41}. \tag{5-45}$$

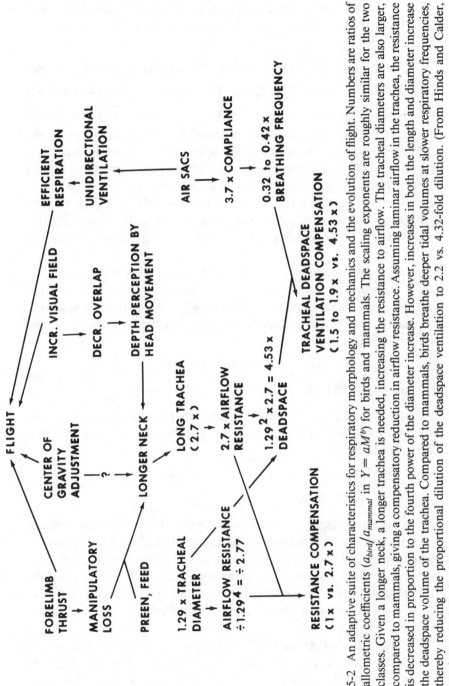

5-2 An adaptive suite of characteristics for respiratory morphology and mechanics and the evolution of flight. Numbers are ratios of allometric coefficients (a_{bird}/a_{mammal} in $Y = aM^b$) for birds and mammals. The scaling exponents are roughly similar for the two classes. Given a longer neck, a longer trachea is needed, increasing the resistance to airflow. The tracheal diameters are also larger, compared to mammals, giving a compensatory reduction in airflow resistance. Assuming laminar airflow in the trachea, the resistance is decreased in proportion to the fourth power of the diameter increase. However, increases in both the length and diameter increase the deadspace volume of the trachea. Compared to mammals, birds breathe deeper tidal volumes at slower respiratory frequencies, thereby reducing the proportional dilution of the deadspace ventilation to 2.2 vs. 4.32-fold dilution. (From Hinds and Calder, *Evolution* 25:347, © 1971).

The intrapulmonary pressure required to ventilate these volumes and velocities can be calculated as the tidal volume (V_t) divided by the compliance ($\Delta V/\Delta p$):

$$\text{mammals:} \quad 7.69\ M^{1.04}/1.56\ M^{1.04} = 4.93\ M^0 = 483\ \text{Pa}, \quad (5\text{-}46)$$

$$\text{birds:} \quad 16.93\ M^{1.05}/5.86\ M^{1.03} = 2.89\ M^{0.02} = 283\ \text{Pa}. \quad (5\text{-}47)$$

The pressures involved are size independent, and lower for the more compliant avian system. Furthermore, since the pressures developed by the respiratory (thoracic) pump are a size-independent constant, respiratory resistance should scale as the reciprocal of the flow rate (expressed as minute volume, $\dot{V}_e$). Stahl (1967) found that $\dot{V}_e$ for mammals was proportional to $M^{0.80}$ ($s_b = 0.01$), and that the airway resistance scaled as $M^{-0.70}$ ($s_b = 0.04$). Bennett and Tenney (1982) determined that the upper airway resistance for the mouse-dog series scaled as $M^{-0.70}$, and total respiratory resistance as $M^{-0.82}$.

The respiratory work (ml^2t^{-2}) involved in the pumping of resting respiration is a product of the pressure ($ml^{-1}t^{-2}$) and the volume (l^3) involved. Since the pressures are size independent, the work should be proportional to the volume. Stahl (1967) analyzed the mammalian data as follows (with the work unit g · cm):

$$\text{work/breath} = 17.1\ M^{1.08}; \quad r = 0.99; \quad n = 68; \quad s_b = 0.04;$$
$$\text{joules} = 0.0017\ M^{1.08}. \quad (5\text{-}48)$$

The power of breathing (P_{resp}), or work rate, is of course the product of work per breath and breaths per minute. Stahl's equation for this, in g · cm/min for mammals, is

$$P_{resp} = 962\ M^{0.78}; \quad r = 0.98; \quad n = 89; \quad s_b = 0.03;$$
$$\text{watts} = 0.00157\ M^{0.78}. \quad (5\text{-}49)$$

Resting breathing frequencies have been correlated with minimum ventilatory work rates for a series of mammals ranging in size from mouse to man (Agostini et al., 1959; Crosfill and Widdicombe, 1961). Note the similarity of the exponent to that for the basal metabolic rate. This indicates that the fraction of the metabolic rate necessary for resting respiration is independent of size (assuming that the efficiency of this effort is size independent, as well), as anticipated.

The Challenges of Activity. Most of the information to date on the allometry of oxygen delivery has been obtained from resting animals, just as most of the information on metabolic rates is at the basal level. While this information provides a useful baseline, a major reason for the emphasis on the inactive state is that it is technically easier to measure. This is rather like the story of the man who lost a key while walking at night. A friend who chanced by and saw the man on hands and knees beneath a street light, asked what was missing and joined to help him find the key. When the object could not be located, the usual questions ensued in an attempt to reconstruct the exact time and place of the loss. At this point the man acknowledged that he had actually lost the key down the street where it was dark, but had come up the street to search where the lighting was better.

The design of the respiratory system is probably adapted to the maximum challenge, not the minimum demand, that will be placed upon it. Or, as Gans (1979) suggested, "Most aspects of phenotypes will . . . be capable of fulfilling demands much greater than those routinely encountered." Fortunately, the design of the mammalian (eutherian) respiratory system has been examined recently and very systematically in relation to $\dot{V}_{O_{2max}}$ (Weibel and Taylor, 1981). I recommend this paper as a model study in allometry.

We can safely assume, with Taylor and Weibel (1981), that "animals are built reasonably," with enough alveolar surfaces for a rate of diffusion that will match the maximal rate of oxygen consumption of the running animal, and that they will have enough mitochondria to provide the chemical energy from oxidative metabolism for maximal work performance. Thus the structural design is adequate for peak demands with no costly excess of construction or maintenance of tissues beyond those demanded functionally. The matching of quantity of structure to functional needs is accomplished by a regulated morphogenesis called *symmorphosis,* which relates to anatomy as *homeostasis* does to physiology.

Taylor and Weibel (1981) list three basic hypotheses which follow from the principle of symmorphosis:

1. The structural design is a rate-limiting factor for O_2 flow at each level [as in Figure 5-3]. . .
2. The structural design is optimized . . . just enough to support the maximal O_2 flow rate. . .
3. The structural design is adaptable, at least within certain limits.

One can only be impressed by the clarity of these hypotheses and by the orderly and resourceful plan upon which were based the extensive experimental efforts. According to this principle of symmorphosis, we would expect the respiratory system to be scaled in proportion to the oxygen requirements that could be dependent upon it, specifically that the diffusing capacity of the lungs and maximum oxygen consumption rate would be

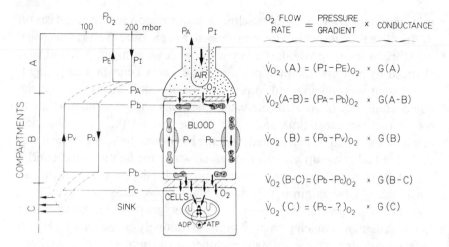

5-3 A model of oxygen delivery in mammals. Oxygen moves from higher to lower partial pressures (P) as indicated in the "respiratory cascade" graph at the left. The diffusional gradients (ΔP_{O_2}/distance) between the alveolar air and blood, and between the systemic capillaries and the metabolizing cells, are kept high to expedite the flow by two convectional loops, the respiratory (lungs, thoracic muscles, ribs, and diaphragm) and circulatory (heart) pumps. Flows are characterized by their rates, the driving pressure differences, and the conductances (relative ease of passage). The conductances via air (A) and blood (B) are related to bulk flow rates ($\dot{V}$) times capacitance (volume of oxygen per volume of air or blood). Conductances at the diffusional transfer steps (air-blood, A-B; blood-cells, B-C) are diffusing capacities, determined in part by the morphometry. The mitochondrial conductance is related to the respiratory chains of the mitochondrial membranes. (From Taylor and Weibel, 1981, $Respir. Physiol.$ 44:2.)

similarly scaled. The diffusing capacity (or diffusive conductance, $G_{L_{O_2}}$) is merely a collection of the diffusion coefficient (D), surface area for diffusion (A), and thickness (x) of the interface or membrane across which diffusion must occur:

$$G_{L_{O_2}} = -DA/x. \tag{5-50}$$

This can be substituted into Eq. (5-16) as:

$$\dot{V}_{O_2} = (G_{L_{O_2}})(\Delta p_{O_2}). \tag{5-51}$$

When $\dot{V}_{O_2}$ is measured in ml O_2/s and Δp_{O_2} is in mbar, $G_{L_{O_2}}$ has the units of ml O_2/(s · mbar) for dimensions of ml/(time · Δ pressure).

An oxygen molecule that diffuses from the alveolar air to a binding site on a hemoglobin molecule within an erythrocyte encounters a series of resistances to diffusion — resistances of the alveolar surface lining (R_s), alveolar and capillary wall tissues (R_t), plasma (R_p), and erythrocyte membrane (R_e). Resistances in series are added:

$$R_{total} = R_s + R_t + R_p + R_e. \tag{5-52}$$

Diffusive capacities or conductances are reciprocals of the respective resistances:

$$1/G_{L_{O_2}} = 1/G_s + 1/G_t + 1/G_p + 1/G_e \tag{5-53}$$

(Weibel, 1970/71; Weibel et al., 1981a). As is seen from Eq. (5-50), a diffusive conductance is the combined effect of the diffusion coefficient for a particular material (membranes, protoplasm, surfactant secretion) and the dimensions of exposed area and the thickness of the material. The thickness of the surfactant lining of the alveoli appears to be independent of body size, since the amounts extracted cover an area directly proportional to pulmonary surface area when submitted to a surface tension of 12 dyne/cm (Clements et al., 1970). The distances through the plasma layer and the erythrocyte must also be independent of animal size (see Eq. 5-101 and Table 5-6). The mean harmonic barrier thickness, $\tau_{h(t)}$, of the alveolar plus capillary tissues is size independent:

$$\tau_{h(t)} = 0.444\ M^{0.028}, \tag{5-54}$$

the 95% confidence interval of the exponents being -0.001 to $+0.057$ (Gehr et al., 1981a). From a larger sample of 32 species of mammals a significant, though slight, size dependence appears:

$$\tau_{h(t)} = 0.416\ M^{0.05}. \tag{5-54a}$$

Thus, if we assume that the diffusion coefficients of membranes, protoplasm, and body fluids are size independent as well, the component diffusive conductances must scale in direct proportion to the surface areas involved.

The total alveolar surface area $(A_A$, in square meters) approaches but does not quite reach a linear scaling to body mass for 27 African mammals (95% confidence interval of b, 0.87 to 0.97):

$$A_A = 3.58\ M^{0.92}, \tag{5-55}$$

and for the larger sample of 32 species (95% confidence interval of b, 0.92 to 0.98):

$$A_A = 3.34\ M^{0.95}. \tag{5-55a}$$

From Eqs. (5-54) and (5-55), or (5-54a) and (5-55a), the scaling of G_t can be approximated:

$$G_t \propto M^{0.92}/M^{0.03} \propto M^{0.89}, \tag{5-56}$$

or from the larger collection of data,

$$G_t \propto M^{0.95}/M^{0.05} \propto M^{0.90}. \tag{5-56a}$$

The scaling of total diffusive conductance is closer to and statistically indistinguishable from a linear relationship:

$$G_{L_{O_2}} = 0.055\, M^{0.95}, \qquad n = 27, \tag{5-57}$$

or for the 32 species (Gehr et al., 1981a):

$$G_{L_{O_2}} = 0.049\, M^{0.99}. \tag{5-57a}$$

The maximum oxygen consumption rate of these mammals when running on treadmills was about 11 times the basal metabolic rate for all sizes with confidence intervals of the scaling exponents which include the $M^{3/4}$ of the basal rate:

$$\dot{V}_{O_2 max} = 1.84\, M^{0.77}, \qquad n = 27, \tag{5-58}$$

or for the 32 species:

$$\dot{V}_{O_2 max} = 1.94\, M^{0.79}. \tag{5-58a}$$

A paradox appeared when Weibel and colleagues (1981b) combined expressions for $G_{L_{O_2}}$ and $\dot{V}_{O_2 max}$, substituting Eqs. (5-57) and (5-58) into (5-51) and solving for the necessary p_{O_2} needed to diffuse the oxygen from alveolus to erythrocyte:

$$\Delta p_{O_2} = \frac{\dot{V}_{O_2 max}}{G_{L_{O_2}}} = \frac{1.84\, M^{0.77}}{0.055\, M^{0.95}} = 33.5\, M^{-0.18}. \tag{5-59}$$

According to this result, a 20-g mouse or shrew would need a Δp_{O_2} of 68 mbar, 32% of the difference between atmospheric p_{O_2} and complete anoxia, just to accomplish the necessary diffusion to support $\dot{V}_{O_2 max}$, while a 500-kg galloping horse would have a drop of only 11 mbar for the same segment of the oxygen delivery process. Recall that this process, or respiratory cascade, consists of a sequence of drops in p_{O_2} due to humidification and dilution with dead-space and residual air left from the previous expiration, the drop across the alveolar-capillary barrier, and the drop from the capillary blood to the metabolizing cells in the metabolizing tissues. At least in resting mammals, the p_{O_2} and p_{CO_2} values are size independent (Table 5-4) and nothing

Table 5-4 Gas tensions within tissues of resting mammals.

Species	Body mass (g)	p_{O_2} (mbar)	p_{CO_2} (mbar)	Sample site	References
Bat	6	60	51	Subcutaneous gas pocket	Tenney and Morrison, 1967
Pocket mouse	18	59	51	Subcutaneous gas pocket	Tenney and Morrison, 1967
Shrew	20	49	71	Subcutaneous gas pocket	Tenney and Morrison, 1967
Ground squirrel	220	45	—	Venous	Parer and Metcalfe, 1967
Rat	400	53	60	Venous	Parer and Metcalfe, 1967
Rabbit	1,900	51, 65	31	Venous	Parer and Metcalfe, 1967
Cat	2,000	57	29	Venous	Parer and Metcalfe, 1967
Dog	22,000	48	—	Venous	Parer and Metcalfe, 1967
Sheep	62,000	54	—	Venous	Parer and Metcalfe, 1967
Man	75,000	55	61	Venous	Parer and Metcalfe, 1967
Cattle	414,000	53	—	Venous	Parer and Metcalfe, 1967
Allometry		$55\,M^{-0.005}$	$56\,M^{-0.03}$		
r^2		0.04	0.07		
n		11	7		

like the $M^{-0.18}$ scaling of Eq. (5-59). By creating subcutaneous gas pockets of inert SiF_6 gas in a small bat, a mouse, and a shrew, then waiting two hours or more for equilibration, Tenney and Morrison (1967) found that tissue partial pressures of oxygen were similar to those recorded from large mammals.

The scaling of A_A and $G_{L_{O_2}}$ obtained by Gehr and associates suggests that either the smaller mammals are inadequately endowed with alveolar surface area or the large mammals are oversupplied, a paradox that does not fit the idea of symmorphosis, and a paradox that has the shrew or mouse spending a third of the available Δp_{O_2} for a tiny segment of the total journey from *milieu extérieur* to *milieu intérieur*. These small mammals must have to tolerate much lower tissue p_{O_2} when running, and perhaps hyperventilation as well, to enhance the Δp_{O_2}.

With this $G_{L_{O_2}}$ scaling, the smaller mammal should have a somewhat greater Δp_{O_2}, either because of or to permit a relatively smaller diffusion surface area. Lindstedt (MS) explains this as a requirement of the shorter transit time for blood through the pulmonary capillaries of the smaller mammal. The transit time can be expected to be proportional to $M^{-1/4}$:

$$t_{transit} = \frac{\text{capillary blood volume}}{\text{blood volume flow rate}} = \frac{3.2 \, M^{1.0}}{2.77 \, M^{0.79}} = 1.16 \, M^{0.21}. \quad (5\text{-}60)$$

If the rate of oxygen binding by the hemoglobin in the pulmonary capillary bed is size independent, then Δp_{O_2} would have to scale as the reciprocal of transit time; that is, the shorter the time that an individual red blood cell is exposed, the higher the Δp_{O_2} must be to saturate it with oxygen. Lindstedt's reasoning yields Δp_{O_2} scaling similar to that in Eq. (5-59):

$$\text{at } \dot{V}_{O_2rest}: \quad \Delta p_{O_2} = 6.28 \, M^{-0.21}, \quad\quad\quad (5\text{-}60a)$$

$$\text{at } \dot{V}_{O_2max}: \quad \Delta p_{O_2} = 31.4 \, M^{-0.20}. \quad\quad\quad (5\text{-}60b)$$

A recent analysis of $\dot{V}_{O_2max}$ in reptiles makes it possible to compare their maximum aerobic capacity with that of mammals (Table 5-5). At body temperatures of 30° C to 35° C, the scaling exponents of lizards and of reptiles in general are similar to those for $\dot{V}_{O_2max}$ of mammals, but these exponents are temperature dependent, as are the 1-g intercept values. At body temperatures similar to those of active mammals, the $\dot{V}_{O_2max}$ of reptiles is only 10% or less of what the mammalian system can supply.

A little later I shall examine the growing body of evidence for small inverse exponents in blood and enzyme properties that seem to support the fast pace of metabolic physiology in small mammals. First, however, we

Table 5-5   Maximum oxygen consumption of mammals and reptiles as a function of body mass (kg).

Animals	Body temperature (°C)	Maximum oxygen consumption (ml/min)	Number of species	Correlation coefficient
Wild mammals[a]	≥ 37	116 $M^{0.79}$	14	0.995
Domestic mammals	≥ 37	101 $M^{0.86}$	8	0.961
Reptiles[b]	40	8.1 $M^{0.64}$	17	0.85
Lizards[b]	40	8.1 $M^{0.64}$	14	0.86
Reptiles	35	8.1 $M^{0.71}$	20	0.66
Lizards	35	9.5 $M^{0.76}$	13	0.76
Reptiles	30	6.1 $M^{0.77}$	21	0.71
Lizards	30	6.2 $M^{0.76}$	15	0.82
Reptiles	20	3.3 $M^{0.83}$	18	0.72
Lizards	20	3.1 $M^{0.81}$	12	0.78

a. Data on mammals from Taylor et al., 1981.
b. Data on reptiles and lizards from Bennett, 1982.

need an allometric view of the circulatory system which transports oxygen from the respiratory system to the metabolizing tissues, and of the size-dependent aspects of respiratory mechanics that make oxygen delivery different in small and large animals.

Respiratory Mechanics during Activity. The ventilation during running and flying has not been measured to any great extent, but allometry can be used for some preliminary projections. This attempt was stimulated by a recent study of running and breathing in mammals (Bramble and Carrier, 1983), which reported a phase-locking of limb and respiratory frequency for jackrabbits, dogs, men, and horses. The quadrupeds synchronized their breathing and stepping at a constant ratio of 1 : 1, trotting and galloping. Bramble and Carrier did not go into the allometry, but we can borrow from our own study of locomotion allometry in Chapter 7. At the speed of transition from trotting to galloping, the stepping frequency (f_{t-g}), in cycles per minute, for mouse, rat, dog, and horse, as we shall see, scales as

$$f_{t-g} = 269 \, M^{-0.14}. \tag{7-1}$$

With 1 : 1 phase-locking, the respiratory rate would be predicted by the same equation. In comparison, the resting f_r (Table 5-3) has a steeper inverse scaling, so that the proportional change in breathing rate going from rest to a trot or gallop is the ratio

$$f_{r,t-g}/f_{r,rest} = 269 \, M^{-0.14}/53.5 \, M^{-0.26} = 5.02 \, M^{0.12}. \tag{5-61}$$

On this basis a 25-g "mouse" would increase its f_r by a factor of 3.2, a 15-kg "dog" by a factor of 7, and a horse-sized mammal by a factor of 10.

Respiratory rates can be measured simply with small microphones (Bramble and Carrier, 1983) or thermocouples to detect the alternation between inspiratory and expiratory airflow. The measurement of tidal volumes is not so easy, and there are correspondingly fewer data. Perhaps they can be obtained indirectly. The ratio of metabolic power while running at the trot-gallop transition to standard metabolic power (derived from table 1 of Taylor, 1977c) is

$$\dot{E}_{t\text{-}g}/\dot{E}_{std} = 6.46 \; M^{0.07 \pm 0.002}. \tag{5-62}$$

As oxygen consumption, these metabolic rates are the product of f_r, V_t, and ΔF_{O_2} (Eq. 5-19), so the oxygen uptake per breath ($V_t \cdot \Delta F_{O_2}$) can be solved by substituting f_r and $\dot{V}_{O_2}$ ratios from Eqs. (5-61) and (5-62) and then solving for V_{O_2}/breath ($= V_t \cdot \Delta F_{O_2}$):

$$(V_t \cdot \Delta F_{O_2})_{t\text{-}g}/(V_t \cdot \Delta F_{O_2})_{rest} \propto M^{0.07}/M^{0.20} \propto M^{-0.13}. \tag{5-63}$$

To break the oxygen uptake per breath into separate ratios for V_t and ΔF_{O_2} it would be helpful to have ΔF_{O_2} regressions (inspired-expired); but if they exist, I have not seen them. At least at the alveolar-capillary level interface, there is no size dependence for the ratio of Δp_{O_2} (Lindstedt, MS) going from resting to maximum $\dot{V}_{O_2}$:

$$\Delta p_{O_2,max}/\Delta p_{O_2,rest} = 31.4 \; M^{-0.20}/6.28 \; M^{-0.21} = 5 \; M^{0.01}. \tag{5-64}$$

This suggests that the extraction ratio is size independent. If so, Eq. (5-63) indicates that V_t does not have as great an expansion factor for exercise in large mammals as in small ones. Direct information confirming or rejecting this thesis should come soon, to extrapolate from the pace of recent progress in the comparative physiology of exercise.

The patterns of change from resting to flying respiration in birds are, in general, similar to the change to running in mammals. The increase in f_r is greater for larger birds than for smaller ones. The extraction ratio appears to be independent of size, and the V_t increase is less for larger birds than for smaller birds (Berger and Hart, 1974):

$$f_{r,flight}/f_{r,rest} = 414 \; m^{-0.17}/182^{-0.33} = 2.27 \; m^{0.16}, \tag{5-65}$$

$$V_{t,flight}/V_{t,rest} = 0.10 \; m^{0.90}/0.0058 \; m^{1.04} = 17.4 \; m^{-0.14}, \tag{5-66}$$

$$\dot{V}_{e,flight}/\dot{V}_{e,rest} = 41.4 \; m^{0.73}/1.056 \; m^{0.71} = 39.2 \; m^{0.02}. \tag{5-67}$$

The $f_{r,flight}$ came from five species ranging from hummingbird to duck (table 2 of Berger and Hart, 1974) and from the 1,600-g barnacle goose *Branta leucopis* (Butler and Woakes, 1980) to obtain $r^2 = 0.727$; the $V_{t,flight}$ was derived by Berger and Hart, 1974; the resting expressions came from Lasiewski and Calder, 1971; and the $\dot{V}_e$ was calculated from the product $f_r V_t$.

The coordination between respiration and wingbeat frequencies in flying birds is not like the simple 1:1 synchrony of breathing and stepping of running mammals, partly because the scaling of wingbeat frequency varies with wing-loading type (see Table 7-1). For the lightest wing loading, like that of passerine birds, the allometry is

$$f_{wing}/f_{r,flight} = 2{,}188 \ m^{-0.36}/414 \ m^{-0.17} = 5.29 \ m^{-0.19}. \qquad (5\text{-}68)$$

For heavily loaded ducks and fowl it is

$$\begin{aligned}
f_{wing}/f_{r,flight} &= 352.8 \ M^{-0.24}/127.9 \ M^{-0.17} \\
&= 2.76 \ M^{-0.07} = 4.48 \ m^{-0.07}.
\end{aligned} \qquad (5\text{-}69)$$

These equations predict a gradual decrease in the ratio of wingbeats to breaths with an increase in bird size. This trend is indeed confirmed, but the observed coordination proceeds as a series of steps, and predictions from Eqs. (5-68) and (5-69) averaged $80 \pm 28.8\%$ of the observed range in $f_{wing}/f_{r,flight}$ ratios. (This 20% error could be smoothed out by rounding the decimal ratios to the next highest integer.)

It seems reasonable to expect that natural selection for conservation of effort or achievement of greater speeds or endurance would necessitate simultaneous modifications of more than one functional variable. Thus in our attempts to evaluate efficiencies and capacities, we must consider more than one function at a time. Initial explorations of the relations between respiratory and locomotory mechanics are encouraging. With the refinements that will be possible from a larger selection of data, our knowledge of functional interactions could advance significantly.

CIRCULATION

In macroscopic animals the circulatory system serves as an intermediary between diffusion of oxygen from the alveoli and diffusion of oxygen into the working tissues, an express route or bypass of the slowness of diffusion over long (> 1 mm) distances (shown by Eq. 5-17). There are two phases of scaling to be considered with respect to circulatory logistics, the scaling of the oxygen delivery, which is the primary function of this system, and the scaling of the fluid movement, which is the means by which the function is accomplished. The rate of oxygen delivery to the tissues by the heart and circulatory system is also described by Fick's Principle:

$$\left(\begin{array}{c}\text{heart}\\ \text{rate or}\\ \text{frequency}\end{array}\right) \times \left(\begin{array}{c}\text{blood volume}\\ \text{per beat or}\\ \text{stroke volume}\end{array}\right) \times \left(\begin{array}{c}\text{concentration change,}\\ \text{arterial to mean}\\ \text{venous blood}\end{array}\right) = \begin{array}{c}\text{oxygen}\\ \text{uptake}\end{array} \tag{5-19a}$$

$$\begin{array}{ccccccc} \text{(per min)} & & \text{(ml)} & & (\text{ml }O_2/\text{ml blood}) & & (\text{ml }O_2/\text{min}) \\ f_h & \times & V_s & \times & (C_{a_{O_2}} - C_{\bar{v}_{O_2}}) & = & \dot{V}_{O_2} \end{array}$$

where $(C_{a_{O_2}} - C_{\bar{v}_{O_2}})$ is the difference in oxygen concentration of arterial blood and the average for returning venous blood. We shall examine each of these allometrically. A summary is found in Table 5-3.

Resting Heart Rate. Depending on which ornithology book you read, you may find "The rate of the heart beat, long considered an indicator of the physiological activity in mammals, is higher in most birds than in mammals of equivalent size" (Pettingill, 1970), or "As a rule, the heart beats less rapidly in birds than in mammals of comparable size" (Welty, 1982). The allometry supports Welty. For resting mammals Stahl (1967) reported:

$$f_h = 241 \ M^{-0.25}, \tag{5-70}$$

and for resting birds Calder (1968) derived:

$$f_h = 156 \ M^{-0.23}. \tag{5-71}$$

Resting marsupial mammals do have lower heartbeat frequencies than birds (Kinnear and Brown, 1967):

$$f_h = 106 \ M^{-0.27}. \tag{5-72}$$

Bird hearts are larger than those of equal-sized mammals, have lower resting heartbeat frequencies, and develop higher mean blood pressures ($kPa = 21 \ M^{0.036}$ for birds vs. $12 \ M^{0.032}$ for mammals). Although the regression for avian mean arterial pressure is not significant ($n = 11, r = 0.43, p > 0.10$), it appears to be size independent and consistently above mammalian arterial pressure (Calder, 1981).

Within a class a larger animal's larger heart beats more slowly than that of a smaller animal. Is the difference in rates of birds and mammals caused solely by differences in heart mass (m_h) per unit of body mass? No:

$$\text{mammals:} \quad f_h = 379 \ m_h^{-0.255}, \tag{5-73}$$

$$\text{birds:} \quad f_h = 264 \ m_h^{-0.244}. \tag{5-74}$$

Furthermore, the longer life spans of birds are not a mere result of conservation of heartbeats at a slower frequency, for in a lifetime $2.4 \ M^{-0.01}$ times that of a mammal's, a bird's heart will have 38% more resting beats (Lindstedt and Calder, 1976, 1981), each developing and withstanding a significantly higher blood pressure. The reasons for this superior endurance are not known.

Respiratory frequencies of resting mammals are correlated with a minimum ventilatory work rate. Similarly, ventricular work is frequency dependent, in this case because of aortic impedance. Minimal cardiac work apparently has been selected via a scaling of f_h that preserves a constant ratio of pulse wavelength (λ) to aortic length (l_{aorta}) (Milnor, 1979; Noordergraff et al., 1979). The velocity of this pulse wave is about 6 m/s and it is essentially size independent, like the aortic blood velocity (0.26 $M^{0.07}$), but faster (see Eq. 5-102 and Milnor, 1979). The pulse wave is reflected throughout the arterial tree when the pathway bends, narrows, or splits. This causes an impedance to flow that is frequency dependent (consider the difference in work required to overcome a wave reflected back toward the ejecting ventricle compared to the work required when the wave is still going forward with the kinetic energy in the right direction). The hydraulic power $\dot{W}_h$ is proportional to the impedance modulus and is greatest at very low frequencies. Impedance to flow in the rat, dog, and man studied by Milnor was minimal when λ was approximately 4 l_{aorta}; since velocity (c_p in meters per second) is the product of λ (in meters) and f (in hertz, or cycles per second):

$$\lambda = c_p/f_h = 4 \, l_{aorta}. \tag{5-75}$$

This relationship can also be approached allometrically, using Milnor's data (Calder, 1981):

$$\lambda = 6/4.7 \, M^{-0.32} = k(0.132 \, M^{0.27}), \tag{5-76}$$

in which the proportionality between λ and l_{aorta} is not 4 but 9.7 $M^{0.05}$. If we substitute the allometry for f_h and l_{aorta} from a more extensive mammalian data base (Eq. 5-70 and Table 5-6), this difference is reduced slightly:

$$\lambda = 6/4.02 \, M^{-0.25} = (8.7 \, M^{-0.05})(0.171 \, M^{0.30}). \tag{5-77}$$

While there is a need for c_p data from additional species to fine-tune the λ/l_{aorta} relationship, Milnor has provided an explanation of heart rate scaling. That this was a new hypothesis, or the only hypothesis, was subsequently questioned (Iberall, 1979; O'Rourke, 1981). Iberall favors a geometric similarity model relating the undamped standing wave frequency (f_o) to a quarter-wavelength arterial "terminus" of total length (l_{tot}) where the arterial tree becomes highly resistant to flow:

$$f_o = u_{blood}/(4l_{tot}). \tag{5-78}$$

Table 5-6 Comparison of the dimensions of major circulatory and respiratory vessels by allometry and by support and theoretical scaling.

Structure	Length	Diameter	Volume[a]	References
Mammal				
Left ventricle	—	$M^{0.44}$	$M^{1.02}$	Holt et al., 1968; Martin and Haines, 1970
Aorta	$16.1\ M^{0.32}$	$0.34\ M^{0.36}$	$M^{1.04}$	Holt et al., 1981
Ascending aorta	$1.0\ M^{0.28}$	$0.41\ M^{0.36}$	$M^{1.00}$	Holt et al., 1981
Inferior vena cava	$13.3\ M^{0.33}$	$0.48\ M^{0.41}$	$M^{1.15}$	Holt et al., 1981
Right renal artery	—	$M^{0.30}$	—	Holt et al., 1981
Right iliac artery	—	$M^{0.31}$	—	Holt et al., 1981
Capillary	—	$M^{-0.02}$	—	Assumed from erythrocyte diameters
Capillary density/cm^2				
Masseter, 10 spp.	—	$7.96\ M^{-0.14}$	—	Schmidt-Nielen and Pennycuick, 1961
Semitendinosus, 19 spp.	—	$7.47\ M^{-0.14}$	—	Hoppeler et al., 1981
Longissimus dorsi	—	$7.60\ M^{-0.10}$	—	Hoppeler et al., 1981
Vastus medialis	—	$9.65\ M^{-0.10}$	—	Hoppeler et al., 1981
Diaphragm	—	$10.60\ M^{-0.05}$	—	Hoppeler et al., 1981

Trachea (cm, ml)	$6.2\,M^{0.40}$	$0.41\,M^{0.39}$	$0.82\,M^{1.18}$	Tenney and Bartlett, 1967 and personal communication
Alveolus	—	$M^{0.137}$	—	Spells, 1968
Alveolus	—	$77\,M^{0.13}$	—	Tenney and Remmers, 1963
Alveolus	—	$95\,M^{0.18}$	—	Calculated from Tenney and Remmers $(\dot{V}_{O_2}/m)$, 1963
Bird trachea (cm, ml)	$16.8\,M^{0.39}$	$0.53\,M^{0.35}$	$3.70\,M^{1.09}$	Hinds and Calder, 1971
Elastic similarity	$M^{0.25}$	$M^{0.38}$	$M^{1.01}$	McMahon, 1975
Cattle bodies	$M^{0.34b}$	$M^{0.35}$	$M^{0.96}$	Brody, 1945
Mammal bodies (m)	$0.341\,M^{0.31}$	—	—	Economos 1982, 1983
Small, $M \le 20$ kg	$0.329\,M^{0.34}$	—	—	Economos 1982, 1983
Large, $M > 20$ kg	$0.441\,M^{0.27}$	—	—	Economos 1982, 1983
Limb bones				
Bovidae	$M^{0.25}$	$M^{0.38}$	$M^{1.01}$	McMahon, 1975
Eutheria	$M^{0.35}$	$M^{0.36}$	$M^{1.07}$	Alexander et al., 1979
Horizontal, birds	$M^{0.28}$	$M^{0.40}$	$M^{1.08}$	Maloiy et al., 1979
Vertical, birds	$M^{0.43}$	—	$M^{1.23}$	Maloiy et al., 1979

a. $V = \pi\,(D/2)^2\,L$.

b. Height at withers $\propto M^{0.24}$.

For blood velocity (u_{blood}) he used 0.47 m/s compared to the 0.26 $M^{0.07}$ used above. This model predicted the f_h better at extremes in the mammalian body-size range, coming closer than Milnor's model, and is consistent with the geometric constancy of l_{aorta}/d_{aorta} described by Holt and coworkers (see Table 5-6).

O'Rourke basically favored the hypothesis of Milnor, but added several interesting points. The heart is located near a node of pressure and an antinode of flow. This provides the best impedance match for f_h just above the resting rate and allows performance near the optimal level for demands above the resting level, in activity necessary for survival in fright or flight situations. It is not the rate as such, but rather the ejection duration that is crucial for optimizing the impedance matching. The ratio of durations of systole and diastole is relatively constant among species, but it changes in cardioacceleration. The relationships are more or less the same in the pulmonary circulation, where the vascular beds are shorter, but the pulse wave velocity is also lower; so the two factors offset each other in setting the same optimal frequency or ejection duration. Previous authors had reported similar ideas, with data for species beyond those used by Milnor.

Resting Cardiac Output. The cardiac output (milliliters of blood pumped per minute, $\dot{V}_{blood}$) is the product of heartbeat frequency and volume per heartbeat (stroke volume). The stroke volume is calculated from direct measurement of f_h and $\dot{V}_{blood}$; the latter, in resting mammals (Stahl, 1967), scales approximately the same as $\dot{V}_{O_2}$ ($r = 0.98$; $n = 568$; standard error of coefficients, or s_a, $= \pm 31\%$; $s_b = 0.01$):

$$\dot{V}_{blood} = 187 \ M^{0.81}. \tag{5-79}$$

This is a very important baseline, and we can use it in the estimation of many design features in the scaling of circulatory function to body size. A few allometric equations can help us deduce a great deal—and a lot more easily than if we were to measure everything directly.

First of all, if we divide $\dot{V}_{blood}$ by heart rate, we see that stroke volume (V_s) is approximately a linear function of body mass in resting mammals:

$$V_s = \dot{V}_{blood}/f_h = 187 \ M^{0.81}/241 \ M^{-0.25} = 0.78 \ M^{1.06}. \tag{5-80}$$

Heart mass (M_h) is also scaled almost linearly ($r = 0.99$, $n > 200$, $s_a = 41\%$, $s_b = 0.01$):

$$M_h = 0.058 \ M^{0.98}. \tag{5-81}$$

Thus in a 1-kg mammal the 5.8 g of heart pumps 0.78 ml of blood per contraction during rest. The residual mass exponent of 0.08 is probably not significant; alternatively it suggests that with a larger heart, a slightly greater proportional ejection of blood will be accomplished. The heart mass scales in the same way for marsupial mammals, but is at all body sizes a larger heart than for eutherians (Needham and Dawson, 1982):

$$M_h = 0.077 \, M^{0.98}. \tag{5-82}$$

This is 32% larger than the eutherian heart, and during rest the marsupial heart pumps 56% more blood per stroke (Dawson and Needham, 1981).

The protherians or monotremes have persisted to this day as only three species in Australia and New Guinea, with a body mass range of only 1.0 kg to 16.5 kg, which limits the application of allometry beyond that for standard metabolism (seen in Figure 5-1). Comparison has been made, however, on an equal-size basis by Dawson (1983), who points out the following. Compared to an equivalent placental, say a rabbit, an echidna uses oxygen at one-fourth the rate; this is provided by a cardiac output one-fourth that of a rabbit, so the oxygen extraction per milliliter of blood pumped is the same. The body masses of the aquatic platypus and terrestrial echidna do not overlap, but their basal oxygen consumption can be compared per metabolic unit of kg $^{0.75}$, in effect an extrapolation of the appropriate plots in Figure 5-1. On this basis the platypus has twice the basal oxygen consumption rate of the echidna, but this oxygen is delivered at the same resting heart frequency by a heart proportionately twice as large and presumably delivering a proportionately larger cardiac output.

The data on resting cardiac output in birds are correlated with experimental heart rates that are considerably higher than those observed in truly resting birds, so we have no reliable allometry for this function (Calder, 1981). However, if stroke volume has the same proportionality to heart mass in birds as in mammals (Eqs. 5-80 and 5-81), the V_s for birds would be expected to be about 1.5 times that of mammals:

$$M_{h,birds}/M_{h,mammals} = 0.086 \, M^{0.94}/0.058 \, M^{0.98} = 1.48 \, M^{-0.04}. \tag{5-83}$$

Now if $\dot{V}_{blood}$ is substituted from Eq. (5-79) into (5-19a), we can solve for the difference in oxygen content between arterial and venous blood (as it returns mixed from the various veins to the heart):

$$\dot{V}_{O_2}/\dot{V}_{blood} = 11.6 \, M^{0.76}/187 \, M^{0.81} = 0.06 \, M^{-0.05}. \tag{5-84}$$

Assuming that the residual mass exponent (r.m.e.) of -0.05 is not significant, we see that across the size range, during rest, the volume of oxygen extracted is about 6% of the volume of blood pumped. The oxygen capacity of the blood of mammals varies with habits (highest in diving mammals), but not systematically with size, and is typically about 20 ml per 100 ml of blood. At rest, mammals then typically remove $\%_{20}$, or 30%, of the available oxygen. Mst of the remaining 70% can be used in exercise.

Exercise. Baudinette (1978) summarized the maximum heart rates of running and hopping mammals, both eutherian and marsupial, as $r = 0.967$, $p < 0.01$:

$$f_{h,max} = 375 \, M^{-0.19}. \tag{5-85}$$

This is about 1.6 times resting rates if we ignore the r.m.e. of 0.06, when compared with Eq. (5-70). If available data are compared for different taxonomic groups on a species-by-species basis, the expansion factors from rest to activity appear to be similar (Calder, 1981):

Eutheria: $f_{h,max}/f_h = 2.13 \pm 0.94$ s.d., $n = 6$;   (5-86)

Marsupialia: $f_{h,max}/f_h = 2.24 \pm 0.35$ s.d., $n = 6$;   (5-87)

Aves: $f_{h,max}/f_h = 2.56 \pm 0.73$ s.d., $n = 10$.   (5-88)

The ratio of $\dot{V}_{O_2 max}$ to $\dot{V}_{O_2}$ at rest in wild eutherians (from Stahl, 1967, and Taylor et al., 1981), is

$$\dot{V}_{O_{2,max}}/\dot{V}_{O_{2,rest}} = 116.4 \, M^{0.79}/11.6 \, M^{0.76} = 10 \, M^{0.03}. \tag{5-89}$$

Thus for a tenfold expansion in rate of oxygen consumption, with heart rate increased by a factor of 2 to 2.5, the product of stroke volume and arteriovenous oxygen differential (the "oxygen pulse") must increase by a factor of 4 to 5 (4.7 in Eutheria). In man, the stroke volume can increase by about 25% and the arteriovenous oxygen difference by a factor of 3.4, an oxygen pulse ratio of 4.2 coming reasonably close to our allometric expectations.

Structural Allometry. In view of the approximately linear scaling of heart mass and stroke volume to body mass, ventricular volumes (end of diastole, V_{ed}, and end of systole, V_{es}, in milliliters) would also be expected to scale linearly to body mass—which they do, according to the excellent description of Holt and associates (1968):

$$V_{ed} = 1.76 \, M^{1.02}, \tag{5-90}$$

$$V_{es} = 0.59 \, M^{0.99}. \tag{5-91}$$

This tells us that just about one-third of the blood volume is ejected per heart contraction. This ejected volume, $1.76 - 0.59 = 1.17$ ml, should be the stroke volume, 50% greater than the prediction from Eq. (5-80). This is twice the increase noted above in exercising man, suggesting that there is a need for some refinement in the data. However, one must realize that the allometry is only a generalization over four or more orders of magnitude in body size, and therefore limited in predictive precision.

Having measured the size of the heart, externally and internally, we might wonder about the strength of the walls. The pressure (p) expanding a vessel of circular or oval cross-section is related to the radii (r_1 and r_2, perpendicular minor and major axes) and tension (Fl^{-1}) by Laplace's Law (which is perhaps more familiar as $p = T/R$):

$$p = (Fl^{-1})(r_1^{-1} + r_2^{-1}). \tag{5-92}$$

Martin and Haines (1970) evaluated the dimensions of the left ventricle and found that it conformed to the above relationship for seven species of mammals ranging in size from 310 g to 252 kg. They assumed that the thickness of the ventricular walls (x_h) was proportional to the tension they withstood or could develop. Analyzing the dimensions they measured, we find:

$$r_1 = 4.83 \, M^{0.44}, \tag{5-93}$$

$$r_2 = 13.52 \, M^{0.45}, \tag{5-94}$$

$$x_h = 2.50 \, M^{0.46}. \tag{5-95}$$

Substituting these dimensional allometries (all with $p < 0.001$) into Eq. (5-92) we find:

$$p \propto M^{0.46} / (k_1 \, M^{0.44} + k_2 \, M^{0.45}) \propto M^{-0.01}. \tag{5-96}$$

This not only confirms the conformity to Laplace's Law, but leads us to expect the pressure developed by the heart contraction to be independent of body size. As a check on this, we can run a regression on mean arterial pressures (diastolic plus one-third of pulse difference, in millibars) of mammals from mouse to elephant (excluding the giraffe) calculated from table 20-2 of Prosser (1973) (with typical body masses from other sources):

$$\bar{p} = 116 \, M^{0.026}. \tag{5-97}$$

A similar calculation for six species of birds (whose heart rates were elevated from basal predictions) was

$$\bar{p} = 208 \, M^{0.036}. \tag{5-98}$$

Günther (1975) listed a similar scaling for systemic arterial pressure ($\propto M^{0.032}$). It would appear that blood pressures developed by the larger bird hearts are higher than those of mammals, with a similar size independence —although some of the increase may have been the result of procedures during measurement.

The total peripheral resistance encountered can be appreciated by application of Ohm's Law, which says:

$$\text{flow rate} = \text{potential difference/resistance,}$$
$$\text{or} \quad \dot{V} = \Delta p / R. \tag{5-99}$$

Solving for total peripheral resistance, allometrically, we get:

$$R = \Delta p / \dot{V} = 116 \, M^{0.026} / 187 \, M^{0.81} = 0.62 \, M^{-0.78} \tag{5-100}$$

The units are "peripheral resistance units" of millibars $(ml \cdot min)^{-1}$. Certainly there is less resistance in the bloodstream of an elephant than of a mouse, but that is not quite the proper way to view Eq. (5-100), because the diameters of red blood cells are not scaled to body sizes differently, their diameters in micrometers from Altman and Dittmer (1972) scaling as

$$d_{rbc} = 6.3 \, M^{-0.02}, \tag{5-101}$$

and all erythrocytes just barely pass through their capillaries, which must then have similar diameters. The resistance, proportional to $M^{-3/4}$, must therefore be a consequence of the number of parallel routes from which we can choose.

The problem is that the cross-sectional area (A_x) is proportional to diameter squared. The scaling decreases as we leave the ventricle (Table 5-6), with its $d \propto M^{0.44}$ and $A_x \propto M^{0.88}$, passing through the aorta $(d \propto M^{0.36}; A_x = M^{0.72})$ and on to an artery such as the renal or right iliac where $d \propto M^{0.30}, M^{0.31}; A_x \propto M^{0.60}, M^{0.62}$. This becomes M^0 in the capillaries, but returns to a scaling of $d \propto M^{0.41}, A_x \propto M^{0.82}$, in the inferior vena cava before reentry into the heart.

The principle of continuity (as discussed in Vogel's 1981 gem, *Life in Moving Fluids*) states that the blood is not destroyed or lost from the system (unless the animal is wounded), nor is the fluid compressed. Hence at any fraction of the distance from left ventricle to right atrium the total flow rate, in all parallel vessels including the lymph system, is the same.

The function of the bloodstream is to expedite oxygen transport over longer distances. Logically the stream should be fast in the major vessels. However, at the point of oxygen delivery diffusion must be relied upon. To facilitate this slow process, the blood speed must be drastically reduced to allow time for adequate exchange. How so, if the volume flow rate is unchanged?

Rate of flow has the dimensions $l^3 t^{-1}$, whereas stream velocity is in terms of lt^{-1}. The difference between the two is l^2, or total cross-sectional area. The velocity or speed of blood in the aorta (u_{aorta}, in centimeters per second) would be

$$u_{aorta} = \dot{V}_{blood}/A_x = 3.12\ M^{0.81}/\pi \left(\frac{0.34}{2}\ M^{0.36}\right)^2 = 34.3\ M^{0.09}. \qquad (5\text{-}102)$$

However, cardiac output ($\dot{V}_{blood}$, in milliliters per minute) has been described as 2.70 $M^{0.75}$ (White et al., 1968) and 2.77 $M^{0.79}$ (Holt et al., 1968), predicting a u_{aorta} of 27.7 $M^{0.03}$ to 30.5 $M^{0.07}$. Further exponential variability exists in the derivation of aortic diameters, so presumably the r.m.e.'s are insignificant; hence the suggestion that aortic blood velocity is size independent (Holt et al., 1981). The cross-sectional area of the individual route to one capillary has diminished drastically, being measured in square micrometers instead of square centimeters, but the total cross-sectional area ($A_{x,total\ c}$) is greatly increased by the large number of parallel vessels. At any one level the cross-sectional area is the roduct of vessel size (capillary diameter squared, d_c^2), density of vessels per unit of body or muscle cross-section, and that cross-section (see Figure 5-4).

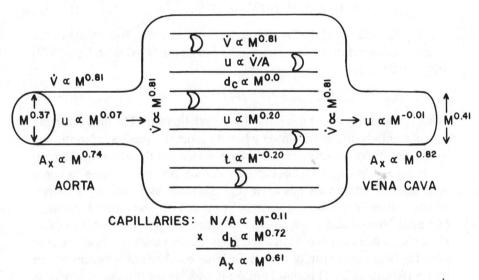

5-4 Circulatory velocities (u) are, on the average, equal to volume flow rates ($\dot{V}$) divided by total cross-sectional area at that level (A_x in aorta, capillaries, or vena cava). The total A_x of the systemic capillary beds is estimated as the product of capillary density (N/A) and A_x of the body. If the scalings substituted herein are correct, capillary transit time is inversely related to body size, giving a better chance for O_2-CO_2 exchange in the intensely metabolizing tissues of mice and shrews.

For the capillaries, we have:

$$A_{x,totalc} \propto (d_c^2)(N_c/A_{tissue})(A_{tissue}). \tag{5-103}$$

Do we have enough information to fill in the exponents? The diameter of the erythrocytes can be substituted from Eq. (5-101) for the capillary diameter (d_c). The capillary density N_c/A_{tissue} can be calculated for masseter muscles from Schmidt-Nielsen and Pennycuick (1961), and has been determined on muscle tissue also in the diaphragm, semitendinosus, longissimus dorsi, and vastus medialis; the scaling averages $M^{-0.11}$ ($M^{-0.14}$, $M^{-0.05}$, $M^{-0.14}$, $M^{-0.10}$, $M^{-0.10}$ respectively; see Table 5-6). The total N_c is the product of density and muscle cross-sectional area. Trunk diameters of cattle and primates scale as $M^{0.35}$ to $M^{0.38}$, as do bovid limbs (Brody, 1945; McMahon, 1973, 1975). Therefore, the scaling of the total cross-sectional A_x of the capillary bed can be estimated:

$$A_{x,totalc} \propto [M^{(-0.02)2}][M^{-0.11}][M^{(0.36)2}] \propto M^{0.57}, \tag{5-103a}$$

or, since the scaling of d_{rbc}^2 is not significant, the first term could be considered to be M^0:

$$A_{x,totalc} \propto (M^0)(M^{-0.11})(M^{0.72}) \propto M^{0.61}. \tag{5-103b}$$

If Eq. (5-103a) or (5-103b) is close to being accurate, the scaling of the mean blood speed in the capillaries (u_{cap}) can be predicted from Eqs. (5-79) and (5-103b):

$$u_{cap} = \dot{V}_{blood}/A_{x,totalc} \propto M^{0.81}/M^{0.61} \propto M^{0.20}. \tag{5-104}$$

What is interesting about this is the inverse relationship: The time taken per unit of capillary distance is slower when the animal is smaller, as contrasted to the essentially size-independent blood velocity in the aorta. In the past, the benefit of having a higher capillary density per unit of cross-sectional area in small mammals has been regarded only as providing a shorter distance from the intensely metabolizing cell to the nearest capillary (Schmidt-Nielsen and Pennycuick, 1961). Just as important, or even more so, it also confers the benefit of a relatively slower capillary blood velocity, allowing better opportunity for a diffusional equilibrium between oxygen and carbon dioxide. This should make it possible for the shrew or mouse muscles to get their oxygen with a Δ_{pO_2} only slightly greater than in the elephant.

SMALL INVERSE EXPONENTS AND INTENSE METABOLISM
OF SMALL HOMEOTHERMS

It appears that the scaling of convective gas transport, both in respiration and circulation, matches up well with metabolic oxygen demand (Table 5-3). This is best expressed through the cancellation of power law formulas in search of the "size-independent constants" and "dimensionless variables" advocated by Stahl (1962, 1963a,b, 1967). So doing, we obtain (from Stahl, 1967; Lasiewski and Calder, 1971):

mammals:

$$\dot{V}_{O_2}/(f_r V_t) = 11.6 \ M^{0.76}/(53.5 \ M^{-0.26} \ 7.69 \ M^{1.04}) = 0.03 \ M^{-0.02}$$
$$95\% \ \text{c.i. on exponents: } 0.01/0.01, \ 0.01; \qquad (5\text{-}105)$$

$$\dot{V}_{O_2}/(f_h V_s) = 11.6 \ M^{0.76}/(241 \ M^{-0.25} \ 0.78 \ M^{1.06}) = 0.06 \ M^{-0.05}$$
$$95\% \ \text{c.i. on exponents: } 0.01/0.02, \ 0.01; \qquad (5\text{-}106)$$

birds (other than Passeriformes):
$$\dot{V}_{O_2}/(f_r V_t) = 11.3 \ M^{0.72}/(17.2 \ M^{-0.31} 16.9 \ M^{1.05}) = 0.04 \ M^{-0.02}. \qquad (5\text{-}107)$$

Note that the residual mass exponents at the far right above are all well within the range of ± 0.08, which Stahl considered to be experimental variability not distinguishable from $M^{0.0}$. Thus the dimensionless ratios of $\dot{V}_{O_2}/\dot{V}_{air}$ and $\dot{V}_{O_2}/\dot{V}_{blood}$ above seem to meet the challenges of supporting the intense specific metabolic rates, $\dot{V}_{O_2}$ per unit of mass ($M^{3/4}/M = M^{-1/4}$), of small homeotherms.

I find myself in the ambiguous position of suggesting on the one hand that small r.m.e. values are not significant, and of drawing attention on the other hand to several indications of small inverse exponents that appear to be statistically significant and can be rationalized relative to the needs of small-sized homeotherms and their metabolic intensities.

These exponents, collected in Table 5-7, derive much of their meaning through mutual reinforcement of a pattern. There are significant, though small, negative exponents for oxygen affinity*, erythrocyte carbonic anhydrase concentration, oxidative enzyme activity, sugar concentration, liver mitochondrial density, and capillary density, but not for hematocrit.

---

* Note that the significance of the negative regression for avian p_{50} on body mass (Lutz et al., 1974) was disputed by Baumann and Baumann (1977). From their additional data (apparently using hemoglobin solutions rather than whole blood) and a variety of p_{50} values from the literature, they obtained a lower correlation coefficient, $r = -0.587$. Palomeque and colleagues (1980) reported p_{50} values for eight species of Passeriformes, 10.6 g to 86 g in body mass—a range too small to see a size relationship. These values, however, averaged 11% greater than predictions from the Lutz equation (94% to 134% of predicted values).

Table 5-7 Meeting the intense needs of small homeotherms by small inverse exponents.

Variable and taxon	No. of species	Relationship	S_b	S_{yx}	r	p	References
Oxygen affinity of blood: p_{50} = O_2 partial pressure in torr (mm Hg) at which the blood is half-saturated; the higher the p_{50}, the more readily it releases oxygen at the tissues							
Eutherian mammals	17	$p_{50} = 50.34\ M^{-0.054}$	±0.01	1.15	—	<0.001	Schmidt-Nielsen and Larimer, 1958
Marsupial mammals	7	$p_{50} = 32.14\ M^{-0.074}$	—	—	—	—	Bland and Holland, 1977
Eutherian and marsupial mammals	58	$P_{50} = 31.62\ M^{-0.03}$	±0.007	—	−0.45	<0.001	Prothero, 1980
Birds	7	$p_{50} = 32.12\ M^{-0.079}$	±0.026	1.17	−0.809	<0.05	Lutz et al., 1974
Hematocrit							
Mammals	52	$\%_{rbc} = 45.8\ M^{-0.01}$	±0.005	—	−0.169	<0.25	Prothero, 1980
Carbonic anhydrase in erythrocytes (arbitrary enzyme units per μl rbc)							
Mammals		$C = 14.0\ M^{-0.107}$	—	—	—	—	Larimer and Schmidt-Nielsen, 1960
Capillary density (mm^{-2})							
Mammalian masseter muscle		$796\ M^{-0.14}$	—	—	—	—	Calculated from Schmidt-Nielsen and Pennycuick, 1961
Sugar concentration, whole blood (mg/100 ml)							
Mammals		$C = 115\ M^{-0.101}$	—	—	—	—	Umminger, 1975

Parameter	N	Equation					Reference
Blood volume (ml)							
Eutherian mammals		$V_{blood} = 65.6\,M^{1.02}$	±0.01	3.3	0.995	—	Stahl, 1967
Mammals (including marsupial), terrestrial	28	$V_{blood} = 81.3\,M^{0.99}$	±0.01	—	0.997	—	Prothero, 1980
Terrestrial, double-labeled	28	$V_{blood} = M^{1.04}$	±0.02	—	0.996	—	Prothero, 1980
Terrestrial and aquatic	34	$V_{blood} = 77.6\,M^{1.00}$	±0.01	—	0.994	—	Prothero, 1980
Oxidative enzyme activity (μmol/[min·g wet tissue mass])							
Eutherian mammals	10						Emmett and Hochachka, 1981
β-OH butyrlCoA dehydrogenase		$= 14.8\,M^{-0.21}$	0.009	—	0.72	—	
Citrate synthase		$= 24.1\,M^{-0.106}$	0.032	—	0.61	—	
Malate dehydrogenase		$= 629\,M^{-0.065}$	0.017	—	0.65	—	
Glycolytic enzyme activity, μmol/(min·g wet tissue mass)							
Eutherian mammals	10						Emmett and Hochachka, 1981
Pyruvate kinase		$= 631\,M^{0.14}$	0.026	—	0.80	—	
Lactate dehydrogenase		$= 978\,M^{0.15}$	0.027	—	0.77	—	
Phosphorylase		$= 67.4\,M^{0.09}$	0.021	—	0.71	—	
Oxygen handling capacity, eutherian mammals							
Cytochrome oxidase activity	4	activity $= 208\,M^{-0.239}$	0.132	—	−0.89	—	Kunkel et al., 1956
Cytochrome c concentration	5	g/kg $= 0.036\,M^{-0.16}$	—	—	−0.958	—	Drabkin, 1950
Myoglobin concentration	5	g/kg $= 0.52\,M^{0.22}$	—	—	0.987	—	Drabkin, 1950
Hemoglobin concentration	5	g/kg $= 13.2\,M^{-0.01}$	—	—	−0.999	—	Drabkin, 1950
Mitochondrial density, liver	4	N/g tissue $= 37.6\,M^{-0.099}$	0.012	0.085	—	<0.02	Smith, 1956
Mitochondrial volume density, muscle:			95% c.i., b		95% c.i., a	r	Mathieu et al., 1981
Semitendinosus	21	V_v/g tissue $= 0.065\,M^{-0.231}$	−0.281, −0.182		0.055, 0.076	−0.92	
Longissimus dorsi	21	V_v/g tissue $= 0.066\,M^{-0.163}$	−0.210, −0.116		0.056, 0.077	−0.85	
Vastus medialis	21	V_v/g tissue $= 0.071\,M^{-0.139}$	−0.196, −0.083		0.059, 0.085	−0.77	
Diaphragm	21	V_v/g tissue $= 0.131\,M^{-0.055}$	−0.094, −0.017		0.116, 0.148	−0.57	

Concentrations of a substance such as blood sugar or carbonic anhydrase are only one factor in the total body capacity for transport by the blood. Total capacity is concentration times blood volume. Table 5-7 includes equations for blood volume allometry, which are close enough to linearity that it seems safe to conclude that the concentrations are not merely the result of more or less dilution into a heterogonic blood volume, and that the actual amounts of these substances transported per gram of body mass are greater in smaller mammals.

With convective transport scaled to metabolic needs, partial-pressure differentials for gas diffusion and nutrient concentrations are maintained at the level of the capillaries. From this point the needs of the tissues are met by diffusion, as described by the Fick equation (Eq. 5-16).

Each gram of flesh is metabolizing according to an inverse relationship with the total size of its owner, so it is reasonable to expect that other variables in the diffusion equation might be exploited to keep up with the needs. By increasing capillary density with small body size, x is decreased, increasing the rate of diffusion. The higher p_{50} values of smaller animals facilitate off-loading of oxygen from the erythrocytes, at p_{O_2} values that are higher for a given percentage of oxygen saturation. Carbonic anhydrase speeds up the equilibration of $CO_2 + H_2O \rightarrow H_2CO_3$, so a higher concentration of this enzyme appears to be necessary to ensure rapid carbon dioxide transport from the tissues of the small homeotherms. Just as oxygen is used more rapidly and carbon dioxide is produced more rapidly per gram of smaller beast, blood glucose should also be supplied at a higher rate, as manifested in the higher blood glucose concentrations in smaller mammals.

Fuel Supply

GUT CAPACITY

Rates of loss of heat, water, salts, and nitrogen, and rates of consumption of oxygen and energy reserves are proportional to the metabolic rate. It is thus to be expected that rates of feeding and digestion would follow the metabolic scaling in proportion to $M^{3/4}$.

Empirical verification of this expectation would be especially important for those ecologists who deal with optimality models of body size and foraging strategy. These models seek to show a net gain of energy intake over energy expenditure. The energy expenditure includes metabolic rates during rest and activity, for which there are good allometric descriptions. A comparable allometry of energy intake is not available, so modelers sometimes substitute gut size, which unfortunately omits the dimension of time or feeding frequency and leads to invalid comparisons.

There are equations for gut mass. Table 3-4 gave Brody's (1945) equation for mass of stomach plus intestine of eutherian mammals. Regression of a larger data base gives a larger exponent ($M^{1.02}$ vs. $M^{0.94}$) when ruminants and cetaceans are excluded. Although there was a difference in the state of the animal when total body mass was determined (depilated, gut emptied by Pitts and Bullard, 1968), there was also a difference in the range of body sizes that would have tended to bias toward a smaller exponent. Thanks to Pitts, I have the original data of these researchers on live body mass, pelage, and gut content mass. The pelage contributes an essentially size-independent 3% of body mass (Table 3-4), while the gut contents scale in the same fashion as empty gut mass (the latter from line 7, Table 5-8):

$$\text{gut contents/gut mass} = 0.078\ M^{0.87}/0.068\ M^{0.89} = 0.56\ M^{-0.02}. \quad (5\text{-}108)$$

The variability in both exponents and coefficients is damped somewhat when groups with different diets or digestive tracts are treated separately (Table 5-8), and we see that the mass of the gut contributes an essentially size-independent fraction slightly more than 5% of body mass in mammals. The linearly scaled 9% of avian body mass devoted to the gut probably reflects the addition of crop and gizzard, the latter a functional analogue to teeth (which have not been allometrized, but amount to 0.06% of human body mass; Kleiber, 1961) and jaw muscles.

The volume of the gut appears to be proportional to gut mass and therefore to body mass (Eq. 5-108). Do the length and diameter scale as geometrically similar cylinders, or might they scale according to McMahon's elastic similarity model? Preliminary indications are that neither is the case. Davis (1962) measured the length of the digestive tube (gut minus stomach) of four cats (male and female housecats at 2.8 kg body mass, male leopard at 45 kg, and male and female lions at 190 and 131 kg). While body length (in centimeters) scaled in such a way that a clear distinction between geometric and elastic similarity is not possible,

$$\text{body length} = 38.63\ M^{0.29}.$$

The gut lengths (in centimeters) were consistently of a higher allometry:

$$\text{digestive tube} = 16.29\ M^{0.84}\ (r^2 = 0.99), \quad (5\text{-}109)$$

$$\text{small intestine} = 11.12\ M^{0.83}\ (r^2 = 0.99), \quad (5\text{-}110)$$

$$\text{colon and rectum} = 1.865\ M^{0.85}\ (r^2 = 0.98). \quad (5\text{-}111)$$

Table 5-8 The scaling of gut mass to body mass in mature homeothermic animals.

Group	Gut mass (kg)	Body = size range (kg)	No. of species (n)	r^2	References
Mammals	$0.075 \, M^{0.94}$	0.02 – 700	41	0.925	Brody, 1945
Eutheria	$0.063 \, M^{1.07}$	0.007–6,654	78	0.974	(Q,P)[a]
Carnivora	$0.056 \, M^{1.01}$	0.183– 200	23	0.930	(Q,P)
Rodentia	$0.053 \, M^{0.95}$	0.018– 17.6	30	0.950	(Q,P)
Ruminants	$0.059 \, M^{1.16}$	8.62 – 665	11	0.914	(Q)
Eutheria, except ruminants and whales	$0.053 \, M^{1.02}$	0.007–6,654	67	0.926	(Q,P)
	$0.068 \, M^{0.89b}$	0.007– 17.6	42	0.934	(P)
Birds	$0.090 \, M^{0.99}$	—	—	—	Brody, 1945
	$0.098 \, M^{1.00}$	0.016– 123	31	0.951	(Q)

a. Parentheses indicate sources of data used to derive the new equations: Q = Quiring, 1950; P = Pitts and Bullard, 1968, and Pitts, personal communication.

b. Derived as a function of body mass minus pelage and gut contents; if pelage is included in body mass, the value becomes $0.066 \, M^{0.89}$.

If volume is indeed directly proportional to mass, the cylindrical volume would be the product of length and cross-sectional area. Substituting the allometry of the cats and the carnivore gut mass equation from Table 5-8, we obtain:

$$V \propto M^{1.01} \propto lA_x \propto ld^2 \propto (M^{0.84})(d^2), \tag{5-112}$$

and the diameter would therefore be

$$d \propto (M^{0.17})^{1/2} \propto M^{0.08}, \tag{5-113}$$

which according to Appendix Table 1 would not allow for a very sizable increase in gut diameter over a large body-size range, a 36% increase for the cat sizes involved. This scaling is similar to that of intercapillary distance in the circulatory system and contrasts with the size independence of capillary diameter and nephron (proximal tubule) diameter.

Let us return to the matter of turnover time for gut contents (t_{gut}), an important measurement for estimating digestive capacity. The many variations of diet and digestive system types (carnivorous, herbivorous, omnivorous, frugivorous, nectivorous, granivorous, rumination, coprophagy, and so on) have no doubt impeded empirical quantification of t and V. How much can a gut hold? Schmidt-Nielsen (1964) relates an incident about a dehydrated donkey that drank to replace its water loss until water spurted out the anus, taking 20.3 l of water in a few minutes. The normal (hydrated) body mass of this donkey was 104 kg. Using the equation for nonruminant eutherians (Table 5-8, line 6), we get a gut mass (in kilograms) of

$$M_{gut} = 0.053 \, M^{1.02} = 6.05.$$

If that mass contained 20.3 l, a factor of 3.4 times M_{gut} might predict full-gut capacity. Assume that the contents are carbohydrate in an aqueous slurry — the digestive contents may be viewed as a slurry with a fringe (of intestinal villi) on top. Carbohydrates yield about 19 kJ/g, but in animal and plant tissue water this is diluted to about 4.2 kJ/g, or 4,200 kJ/kg, assuming a density of 1 kg/l.

The energy content of a full gut (in kilojoules) would then scale with body mass as

$$3.4 \, kg \, (M_{gut})^{-1} \cdot 4,200 \, kJ/kg^{-1} \cdot 0.053 \, M^{1.02} = 757 \, M^{1.02}. \tag{5-114}$$

If the average rate of metabolic demand (in kilojoules per day) is 1.5 times the resting metabolic rate of mammals from the very large ($n = 349$; $r = 0.98$) data base summarized by Stahl (1967), we have:

$$(1.5)(11.6 \text{ ml } O_2/\text{min})(M^{0.76})(20.1 \text{ J/ml } O_2)(1,440 \text{ min/day})$$
$$= 504 \ M^{0.76}. \qquad (5\text{-}115)$$

The turnover time in days for the contents of a full gut would then be

$$\text{time} = \frac{\text{amount}}{\text{rate}} = \frac{757 \ M^{1.02}}{504 \ M^{0.76}} = 1.5 \ M^{0.26}, \qquad (5\text{-}116)$$

or 36 $M^{0.26}$ in hours. For a hypothetical 1-g mammal this would be 6 hours (filling 4 times per day); for 10 g, 10.8 hours (filling 2.2 times a day); for 100 g, 19.7 hours (filling 1.2 times per day); and for 10 kg, 2.73 day (filling one-third per day).

Is this theoretical scaling of gut turnover time reasonable? Adolph (1949) calculated the relationship of peristaltic gut beat duration, in hours, as

$$0.000093 \ m^{0.31} = 7.92 \times 10^{-4} \ M^{0.31}. \qquad (5\text{-}117)$$

The gut contents are the product of several meals or feeding bouts, so if enough field ecologists obtain data on feeding frequency, duration, and meal size, several aspects of life support could be tied together without the allometric and dimensional errors that may exist in some current optimality models.

DAILY FOOD INTAKE

From Table 5-9, which summarizes data on captive animals, it is clear that the food intake scales as one might expect from the allometry of metabolic physiology, approximately as $M^{3/4}$. This information could also be used with the allometry of gut mass (mean exponent for nonruminants, $M^{0.98}$) as an index of gut volume (Table 5-8), to calculate the scaling for gut turnover time, as in Eq. (5-116):

$$\text{time} = \frac{\text{amount}}{\text{rate}} \propto \frac{M^{0.98}}{M^{0.75}} = M^{0.23}. \qquad (5\text{-}116a)$$

Shrews, known for their apparently voracious appetites, are among the very small mammals that eat their own mass in food per day. The least shrew *(Cryptotis parva)* eats 1.1 g of baby mouse meat per gram of its own mass per day (Barrett, 1969). The water shrew *(Sorex palustris)* has been known to take 1.29 g per gram of body mass of a prepared diet of beef liver, brain, and suet (Sorenson, 1962). Some of this was hoarded, but about 0.95 g/g was utilized. In comparison with our own proportionate food consumption, this seems amazing; but is the comparison valid, when the shrew is living on a

Table 5-9 Daily food intake of mammals and birds in captivity.

Group	Food intake	r^2	s_b	No. of species (n)	References
Mammals					
Vegetable and seed eating	kg food/day = $0.157\ M^{0.84}$	0.971	0.046	12	Bourlière, 1969, table 9 (except bats and whales)
Animal-food eating	kg food/day = $0.234\ M^{0.72}$	0.982	0.044	12	Bourlière, 1969, table 9 (except bats and whales)
Herbivores	kJ/da = $971\ M^{0.73}$	0.942	0.020	$\dfrac{df}{1,083}$	Farlow, 1976
Carnivores	kJ/da = $975\ M^{0.70}$	0.968	0.013	1,100	Farlow, 1976
Mean	$\propto M^{0.75}$				
Birds and mammals					
Herbivores	kJ/da = $1,006\ M^{0.72}$	0.948	0.015	1,118	Farlow, 1976
Carnivores	kJ/da = $917\ M^{0.69}$	0.958	0.012	1,148	Farlow, 1976
Mean	$\propto M^{0.71}$				
Maximum	kJ/da = $1,713\ M^{0.72}$	0.998	0.008	8 spp. mammals, 11 spp. birds	Kirkwood, 1983

faster scale and the day is a much greater portion of the shrew's life than of a human's? If the 4.2-g body mass is used in the second equation of Table 5-9, a daily food consumption of 4.55 g, or 1.08 g/g, is predicted. Only in the allometric context can we correctly evaluate such an astonishing figure. Is the shrew living a frantic life, in which it must catch its own mass in food each day? No more so than the puma (*Puma* or *Felis concolor*, 43.7 kg), which must capture its mass in 12.3 days.

Fluid, Osmotic, and Ionic Homeostasis

For the most part, the responsibility for maintaining the proper amount of fluid in the body, at the proper osmotic pressure and ionic concentration, belongs to the kidneys. Some salts, water, and metabolites exit via the skin, sweat, and salt glands, while volatile metabolites such as carbon dioxide are eliminated by the respiratory system. Nevertheless, as beautifully stated by Homer Smith (1959):

> It is no exaggeration to say that the composition of body fluids is determined not by what the mouth takes in, but by what the kidneys keep: they are the master chemists of our internal environment, which, so to speak, they manufacture in reverse by working it over completely some fifteen times a day . . . Recognizing that we have the kind of internal environment we have because we have the kind of kidneys that we have, we must acknowledge that our kidneys constitute the major foundation of our physiological freedom.

That is the situation for man. What is it like for a mouse or a bird? The preceding sections have had brief references to the faster turnover times, more intense metabolic activity, and greater surface/volume ratios of smaller animals, which might lead us to expect that the kidneys of a mouse work over its body fluids more often than the 15 times a day required for man's 70-kg metabolism. In fact, the body water of a mouse appears to be filtered through the glomeruli of the kidneys at least 31 times per day. Even so, this is a small fraction of the 2,390 times per day that the mouse's total blood volume is pumped through the kidneys.

One of the substantial problems of terrestriality, common to birds and mammals, is the threat of desiccation. In the renal department water must be conserved, but the mechanisms differ in these classes in two ways, the waste product for nitrogen excretion and the action of the antidiuretic hormones. In order to reduce urinary water loss, mammals convert waste nitrogen into dissolved urea, which necessitates the production of urine with a high osmotic concentration. Birds, on the other hand, produce essentially insoluble uric acid, which precipitates without generating an osmotic pressure. The antidiuretic hormone (ADH) of mammals increases the permeability of the collecting ducts for reabsorption of water, whereas the ADH of birds simply shuts down the filtration into nephrons that cannot concentrate their contents. Both classes survive well in the desert despite the differences in renal strategies.

Allometric analysis is helpful for examining kidney function and water balance in animals of different sizes, for making interclass comparisons, and for evaluating adaptations to environmental stress. The material that follows is drawn from a recent review by Calder and Braun (1983).

KIDNEY DESIGN

Despite their crucial role, the kidneys add up to slightly less than 1% of the body mass of a 1-kg bird or mammal:

$$\text{mammalian kidney mass} = 7.32 \, M^{0.85}, \qquad (5\text{-}118)$$

$$\text{avian kidney mass} = 8.68 \, M^{0.91}, \qquad (5\text{-}119)$$

$$\text{in birds without salt glands} = 7.30 \, M^{0.93}, \qquad (5\text{-}120)$$

$$\text{in birds with salt glands} = 11.27 \, M^{0.88}. \qquad (5\text{-}121)$$

The salt glands themselves, at least in marine birds, scale with a similar exponent ($M^{0.92}$) (Willoughby and Peaker, 1979).

Three points emerge from the above relationships. First, it takes roughly the same amount of kidney for a 1-kg animal, whether urea or uric acid and water are excreted. Excretion of osmotically active urea makes mammals rely on greater osmotic concentrating ability, while the uric acid and urate precipitates of birds allow less spectacular concentrating ability. Second, although possession of extrarenal salt glands might seem to reduce the need for renal excretory capacity, birds with salt glands tend to have larger kidneys, perhaps because they live in environments of considerable osmotic or ionic stress. Finally, the mass exponents are consistently less than the essentially linear scaling of the lungs and heart. However, these last two homeostatic organs function in a cyclic or pulsatile action; the product of capacity (at $M^{1.0}$) and frequency (at $M^{-1/4}$) fits the metabolic needs (at $M^{3/4}$). Renal function, being continuous rather than oscillatory, might be accomplished adequately by a kidney mass scaled like metabolic rate and the rate of filtration through the glomeruli ($\dot{V}_{gf}$), which is also scaled as $M^{-3/4}$ (see Eq. 5-122 below).

Further details of the scaling of the mammalian kidneys emerge when we analyze the measurements by Rytand (1938), Sperber (1944), and Oliver (1968), cited by Calder and Braun, 1983 (see Table 5-10). We see that the total glomerular surface area and total volume of the proximal tubules are scaled in approximately the same way as the glomerular filtration and urine flow rates, which in turn are scaled like metabolic rates ($M^{0.68}$ and $M^{0.73}$ vs. $M^{0.72}$ and $M^{0.75}$, respectively). Since there is an increase with body size in the ratio of kidney mass to glomerular surface area, or to nephron number, proximal tubule volume, and glomerular filtration and urine production rates, there must be a proportional increase in the amount of supportive tissue (such as connective and vascular tissues and intertubular stroma) around the working nephrons. This could explain why the kidneys scale with a larger mass exponent than the metabolic functions they provide.

KIDNEY FUNCTION

In a sense, the kidney is a part of the circulatory system—a filter and a regulator of the plasma flowing in that system. The circulation begins with a pulsatile flow of marked amplitude in pressure from systole to diastole. This is progressively reduced in both amplitude and mean pressure as the blood flows to the capillaries. Some of the fluid is forced out of the capillaries, to return to the bloodstream, either as the difference between hydrostatic and colloidal osmotic (oncotic) pressure is reversed in sign along the capillary length, or via the separate lymph system which rejoins the vascular system proper just before the blood reenters the heart.

In the kidney hydrostatic pressure generated by the heart causes plasma fluid to leave the glomerular capillaries surrounded by the Bowman's capsule at the proximal end of the nephron. Over 99% of the filtrate will be reclaimed, ultimately to the circulatory system. What makes the kidney different from the rest of the system is that the unwanted, molecularly unrecognized, exogenous waste that got into the system (either accidentally or having been produced as useless metabolites) will pass out. Other more specifically recognized undesirables are actively pumped out of the blood and into the urine, while valuable molecules and water are reclaimed from the filtrate also by active transport "pumps". This is, in essence, how the "master chemists" known as kidneys work over the body fluids many times a day.

We would logically anticipate that renal design would be scaled to pre-

Table 5-10 The allometry of the mammalian kidney. (From Calder and Braun, 1983.)

Morphometric	Scaling	r^2	s_a	s_b	n	References
Kidney mass (one kidney)	$2.98\ M^{0.88}$	0.99	0.053	0.024	7	Oliver, 1968[a]
Kidney mass (both kidneys)	$7.32\ M^{0.85}$	0.98	—	0.010	111	Brody, 1945
Number of glomeruli	$9.53 \times 10^4\ M^{0.62}$	0.98	0.029	0.06	16	Holt and Rhode, 1976
Surface area per glomerulus (mm²)	$0.086\ M^{0.18}$	0.76	0.095	0.044	7	Oliver, 1968[a]
Total glomerular surface (mm²)	$8,371\ M^{0.73}$	0.97	0.121	0.056	7	Holt and Rhode, 1976
Total glomerular volume (ml)	$0.137\ M^{0.85}$	0.99	0.016	0.05	14	Oliver, 1968[a]
Proximal tubule length (mm)	$14.33\ M^{0.10}$	0.81	0.046	0.021	7	Oliver, 1968[a]
Proximal tubule diameter (mm)	$0.060\ M^{0.02}$	0.56	0.019	0.009	7	Oliver, 1968[a]
Proximal tubule volume (mm³)	$0.046\ M^{0.12}$	0.81	0.056	0.026	7	Oliver, 1968[a]
Total of proximal volumes (mm³)	$4,284\ M^{0.68}$	0.99	0.071	0.033	7	Oliver, 1968[a]
Number of nephrons/g kidney	$3.24 \times 10^4\ M^{-0.32}$	—	—	—	7	
Glomerular surface/g kidney	$2,809\ M^{-0.15}$	—	—	—	7	
Proximal tubular vol/g kidney	$1,472\ M^{-0.20}$	—	—	—	7	
Relative medullary thickness, mesic species	$5.06\ M^{-0.08}$	0.71	—	—	32	Blake, 1977
Relative medullary thickness, xeric species	$6.94\ M^{-0.09}$	0.86	0.014	0.009	18	Gregor, 1975[b]

a. As calculated by Calder and Braun, 1983.
b. Plus *Notiosorex crawfordi* from Lindstedt, 1980a.

serve a relation to metabolic rates, food intake rates, and the rate of blood flow (cardiac output). The expectation is on the mark, as can be seen in Table 5-11. In eutherian mammals, for example, the proportions of the cardiac output flowing through the kidneys, of the renal blood flow actually filtered, and of this filtrate finally released as urine are all size independent. Note that the mean of the small residual exponents from allometric cancellations with the basal metabolic rate is $M^{0.02}$. Scaling in functional design could not be clearer.

There is less information for birds, but the glomerular filtration rates ($\dot{V}_{gf}$) of the two classes scale similarly:

$$\frac{\dot{V}_{gf,birds}}{\dot{V}_{gf,mammals}} = \frac{2.00\ M^{0.73}}{5.36\ M^{0.72}} = 0.37\ M^{0.01}. \qquad (5\text{-}122)$$

For any given size birds filter less than half as much fluid per minute. This reflects the qualitative difference in the mechanism of antidiuresis mentioned earlier. For water conservation to be enhanced, the release of mammalian ADH (arginine vasopressin) increases the permeability of the collecting ducts, facilitating the reabsorption of water from urine forming in the nephron to the surrounding tissues. Bird kidneys, in contrast, have two types of nephrons, some looped and capable of countercurrent multiplication like mammalian nephrons, and others, like reptilian nephrons, unlooped and unable to concentrate. The latter serve only in filtration, during diuresis when there is a fluid excess. Avian ADH (arginine vasotocin) closes down the reptilian-type loops, thereby reducing $\dot{V}_{gf}$ (Braun and Dantzler, 1972, 1974a). Antidiuresis being the normal state in the terrestrial environment, the $\dot{V}_{gf}$ of birds should be lower, as it is. This lower $\dot{V}_{gf}$ in birds is also appropriate to the nature of the nitrogenous wastes, uric acid or urate precipitates that require no large aqueous volume, only enough moisture to be moved as a sort of paste.

We have seen that the exponents for kidney mass are greater than ¾, but less than 1 (Eqs. 5-118 to 5-121). Viewing this in terms of the filtration performed per gram of kidney, we find a similar inverse relationship for both birds and mammals (although, as noted before, the bird filters less):

mammals: $\dot{V}_{gf}/M_{kidneys} = 5.36\ M^{0.72}/7.32\ M^{0.85} = 0.73\ M^{-0.13},$ (5-123)

birds: $\dot{V}_{gf}/M_{kidneys} = 2.00\ M^{0.73}/8.68\ M^{0.91} = 0.23\ M^{-0.18}.$ (5-124)

Just as the small animal's kidney filters at a higher rate per gram, it produces urine at a higher rate per gram—although the residual mass exponent is smaller, whether the difference is significant or not.

Table 5-11 The relation of renal and circulatory support of metabolism to body size in eutherian mammals. (From Calder and Braun, 1983.)

Rate	Allometry	Fractional factor[a]	r^2	s_b	References
Basal metabolism (ml O_2/min)	$11.6\ M^{0.76}$	—	0.96	0.02	Stahl, 1967
Cardiac output (ml/min)	$187\ M^{0.81}$	—	0.96	0.01	Stahl, 1967
Renal blood flow (ml/min)	$43.1\ M^{0.77}$	$0.23\ M^{-0.04}$	0.98	0.08	Edwards, 1975
Glomerular filtration (ml/min)	$5.36\ M^{0.72}$	$0.12\ M^{-0.05}$	0.98	0.04	Edwards, 1975
Urine production (ml/min)	$0.042\ M^{0.75}$	$0.008\ M^{0.03}$	0.90	0.10	Edwards, 1975

a. Relative to the preceding equation (that is, 23% of the cardiac output goes to the kidneys; 12% of the renal blood flow is filtered; 0.8% of the filtrate is released as urine).

With this view of the circulation-to-filtration flow, we can now appreciate other allometric observations about kidney design. The diameter of the proximal tubule is essentially size independent, and the length of this portion of the nephron scales only slightly, $M^{0.10}$, in such a way that the length of the proximal tubule increases by a factor of only 2.5 from rat to horse. This lack of stronger size dependence can be rationalized easily. If the diameter were to increase, diffusion, distances would become limiting in the movement of materials to or from the active transport sites on the tubular membranes. The flow of filtrate along the length of the tubule is initiated by the filtration pressure. We can assume that the filtration pressure is size independent, since it is derived from the size-independent arterial pressure (although the arterial blood pressure has been attenuated as the blood moves along the circulatory vessels en route to the kidneys). With the same driving pressure the flow rate of the filtrate, once in the proximal tubule, is inversely proportional to the length of the tubule according to the Poiseuille equation (5-38). This, then, should severely restrict the dimensions of the proximal tubule.

SIZE AND RECLAMATION OF FILTERED WATER

The ancestors of mammals and birds evolved away from the water, but their internal systems retained an aqueous legacy; a portion of the aquatic milieu had to be carried and regulated via fluid exchanges, wherever they went. The voiding of wastes and excess solutes necessitates some loss of water as the solvent or at least as a lubricant for such elimination.

Considerations such as the ratio of evaporative surfaces ($\propto M^{2/3}$) to body water ($\propto M^{1.0}$) and the more intense respiratory demands of small homeotherms suggest that smaller animals need greater renal concentrating powers in order to conserve water. However, it is obvious that the urine-to-plasma osmotic concentration ratio (C_u/C_p) has been so largely an adaptive response that the body-size dependency has been obscured (Edwards, 1975).

The functional significance of renal design is in the context of water conservation, especially for animals that live in arid habitats. The water is conserved by reabsorption from the filtrate as it passes through the mammalian or mammalian-type avian nephrons, which are looped in the renal medulla (toward or into the papilla).

The relative medullary thickness (τ_{RM}) expresses the proportion, relative to kidney size, of the kidney which contains the loops of Henle and collecting ducts, the structures that are primarily responsible for concentrating the urine to osmotic pressures greater than those of the plasma:

$$\tau_{RM} = \frac{\text{longest axis of the medulla}}{(\text{length} \cdot \text{width} \cdot \text{breadth of kidney})^{1/3}}.$$

The τ_{RM} is a morphological index of water-conserving ability, expressed quantitatively as the maximum urine concentration ($C_{u,max}$, in milliosmoles per liter) by Lindstedt (1980a) for Eutheria:

$$C_{u,max} = 580\tau_{RM} - 39. \tag{5-125}$$

This correlation was highly significant ($r^2 = 0.88$; $p < 0.001$).

The τ_{RM} scales with a small negative exponent similar to those noted for metabolic support (Tables 5-7 and 5-10). Small mammals should be able to produce more concentrated urine than large mammals, whether or not the consideration is limited to species of xeric or mesic environments (size shows about the same exponent for the two groups: $M^{-0.09}$ for xeric, $M^{-0.08}$ for mesic). The severe selective pressure for water economy on the desert increases τ_{RM} about 36% for mammals of the same size (Greegor, 1975; Blake, 1977).

We can predict the size dependency of concentrating ability, substituting predictions of τ_{RM} from Eqs. 14 and 15 of Table 5-10 in Eq. (5-125) and running a secondary regression for hypothetical sizes of 10 g, 100 g, 1 kg, 10 kg and 100 kg (Calder and Braun, 1983):

mesic mammals: $C_{u,max} = 2,894\, M^{-0.08}$ (5-126)

xeric mammals: $C_{u,max} = 3,844\, M^{-0.08}$. (5-127)

The amount of water lost in urine is inversely related to C_u, so adaptation to the desert in eutherian mammals appears to have involved, on the average, a 25% water savings for urine production ($2,894/3,844 = 75\%$).

The body-water content and the plasma osmotic concentration appear to be independent of size (see "Fractional factor" column of Table 5-11 and bottom line of Table 5-12). Assuming a plasma osmotic concentration of $300\, M^{0.0}$, the concentration ratio $C_{u,max}/C_p$ has been predicted by Calder and Braun (1983) to be

mesic mammals: $C_{u,max}/C_p = 2,894\, M^{-0.08}/300\, M^{-0.0} = 9.7\, M^{-0.08}$,

(5-128)

xeric mammals: $C_{u,max}/C_p = 3,844\, M^{-0.08}/300\, M^{0.0}$

$= 12.8\, M^{-0.08}$. (5-129)

There is remarkable agreement between the $C_{u,max}/C_p$ derived in this roundabout fashion with that derived directly (Calder and Braun, 1983; $n = 10$, $r^2 = 0.58$, $p < 0.001$):

Table 5-12   Water balance and body size in resting mammals and birds. (Sources of data and original equations given in Calder, 1981; Calder and Braun, 1983.)

Component	Mammals[a]	Birds	Mammals/birds
Losses (ml/day)			
Evaporation	$38.8\ M^{0.88\ b,c}$	$24.2\ M^{0.61\ b}$	$1.6\ M^{0.27\ b}$
Urine	$60.9\ M^{0.75}$	$27.0\ M^{0.86}$	$2.3\ M^{-0.11}$
Feces	$39.6\ M^{0.63}$	—	—
Total for 1-kg animal	139 ml/day	>51.2 ml/day	—
Intake (ml/day)			
Metabolic[d]	$12.6\ M^{0.75}$	$14.1\ M^{0.72}$	$0.9\ M^{0.03}$
Preformed	$?\ M^{0.75?}$	$?\ M^{0.72?}$	—
Drinking	$99.0\ M^{0.90}$	$59.0\ M^{0.67}$	$1.7\ M^{0.23}$
Total for 1-kg animal	>111.6 ml/day	>73.1 ml/day	—
Total turnover (ml/day)[e]	$123\ M^{0.80\ f}$	$115\ M^{0.75\ g}$	$1.06\ M^{0.05}$
Total body water (%)	$60.5\ M^{1.00}$	$60.3\ M^{1.01}$	$1.0\ M^{-0.01}$

a. Eutheria unless otherwise noted.

b. The difference between exponents for mammals (0.88) and birds (0.61) is not statistically significant (Crawford and Lasiewski, 1968).

c. The relationship for bats treated separately, $40.7\ M^{0.67}$, does not differ significantly in exponent from that for all mammals. At the middle of the body-size range, evaporation from bats is 2.5 times that of typical eutherians.

d. Calculations assume a respiratory quotient of 0.8 and 0.578 g H_2O/liter O_2 at 1.5 standard metabolic rate.

e. Streit (1982) derived $M^{0.91}$ scaling for total turnover in monotremes, marsupials, and eutherians combined.

f. Marsupials' turnover is 73% of that of eutherians, but the scaling exponent is the same (Denny and Dawson, 1975).

g. From Pinshow et al., 1983; $n = 19$ species, $r^2 = 0.88$, $s_b = 0.066$, 95% confidence limits of $b = 0.607$ to 0.887.

$$\text{xeric mammals:}\quad C_{u,max}/C_p = 12.4\ M^{-0.10}. \tag{5-130}$$

The form-function correlation thus seems to be very much present. Interestingly, the standard error of estimate on the exponent, ± 0.03, means that the scaling exponent for $C_{u,max}/C_p$ is the same as that for τ_{RM}. Note the similar size dependence of

$$\text{xeric birds:}\quad C_{u,max}/C_p = 1.92\ M^{-0.09}$$
$$(n = 14;\ r^2 = 0.49;\ p < 0.001). \tag{5-131}$$

Birds and mammals seem equally successful in surviving and reproducing in the desert. In general, the renal concentrating ability of desert mammals is nearly an order of magnitude greater than that of desert birds, as seen in the ratio of coefficients above (12.4/1.92), but the birds were "preadapted" by virtue of their osmotically inert nitrogenous waste products, uric acid and urates. An ordinary chicken, for example, uses no more water per gram of nitrogen excretion than does a desert kangaroo rat. In both classes the emphasis on the need to conserve water seems to be at the small end of the size range, in view of the consistent inverse exponents.

WATER BALANCE

Water balance is an extremely important component of the overall homeostasis of animals. If water loss exceeds water income, concentration of body fluids and desiccation result; if intake exceeds loss, dilution and swelling occur. The intake side of the balance sheet includes drinking water, the "preformed" water that occurs as tissue fluid or succulence of food items, and metabolic water formed during oxidation of the food and stored fat, carbohydrate, and protein. Water is lost in the urine and feces, and by evaporation from the respiratory system and skin. Animals adapted to environments where drinking water is scarce often have abilities or morphologies that reduce the fecal and evaporative water losses (Schmidt-Nielsen, 1964, 1979). However, these are not so dramatic as the differences between desert and nondesert homeotherms in ability to concentrate the urine.

The components of the water balance, summarized in Table 5-12, scale in crude approximation to the metabolic $M^{0.75}$. The scaling for evaporative water losses of mammals ($\propto M^{0.88}$) and birds ($\propto M^{0.61}$), and for avian cloacal losses ($\propto M^{0.86}$, $n = 5$) are not statistically different from $M^{0.75}$, and scaling for fecal loss of mammals is based on assumption of a 60% water content, which may not be size independent. The variability in these exponents, from data available now, probably rules out simple addition of line items in the water budget except for specific sizes. For example, the total loss per day for a 1-kg eutherian mammal is predicted to be 139.3 ml, and the intake exclusive of preformed water is 111.6 ml, figures which bracket the total turnover of 123 ml/day, determined with tritiated water by Richmond and associates (1962). The marsupials turn over only 73% as much water as eutherians of the same size; this comes close to the ratio of marsupial standard metabolic rates to those of the eutherians, about 67%.

Birds are characterized by considerably lower urinary and evaporative loss rates, they require less drinking water, and two of the three derivations for total water turnover are about 40% lower than the figure for eutherian mammals. However, total water turnover, determined directly, does not substantiate the impression from separate components that birds can balance with less water. The total body-water content of birds and eutherian mammals is the same, and size independent. Using the equations for total water and its turnover from the last lines of Table 5-12, the turnover times (t_{H_2O} in days) can be estimated:

$$\text{mammals:} \quad t_{H_2O} = 605 \; M^{1.00}/123 \; M^{0.80} = 4.92 \; M^{0.20}, \quad (5\text{-}132)$$

$$\text{birds:} \quad t_{H_2O} = 603 \; M^{1.01}/115 \; M^{0.75} = 5.24 \; M^{0.26}. \quad (5\text{-}133)$$

If abilities to withstand dehydration are the same, say to loss of 10% of body mass (which is loss of 16% of body water), the birds can last insignificantly longer than mammals of the same starting mass unless they have a proportionately greater advantage at higher temperatures.

Endurance time, like other physiological times, is size dependent, and the smaller the animal, the shorter the absolute time it has to regain water balance (see Chapters 6 and 8). Allometry confirms what we already knew, that homeostasis and endurance are intimately related; now we see more clearly the implications of body size. The small animal can survive each day with lower requirements, and can take refuge in microclimates into which a larger beast cannot fit; but if the small animal faces the full blast of environmental extremes, or is cut off from supplies of food or water, it has only a brief time in which to balance its accounts. A full appreciation of the size dependence of physiological time scales may be the single most valuable key to understanding animal functions and life histories.

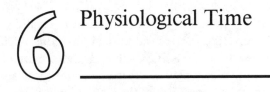

Physiological Time

Time is but the stream I go a-fishing in.
—HENRY DAVID THOREAU, *WALDEN* (1854)

OUR LIVES are set in a precise technology of timekeeping which allows us to predict exactly when the sun will rise or cross the equator. Very small electronic watches and huge clocks on famous towers track the same time scales. However, the life span of an animal and the subunits of its time scale, ticking away as cycles of heartbeat, breathing, metabolic turnover, and reproductive events are all internal matters related to body size. Thus there are two kinds of time scales, absolute time and the size-dependent relative time (or physiological time). Thoreau came close to the biological truth; I would go so far as to say that physiological time is the most important fundamental for understanding life-history comparisons.

Brody (1945) devoted a chapter to "Physiologic Time and Equivalence of Age," in which he introduced the concept that organisms of different sizes have different time scales, although he did not actually reduce his findings to a set of allometric equations. Hill (1950) argued that variables such as power per unit weight, time for growth and maturation, and duration of pregnancy may be constant in all mammals when expressed per unit of size-dependent physiological time; animals similar except in size should "carry out similar movements, not in the same time, but in times directly proportional to their linear dimensions."

This concept of physiological time has persisted through many reexaminations and has proved to have wide application. Adolph (1949) provided some allometric equations for cycle times of heartbeat, breathing, and peristaltic "gut beat." Stahl (1962) derived additional equations of allometric time and observed: "In the analysis of growth and related matters it has become clear that two principles play a key role in biological analysis: conservation of volume and synchronism of times. It appears that the time scale is uniformly that of $M^{0.25}$ to $M^{0.30}$ in mammals and probably in many other organisms."

Adolph's (1949) examination of 34 morphological and physiological variables showed him that there is "quantitative orderliness" in organisms, which can be viewed as "systems of precise multiple interrelationships." Examining those interrelationships, Stahl (1962, 1963a,b, 1967) combined allometric equations to form dimensional and dimensionless "criteria of similarity." For example, he calculated that breath time equaled $4.7 \times 10^{-5} M^{0.28}$ years and pulse time equaled $1.2 \times 10^{-5} M^{0.27}$ years. The ratio of breath to pulse time yields the dimensionless ratio $3.9 M^{0.01}$. Because the residual mass exponent is small, this ratio suggests that there are about 3.9 heartbeats per respiratory cycle in all mammals, regardless of size. From refined equations in the 1967 paper, this ratio was revised to $4.5 M^{0.01}$. By similar analysis, Stahl estimated that all mammals have a basal energy use, or minimum rate of metabolism over a maximum physiological life span, of about 8×10^5 kJ/kg:

$$\frac{(kcal/day)(day/yr)(life\ span,yr/lifetime)}{kg}$$

$$= \frac{(70.5\ M^{0.73})(365)(7.52\ M^{0.29})}{1.0\ M^{1.0}} \tag{6-1}$$

$$= (1.94 \times 10^5)(M^{0.02})kcal/(kg \cdot lifetime)$$
$$= (7.95 \times 10^5)(M^{0.02})kJ/(kg \cdot lifetime).$$

A more widely accepted life-span equation from a larger data base (Sacher, 1959) can be substituted for $7.52 M^{0.29}$ in Eq. (6-1) to give

$$\frac{(70.5\ M^{0.73})(365)(11.6\ M^{0.20})}{1.0\ M^{1.0}}$$

$$= (2.98 \times 10^5)(M^{-0.07})kcal/(kg \cdot lifetime) \tag{6-2}$$
$$= (1.25 \times 10^6)(M^{-0.07})kJ/(kg \cdot lifetime).$$

For passerine birds, although the mass exponent is essentially the same, the coefficient is over three times that for mammals, which means that if we compare the two taxa at any body size, the passerine bird burns three times as much energy at the resting level (Lindstedt and Calder, 1976):

$$\frac{(J/s)(s/yr)(\text{life span,yr/lifetime})}{kg}$$

$$= \frac{(6.25\ M^{0.72})(3.16 \times 10^7)(21.6\ M^{0.26})}{1.0\ M^{1.0}} \qquad (6\text{-}3)$$

$$= (4.26 \times 10^9)(M^{-0.02})J/(kg \cdot \text{lifetime})$$
$$= (4.26 \times 10^6)(M^{-0.02})kJ/(kg \cdot \text{lifetime}).$$

The resting metabolic rate of reptiles at $T_b = 30°C$ can be calculated from Bennett and Dawson (1976) and Calder (1976):

$$\frac{(kJ/day)(365)(yr/\text{lifetime})}{kg} = \frac{(27.38\ M^{0.77})(365)(14.6\ M^{0.23})}{1.0\ M^{1.0}} \qquad (6\text{-}4)$$

$$= 1.46 \times 10^5\ kJ/(kg \cdot \text{lifetime}).$$

In its longer lifetime, then, a passerine bird's basal maintenance uses 3.4 times as much energy as that of a mammal, whereas a reptile at 30° C uses only 12% as much as the mammal.

In like fashion Rahn and Ar (1980) calculated that the total metabolism of a bird's egg, from laying to hatching, is size independent:

$$H_{m,total} = 500\ kcal/(kg \cdot \text{lifetime}) = 2,100\ kJ/(kg \cdot \text{lifetime}), \qquad (6\text{-}5)$$

a small but constant fraction of adult lifetime basal expenditure.

Thus we have reason to expect that consideration of physiological time scales will help us to appreciate some basic features of life's processes. Lindstedt and Calder (1981) compiled a list of allometric expressions relating the time of various life processes to body mass. They either cited or calculated 18 equations for mammals and 9 equations for birds; the life processes ranged from fast muscle contraction time to maximum life span (Table 6-1; Figures 6-1 and 6-2). Although the times for these functions span ten orders of magnitude, from milliseconds to decades of years, the exponents averaged 0.247 ± 0.049, and the 95% confidence intervals of most of the exponents included 0.25. Using maximum life span, rather than absolute time, as the end point of a size-dependent physiological time scale, it appears that each life comprises about the same number of physiological events or actions; in other words, each animal lives its life faster or slower as governed by size, but accomplishes just as much biologically whether large or small.

Where can we get information about physiological times? Any expres-

Table 6-1 Physiological time as a function of body mass (M, in kilograms) of mammals and birds. (Revised from Linstedt and Calder, 1981.)

Period	Units and scaling	C.I.[a]	n	r[b]	Life span/cycle time	References[c]
Mammals (Eutheria)						
Life span in captivity	yr = $11.6\,M^{0.20}$			0.77	1	Sacher, 1959
98% growth time	yr = $1.21\,M^{0.26}$				$9.61\,M^{-0.09}$	Stahl, 1962
Reproductive maturity	yr = $0.75\,M^{0.29}$	0.22–0.36	56	0.76	$15.5\,M^{-0.09}$	Eisenberg, 1981; Economos, 1981
50% growth time	yr = $0.352\,M^{0.25}$				$3.30 \times 10^1\,M^{-0.05}$	Stahl, 1962
Gestation period	da = $65.3\,M^{0.25}$	0.22–0.28		0.85	$6.49 \times 10^1\,M^{-0.05}$	Sacher, 1974
	$66.3\,M^{0.26}$			0.72	$6.39 \times 10^1\,M^{-0.06}$	Blueweiss et al., 1978
Erythrocyte life span	da = $22.6\,M^{0.18}$		7	0.99	$1.87 \times 10^2\,M^{0.02}$	Allison, 1960
	$68.0\,M^{0.13}$		24[d]	0.87	$6.23 \times 10^1\,M^{0.07}$	Vacha and Znojil, 1981
Plasma albumen half-life	da = $4.58\,M^{0.32}$		11	0.96	$9.24 \times 10^2\,M^{-0.12}$	Allison, 1960
	$4.29\,M^{0.31}$		8	0.93	$9.87 \times 10^2\,M^{-0.11}$	Allison, 1960; Munro, 1969
	$3.71\,M^{0.28}$		5	0.99	$1.14 \times 10^3\,M^{-0.08}$	From Munro, 1969
γ-globulin half-life	da = $5.85\,M^{0.26}$		12	0.94	$7.25 \times 10^2\,M^{-0.06}$	From Allison, 1960
Transferrin pool turnover	da = $3.79\,M^{0.24}$		10	0.85	$1.12 \times 10^3\,M^{-0.04}$	From Regoeczi and Hatton, 1980
Metabolism (0.1% of M, fat)	min = $170\,M^{0.26}$	0.24–0.28		1.00	$3.58 \times 10^4\,M^{-0.04}$	Kleiber, 1932
Drug half-life (methotrexate)	min = $58\,M^{0.19}$			0.98	$1.05 \times 10^5 M^{0.01}$	Dedrick et al., 1970
Glomerular filtration time (plasma clearance of insulin)	min = $6.51\,M^{0.27}$	0.23–0.31		0.98	$9.37 \times 10^5\,M^{0.05}$	Stahl, 1967; Edwards, 1975
Total renal plasma clearance (filtration and secretion of PAH)	min = $1.70\,M^{0.22}$	0.14–0.30		0.98	$3.59 \times 10^6\,M^{-0.02}$	Stahl, 1967; Edwards, 1975
Circulation of blood volume	sec = $21.0\,M^{0.21}$			0.98	$1.74 \times 10^7\,M^{-0.01}$	Stahl, 1967

		Range	N		r	Reference
Gut beat duration	$sec = 2.85\,M^{0.31}$			$1.28 \times 10^{8}\,M^{-0.11}$		Adolph, 1949
Respiratory cycle	$sec = 1.12\,M^{0.26}$			$3.26 \times 10^{8}\,M^{-0.06}$	0.91	Stahl, 1967
Time constant for lung filling	$sec = 0.11\,M^{0.30}$		5	$3.32 \times 10^{9}\,M^{-0.10}$	0.86	Bennett and Tenney, 1982
Cardiac cycle	$sec = 0.25\,M^{0.25}$			$1.47 \times 10^{9}\,M^{-0.05}$	0.88	Stahl, 1967
	$0.18\,M^{0.28}$	0.23–0.33		$2.00 \times 10^{9}\,M^{-0.08}$	0.93	Calder, 1968
Twitch contraction cycle, soleus muscle	$sec = 0.064\,M^{0.39}$	0.29–0.49		$5.75 \times 10^{9}\,M^{-0.19}$	0.99	Syrovy and Gutman, 1975
Shiver burst interval (= burst frequency^{-1})	$sec = 0.049\,M^{0.18}$			$7.47 \times 10^{9}\,M^{-0.01}$	0.92	Spaan and Klussman, 1970
Twitch contraction cycle, extensor digitorum longus	$sec = 0.019\,M^{0.21}$	0.12–0.31		$1.94 \times 10^{10}\,M^{-0.01}$	0.99	Syrovy and Gutman, 1975
Intraaural sound arrival difference	$\mu sec = 195\,M^{0.30}$		31	$1.88 \times 10^{12}\,M^{-0.10}$	0.96	Heffner and Heffner, 1980; personal communication
Birds						
Life span in captivity	$yr = 28.3\,M^{0.19}$		58	1	0.70	Lindstedt and Calder, 1976
Life span in the wild	$yr = 17.6\,M^{0.20}$		152	$1.61\,M^{-0.01}$	0.78	Lindstedt and Calder, 1976
First breeding	$yr = 2.33\,M^{0.23}$		40	$1.22 \times 10^{1}\,M^{-0.04}$	0.40	Western and Ssemakula, 1982
Incubation period	$da = 28.9\,M^{0.20}$			$3.58 \times 10^{2}\,M^{0.02}$	0.86	Rahn and Ar, 1974
Respiratory cycle	$sec = 3.22\,M^{0.33}$	0.30–0.36		$2.77 \times 10^{8}\,M^{-0.14}$	0.93	Calder, 1968
Cardiac cycle	$sec = 0.39\,M^{0.23}$	0.17–0.29		$2.32 \times 10^{9}\,M^{-0.04}$	0.85	Calder, 1968
Order Passeriformes						
Life span in the wild	$yr = 21.7\,M^{0.26}$	0.20–0.32	71	$1.31\,M^{-0.07}$	0.76	Lindstedt and Calder, 1976
Fat metabolism, 0.1% of body mass	$min = 1.06\,M^{0.28}$	0.23–0.32		$1.41 \times 10^{5}\,M^{-0.09}$	0.98	Lasiewski and Dawson, 1967
Respiratory cycle	$sec = 2.63\,M^{0.28}$	0.16–0.41		$3.40 \times 10^{8}\,M^{-0.09}$	0.69	Calder, 1968

Table 6-1 (continued)

Period	Units and scaling	C.I.[a]	n	r[b]	Life span/cycle time	References[c]
Non-Passeriformes						
Life span in the wild	yr = 16.6 $M^{0.18}$	0.12–0.24	81	—	1.71 $M^{-0.02}$	Lindstedt and Calder, 1976
First breeding (Procellariiformes)	yr = 4.41 $M^{0.22}$	0.11–0.32	6	0.91	6.42 $M^{-0.05}$	Lack, 1968
Sperm mean fertility, insemination to egg laying (Galliformes and Columbiformes)	da = 11.7 $M^{0.29}$		7	0.82	$8.89 \times 10^2\ M^{-0.10}$	Calculated from Lake, 1975
Reptiles						
Life span	yr = 14.6 $M^{0.23}$	—	8	0.72	—	Calder, 1976a

a. C.I. = 95% confidence interval of slope (exponent).
b. A few equations were derived by combining other regression equations; in those cases r was estimated as the product of the individual correlation coefficients.
c. Source of equation or of data used to calculate equation.
d. Includes six values stated by authors to "have only rough validity."

sions that have been combined, or can be combined, to yield the dimension of time or its reciprocal are useful sources. These may include times for single contractions, durations of repetitive physiological cycles, times between feedings, exponential decay half-life, total pool of some body substance (water, fat, or some protein) divided by rate of turnover, and field and zoo records of incubation, gestation, and maximum life span. When the dimension of time occurs in the denominator (indicating cycles per unit time), the fraction can be inverted to yield time per cycle. For example, heartbeat frequency can be determined with an electrocardiograph. The time between consecutive electrocardiographic (QRS) peaks is the cardiac cycle time, or duration. The reciprocal of cardiac rate over a longer period is the average cardiac cycle time. Tables of undigested facts can be found in the natural history literature (ornithology texts, for instance) and can be analyzed allometrically with body-mass values borrowed from other sources.

Another expression of physiological timing is the half-time. Exponential decay or replacement curves can be used to calculate the time for disappearance of one-half the starting number of labeled molecules or blood cells. Alternatively, we can use the time constant t_c, when $1/e$ of the original number remains, in

$$n = n_o e^{-kt_c}, \tag{6-6}$$

where e = base of natural logarithms, n = number remaining at time t_c when $n = (1/e)(n_o)$, n_o = starting number, and k = fractional disappearance rate.

Endurance or turnover times can be calculated from allometric equations for total body pool and rate as follows:

$$\frac{\text{total body pool} = \text{amount}}{\text{rate} = \text{amount/time}} = \text{time}. \tag{6-7}$$

For example, the total blood volume (in milliliters) of eutherian mammals (Stahl, 1967) is

$$V_{bl} = 65.5 \, M^{1.02}, \tag{6-8}$$

while the cardiac output (in milliliters per minute) is

$$\dot{V}_{bl} = 187 \, M^{0.81}. \tag{6-9}$$

The mean time for circulation of this volume at this rate is

$$t_{circ} = \frac{65.6 \, M^{1.02}}{187 \, M^{0.81}} = 0.35 \, M^{0.21}. \tag{6-10}$$

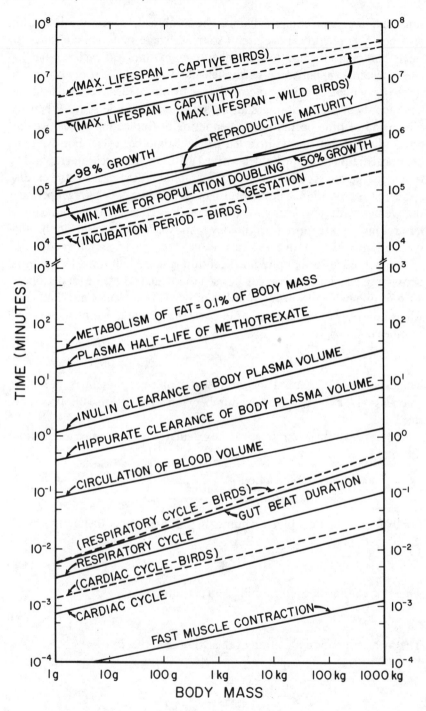

These physiological times and many others are approximately proportional to $M^{1/4}$. Not all come as close as we might like; we may be oversimplifying matters by trying to force all physiological time allometries into an $M^{1/4}$ scaling. The following is a discussion of the allometries in Table 6-1 having body mass exponents that differ from 0.25 by 0.05 or more.

Life Spans ($M^{0.20}$ for mammals; $M^{0.19-0.20}$ for birds). There is a possibility that the data are biased because maximum longevity is more likely to be observed in smaller species. The larger and the longer-lived an animal, the fewer of its kind there are and the less likely it is that they have reached maximum longevity within the period of study and record keeping. The U.S. Fish and Wildlife Service and its predecessor, the U.S. Biological Survey, have supervised a banding program for 62 years, which is 8.2 times the maximum life span of a typical 15-g bird, but only about 2.7 times the maximum life span of a 4-kg bird. Population densities of small birds are greater, as is the feasibility of banding them out of avocational interest. The same principle is in effect in zoological parks, where the basic plan is to keep a pair of each kind of animal on display under conditions adequate for reproduction. Because it takes longer for the larger species to live out their lives, replacement of animals and acquisition of data on their longevity generally emphasize the smaller animals. Furthermore, older records are less reliable because of the turnover of zookeepers. There was, for example, the interesting case of the 125-year-old elephant which, according to the records, had been born an African elephant and died an Asiatic elephant! Jones (1982) has critically reviewed the records on longevity of captive mammals and provided more modest figures of 57 years for *Elephas maximus* and 48 years for *Loxodonta africana*. The St. Louis Zoo in 1953 received a parrot claimed to be 130 years old. A 1913 photograph of this old bird was that of a yellow-headed Amazon, but at the time of donation it was a different species, the greater vasa parrot (William Conway, personal communication).

Another factor that could bias the life span allometry is the differential life span of animals and of the bands or tags used for identification. If a band

---

6-1 Physiological time scales of the warm-blooded animals show size dependencies that consistently are approximately the fourth root of body mass; for both small and large species, these are similar submultiples of the life span (which must be long enough for all of the processes to be completed). This principle, that all animals do the same things about the same number of times in their life spans, extends throughout the life history. (From Lindstedt and Calder, 1981; with permission of the *Quarterly Review of Biology*.)

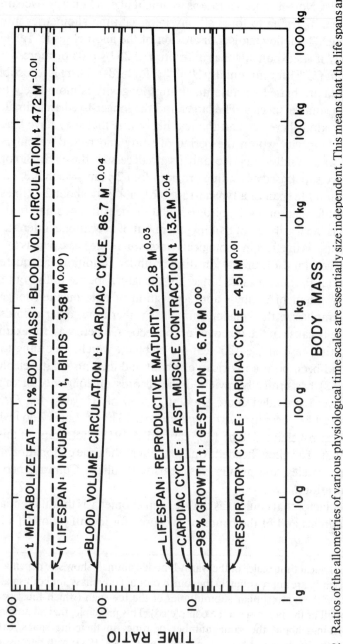

6-2 Ratios of the allometries of various physiological time scales are essentially size independent. This means that the life spans are similarly proportioned, and that physiology can be described by the dimensionless "design criteria" of Stahl (1962). (From Lindstedt and Calder, 1981; with permission of the *Quarterly Review of Biology*.)

wears through and is lost, the bird that wore it will not be accorded proper respect for its old age. But gulls double-banded with a standard aluminum band and a polyvinyl chloride band were later recaptured wearing only the PVC band; seawater had corroded the aluminum, and had it not been for the second band, the life span of these gulls would never have been accurately recorded (Spear, 1980).

Half-Lives and Turnover Times ($M^{0.19}$, $M^{0.18}$, $M^{0.13}$, $M^{0.32}$, $M^{0.34}$). In contrast to the size dependence of some periods or frequencies (such as heart rate and ventilation) that are rigidly determined by mechanical factors, the frequencies of other functions may be more flexible. The rate of a degeneration, for example, is variable; the achievement of a perfect locomotory cadence is not crucial to the survival of the animal, or is lost in transient variations of behavior. Variability can also be affected by sample size and measurement technique; exponents obtained from independent determinations may vary by as much as 0.05 (erythrocyte life span) or 0.04 (half-life of plasma albumen).

One approach to the problem of half-lives and turnover times is to try to explain why some of them vary so widely from $M^{1/4}$. Alternatively, we can search for theoretical reasons that explain why we should expect physiological times to be proportional to $M^{1/4}$. McMahon (1975, 1980) related the physiological "time frame" to the mechanical support of the body frame. Finding statistically significant agreement between the empirically determined exponents for physiological times ($\propto M^{1/4}$) and those predicted from the elastic similarity model, he proposed that mechanical considerations are responsible for the $M^{1/4}$ scaling.

The energy required to maintain a cyclic movement is used to overcome the forces of friction, elasticity, and inertia. At the natural frequency of oscillation, cycles of energy requirement and release of the elastic and inertial components are equal in magnitude, but exactly out of phase. The energy released by elastic recoil is used to accelerate the inertial component, and the energy required to restretch the elastic component is released during deceleration of the inertial component. Thus this portion of the total energy is merely transferred back and forth; once it is operating, an oscillator requires only enough additional energy to overcome friction. For this reason vibrating strings, pendulums, and the like tend to oscillate at their natural frequencies when disturbed.

McMahon describes three basic mechanical geometries that could be involved in a cyclic or oscillatory movement. The first is analogous to a mass suspended by a spring. It has a natural frequency of oscillation (f_{n_1}) of

$$f_{n_1} \propto \left(\frac{1}{l}\right) (E/\rho)^{1/2}, \tag{6-11}$$

where l is the length of the spring, E is the elastic (Young's) modulus, or the slope of its stress/strain curve,* and ρ is the density of the suspended mass. The ratio E/ρ is assumed to be constant within a class of animals. If elastic similarity is maintained, the length l would be proportional to $M^{1/4}$, and a frequency proportional to $M^{-1/4}$ or a time proportional to $M^{1/4}$ would be expected. If geometric similarity applies, on the other hand, the frequency would be proportional to $M^{-1/3}$. For the most part, observed frequencies support the elastic similarity model, but a dynamic similarity would predict exponents that cannot be distinguished from those of elastic similarity (McMahon, 1975, 1980; Alexander, 1982).

The second and third geometries are those that exist when one or more joints are involved in the movement. Scale-dependent mechanical advantages add a factor of d/l to Eq. (6-10) such that the natural frequency in both cases becomes

$$f_{n_2}, f_{n_3} \propto \left(\frac{d}{l^2}\right) (E/\rho)^{1/2}. \tag{6-12}$$

If elastic similarity applies, Eqs. (4-7) and (5-3) can be substituted into Eq. (6-12) to produce

$$f_{n_{2,3}} \propto \left(\frac{M^{3/8}}{M^{2/4}}\right) (E/\rho)^{1/2} \propto (M^{-1/8})(E/\rho)^{1/2}. \tag{6-12a}$$

According to McMahon's model of elastic similarity, then, stride frequencies should be proportional to $M^{-1/8}$ and stride times to $M^{1/8}$. Elastic similarity does not seem to apply to avian incubation periods or mammalian erythrocyte life spans, but there is some correlation between observed and predicted values for running locomotion.

Thus the same biological cycles are about seven times longer in man than in mice, and twenty times longer in elephants than in mice. While few of the lines in Figure 6-1 have a slope of exactly 0.25, they are all very nearly parallel. In fact, the mean of all the body mass exponents is 0.247. Unfortunately the amount of statistical detail provided in the original reports varies considerably. However, where available, the 95% confidence intervals of

---

* Stress is force per unit of cross-sectional area; strain is change in length divided by resting length. Two strips of rubber equal except in length are equally stressed when stretched to the same proportion of starting length; thus a 10-cm strip stretched to 15 cm and a 20-cm strip stretched to 30 cm are equally stressed.

most of the slopes include 0.25 (Table 6-1). The variability in the exponents may represent the variance in the biological periods themselves, though it must also reflect the difficulty of accurate measurement: the end point of the longer cycles, such as growth to maturity or life span, cannot be determined with the same accuracy as cardiac cycle—and even heart rate has been variously reported to be scaled from $M^{0.25}$ to $M^{0.28}$ (Adolph, 1949; Calder, 1968). The observed difference in exponents therefore may not be statistically meaningful. Whether or not the variance in exponents is artifact, small residual mass exponents have relatively little effect over a wide range of bird and mammal body sizes (see Appendix Table 1). According to the equations in Table 6-1, 100 heartbeats occur in 6 seconds for a shrew and in 2.5 minutes for an elephant. The shrew's total blood volume should circulate once in 100 cardiac cycles, whereas the elephant's should circulate 1.5 times in 100 cycles. Although their body sizes differ by a factor of one million, the ratio of blood circulation time to heart contraction time varies only slightly, and the apparent difference may only be "noise" in the measurements or in the power-law equations themselves.

Evolution of Function

It appears that the many biological events and cycles of events in the body occur as relatively constant multiples of one another and in times proportional to (body mass)$^{1/4}$. How are these scalings selected in the evolutionary process? Certainly if one function is designed in proportion to $M^{1/4}$, other functions tied to it will be affected by that scaling, for characteristics as diverse as geometry and ionic conductance of membranes had to evolve together. For example, if the mechanical efficiency of the heart necessitates a size-dependent natural frequency, then the endogenous depolarization-repolarization process of the sinoatrial "pacemaker" must match that natural frequency. The rhythmic timing of cyclic movement in walking, breathing, and swimming are thought to be intrinsic properties of the central nervous system (Delcomyn, 1980). These properties must have been selected to synchronize with the natural frequencies of the oscillating structure.

Are these scales of physiological time controlled separately? Does natural selection occur one feature at a time, or do growth fields exert such effect that once size is settled by natural selection, the component dimensions, rates, and times follow as emergent properties?

The same sort of questions surround the $M^{1/4}$ scaling of physiological time that surround the $M^{3/4}$ scaling of metabolic rate, which is only one of a large number of physiological variables with the dimensions of volume per unit time. As discussed in Chapter 5, supporting physiological variables, such as glomerular filtration rate, respiratory volume per minute, and cardiac output, also scale near $M^{3/4}$. At least in the primary support functions of circulation and respiration, volumes or capacities are scaled in direct proportion to body mass among animals of the same design. Hence, if volume rates are proportional to $M^{3/4}$, physiological time must be proportional to $M^{1/4}$:

$$\frac{\text{volume}}{\text{time}} = \frac{M^{1.0}}{M^{1/4}} = M^{3/4}. \tag{6-13}$$

Explanations of the ¾ exponent have focused on the exponent itself, as if nature had selected the scaling of volume rate. Lindstedt and Calder (1981) have suggested an alternative: the scaling of biological volume rate could just as well be the consequence of physiological time scale constrained by physical factors (limitations of nerve conduction or muscle contraction, for example) to an $M^{1/4}$ scaling.

Consider muscle contraction first. The cross-sectional area of a muscle is proportional to d^2, and the mass of each muscle, limb, or entire animal is proportional to ld^2. Since the maximum stress (force divided by cross-sectional area) generated in homologous muscles is roughly constant (Hill, 1950),

$$\frac{(\text{mass})(\text{acceleration})}{\text{area}} \propto \frac{(ld^2)(l/t^2)}{d^2} = \frac{l^2}{t^2} = \frac{M^{1/2}}{t^2} = \text{constant}. \tag{6-14}$$

This can be solved for time, to get the exponent that fits the empirical description (Lindstedt and Calder, 1981):

$$t^2 = M^{1/2}/\text{constant}; \qquad t \propto M^{1/4}. \tag{6-15}$$

Nerve conduction velocity appears to be independent of body size (M^0). Lindstedt and Calder (1981) reviewed evidence that conduction velocities are correlated with behavior and ecological niches; that is, conduction is slow in sloths and skunks, average in primates, and fast in cats. They were unable, however, to find any reports of systematic relation to body size.

The time required for an impulse to travel the length of a sensory or motor nerve axon is a product of the velocity (u) and the distance traveled. The distance is proportional to a characteristic linear measurement l, which, according to the elastic similarity model, is in turn related to $M^{1/4}$. Therefore:

$$t_{conduction} = ul \propto M^0 M^{1/4}. \tag{6-16}$$

Hormonal transmission is dependent upon circulation of the blood. The velocity of blood flow also approaches size independence (Chapter 5), so that the time required for a hormone to reach the target organ would be simply proportional to the linear distance or to $M^{1/4}$ (Lindstedt and Calder, 1981; Calder, 1981). We can therefore predict an $M^{3/4}$ scaling like that of metabolic rate if we start from a scaling of physiological time made inevitable by the evolution of body size, applied to organ dimensions governed by the principle of symmorphosis (Taylor and Weibel, 1981). Conversely, we could start at the other end with an $M^{3/4}$ scaling from elastic similarity to decide that $M^{1/4}$ timing was the inevitable consequence of metabolic needs in body space limited by symmorphosis. Correlation cannot tell us whether the chicken or the egg came first, only that they could be related.

Life Span

Our species is probably the only one in which the individual lives with an awareness of his or her mortality. Perhaps we would feel better about the inevitability of death if we compared the lengths of our lives with the allometric prediction for average animals of our body mass:

$$t_{ls} = 11.6(70 \text{ kg})^{0.20} = 27 \text{ yr.} \tag{6-17}$$

On this scale the person who attains age twenty-seven has lived a full life; when I reach fifty-four, I will have had twice my share; and if I live to see eighty-one, I will have tripled my size-dependent span.

If an animal succeeds in surviving the threats of disease, predation, and accident, it eventually dies of "old age." There are a number of factors that could determine the timing of death for such an individual (Lindstedt and Calder, 1981; Rosen et al., 1981). Among them are (1) design margin, or safety factor; (2) failure of a crucial component; (3) somatic mutation; (4) degree of protection from toxic metabolites; and (5) brain factors hypothesized from correlation of life span to brain mass.

Safety Factor. The evolutionary survival of the fittest hinges on reproductive success or fitness. Unless an animal lives long enough to reproduce successfully, its genes will be lost forever. Just as the organism has been designed through natural selection to have sufficient strength to resist mechanical stresses that could crush, buckle, shear, or twist the supporting limbs or trunk, and to have adequate respiratory reserve to handle peak demands during sustained activity, life span must be long enough to allow for successful breeding and the rearing of replacement offspring, perhaps controlled by what Sacher (1978) postulated as "longevity-assurance genes."

The existence of these safety factors can be inferred from the ratios of cycle times in col. 6 of Table 6-1, which express life span as multiples of shorter physiological times (for example, 9.61 times the time for mammals to attain 98% of adult mass, 15.5 times that needed to reach reproductive maturity, 64.9 times the gestation period. The table exaggerates the safety factors, however, for the multiples are based on maximum life span. Use of the time for replacement of a population (standing crop turnover time; see Chapter 11) as an expression of average life span would be more appropriate, although the allometry is based on a small sample:

$$\frac{t_{standing\ crop}}{t_{reproductive\ maturity}} = \frac{1.02\ M^{0.29}}{0.75\ M^{0.29}} = 1.36\ M^{0.00}; \tag{6-18}$$

$$\frac{t_{standing\ crop}}{t_{50\%\ growth}} = \frac{1.02\ M^{0.29}}{0.352\ M^{0.25}} = 2.90\ M^{0.04}; \tag{6-19}$$

$$\frac{t_{standing\ crop}}{t_{gestation\ period}} = \frac{1.02\ M^{0.29}}{(66.3/365)M^{0.26}} = 5.62\ M^{0.03}. \tag{6-20}$$

Even when standing crop turnover times are used, there is some evidence for the existence of safety factors, but how this information is stored and how it exerts its effects are not clear.

Failure of a Crucial Component. Physiological homeostasis is dependent upon delivery of oxygen and nutrients to the tissues and removal of waste metabolites via respiratory, circulatory, and excretory systems. If the lungs, heart, or kidney fail, life will end. Since respiratory and heart frequencies are proportional to $M^{-1/4}$, the total number of contractions (at basal rate) is more or less size independent:

$$ft \propto M^{-1/4}M^{-1/4} = M^0. \tag{6-21}$$

Metabolic and renal turnover times have body mass exponents similar to those for life span, so a lifetime can also be viewed as a relatively size-independent multiple of these metabolic events:

$$t_{ls}/t_{metabolic} \propto M^{-1/4}/M^{-1/4} \propto M^0, \tag{6-22}$$

$$t_{ls}/t_{plasma\ clearance} \propto M^{-1/4}/M^{-1/4} \propto M^0. \tag{6-23}$$

These size-independent ratios suggest that the heart, lungs, kidneys, and perhaps other parts of the body are capable of a finite number of operations or contractions, or a finite accumulated level of work, before normal wear and tear reaches the point of impaired function. Cardiac failure and pulmonary emphysema are certainly malfunctions associated with aging, as well as with abuse or inadequate exercise. Humans who exercise have lower basal heart rates and greater life expectancy (Corrsin, 1982). Corrsin suggests extension of the analysis to small mammals, to which an allometrist can only say "Amen!" When oxygen delivery is impaired, other malfunctions are inevitable.

The kidney does a disproportionately large share of the physiochemical maintenance of the body. In man the kidneys are responsible for 7.7% of total basal heat production, but constitute only 0.5% of body mass. The metabolic rate per kilogram of kidney tissue is 17 times the average for the whole body, second only to the cardiac muscles. The kidneys process one-fifth of the entire cardiac output:

$$\frac{(\text{PAH plasma clearance})(1 - \text{hematocrit})}{\text{cardiac output}}$$

$$= \frac{(21.79 \ M^{0.77})(-0.458)}{187 \ M^{0.81}} = 0.21 \ M^{-0.04}. \qquad (6\text{-}24)$$

When the kidney malfunctions, life is threatened. If the concept of a "metabolic life span" is valid at the organismic level, would it not also apply to such a metabolically busy organ? It is not surprising that the onset of decline in renal function occurs almost as early as in cardiovascular or pulmonary function (Strehler, 1959; Epstein, 1979; Lindstedt and Calder, 1981).

Somatic Mutation. One prominent theory of aging holds that as the animal grows older, there is a gradual accumulation of somatic mutations or impairments of cellular proliferation and repair (Timiras, 1978; Harrison, 1978). While this theory has not been submitted directly to allometric analysis, aging is now considered to be a stage in the continuum from fertilization to death—one that includes growth and development. In fact, Timiras (1978) has defined aging as "the sum total of all changes that occur in a living organism with the passage of time." This view is compatible with the empirical expressions, the inverse of the allometric ratios, of Table 6-1. If $10(M^{-0.06})\%$ or $16(M^{-0.09})\%$ of maximum life span is spent respectively for growth to adult size and for attainment of reproductive maturity, the *proportion* of a lifetime beyond these attainments, in which aging occurs, must also be essentially size independent. Thus species within a class are built on a common plan with a program that is expanded or collapsed as appropriate to fit the size-dependent life span.

Aging time, as approached above, is a proportion of the time remaining after growth and development have been completed. This can also be approached directly in the form of an "actuarial aging rate" or annual increase in mortality rate, to be discussed in Chapter 11. This characteristic is scaled as $M^{-0.27}$, so the time for doubling of mortality would scale as $M^{0.27}$ (Eq. 11-53). This exponent is so close to those for growth (Chapter 10) that the coefficients can be compared directly: aging appears to proceed on the same physiological time scale as the earlier chapters of the life history.

Degree of Protection from Toxic Metabolites. Rosen and colleagues (1981) have suggested that somatic mutations have less effect on aging than metabolism. Their conclusion was based upon an exploration of the relation between the maximum potential life span (based on observed record) and the Gompertz constant. The surviving fraction *(n)* of a population in a Gompertz decay curve is

$$n = n_o \exp \frac{q_o}{\alpha} (1 - e^{\alpha t}), \tag{6-25}$$

when n_o = initial population; q_o = initial mortality rate; α = Gompertz constant for increase in mortality rate with time t. From a list of 23 species of animals, calculated values range from 2.2 to 12.8 for the product αt_{ls}, where t_{ls} is the maximum life span observed, usually in the protected environment of captivity, taken as an index of maximum potential life span.

The Rosen group found that the product of maximum potential life span and the Gompertz α, characterizing the age dependence of mortality rate, is almost constant over a wide range of animal sizes. The identification of such a pattern in life-history characteristics encourages those of us who seek to understand the mechanisms of life processes.

The allometric approach seems to me to create an even stronger tie between t_{ls} and the Gompertz α. The α values from table 1 of Rosen et al. (1981) and ten additional points converted from mortality-rate doubling times, t_D, in Sacher's (1978) table 1 (according to the relationship $\alpha t_D = 0.693$), yielded the following relationship:

$$\alpha = 0.71 M^{-0.266}; \qquad r^2 = 0.622. \tag{6-26}$$

The standard error of estimate for the exponent is ± 0.04, which comes amazingly close to the reciprocal of the exponent's ± 1 standard error (s.e.) for the relation of maximum recorded life span to body mass.

The product αt_{ls} for eutherian mammals is therefore

$$(0.71 M^{-0.266})(11.6 M^{0.20}) = 8.24 M^{-0.07}. \tag{6-27}$$

Furthermore, with mass-specific metabolic rate bearing an exponential relation to body mass similar to that for α, Rosen and colleagues suggested that metabolic rate is "a major factor affecting α." They could then associate that suggestion with evidence that aging results from damage inflicted by metabolically produced free radicals in the cell. The idea is intriguing — but we are still in the realm of correlation, not necessarily of causation.

The enzyme superoxide dismutase, otherwise known as superoxide: superoxide oxidoreductase and happily abbreviated SODase, catalyzes the destruction of the toxic free-radical superoxide ions (O_2^-) which are produced in aerobic metabolism. Conceivably, the evolution of longevity could have included genetic regulation of the production of such a protective enzyme, leading Tolmasoff and associates (1980) to measure SODase activity levels from a selection of mammals that included two species of mice and twelve species of primates that have been important in longevity research. These animals ranged in size from 20.7 g to 69 kg. A correlation between these SODase levels and maximum potential life span t_{ls} was then sought.

The Tolmasoff group (1980) found a striking parallel in the irregularities of curves relating both liver SODase and "maximum lifespan calories consumption (MCC)" to t_{ls}; here MCC is the product of $(\dot{E}_{sm}/m)$ and t_{ls}. Regressions of mass-specific metabolic rate $(\dot{E}_{sm}/m_{tissue})$ of liver and brain tissue were closely parallel to the Kleiber relationship, $\dot{E}_{sm}/m \propto m^{-1/4}$. However, there was no general correlation between SODase levels and $t_{ls,max}$.

The authors proceeded to correlate the ratios of SODase:$\dot{E}/m$ with $t_{ls,max}$ for liver, brain, and heart, finding an excellent correlation for all three. This was "the first time an enzyme specific activity level has been correlated with a longevity parameter of a mammal." Further, "this correlation suggests that longer-lived species have a higher degree of protection against by-products of oxygen metabolism."

As an alternative to this explanation, I submit that the correlations of highest significance are those of life span and metabolic rate with body size. Together, they in fact constitute Stahl's (1962) dimensional constant for mammalian lifetime basal power. The ratios of SODase levels (which had "*no* general correlation with life span") to $\dot{E}/m$ (which *did* show a good inverse correlation with life span), therefore have an inevitable correlation with life span. However, this could be due not to the suggested higher degree of protection by SODase, but to the metabolic correlation. Both $t_{ls,max}$ and $\dot{E}/m$ have high correlations with body mass (Eqs. 3-13, 3-13a, and 3-13b and Table 6-1); they should therefore have high correlations with each other.

To demonstrate the spurious outcome of this procedure, we can correlate the ratio of the number of heart chambers in mammals (N_c) to specific metabolic rate with maximum life span:

$$N_c/(\dot{E}/m) = 0.024 + 0.0017 \, t_{ls,max}. \tag{6-28}$$

This shows a higher correlation ($r^2 = 0.91$, $p < 0.001$); still, we would not want to conclude that longer-lived species have a higher degree of physiological protection because of the number of chambers in their hearts, for all mammals have four-chambered hearts. This false correlation does not invalidate the hypothesized relationship between SODase-protection and longevity. Could we utilize allometry in another approach to the hypothesis?

Allometric regression equations serve a useful function as baselines for comparison. Not all animals have values that fall right on the line. Some are higher, some lower, and these may be either adaptations to special circumstances or fortuitous inheritances from an unusual lineage. If there is a departure in one variable, it may be correlated with other, possibly related variables and compensations for their effects. Of course, the larger the sample size, the more "typical" our allometric generalization.

The factors which showed the most strikingly similar patterns in the study by Tolmasoff and coworkers are the curves for SODase vs. t_{ls} and MCC vs. t_{ls}. A straight-line eye fit to these data shows that two species, 6 (the mustached tamarin) and 7 (the lemur), are conspicuously high in SODase and in calculated lifetime metabolism, whereas the gorilla is considerably below the eye fit in both graphs. Two species, the deer mouse and the common tree shrew, are slightly high. The correlation should be expressed mathematically to check this optical perception.

Our baseline for MCC is derived from Eqs. (3-13), (3-13a), and (3-13b) and the equation for t_{ls} in captivity (Table 6-1), converted to the units (kilocalories per gram) used in the Tolmasoff study.

$$\begin{aligned} \text{MCC} &= (\dot{E}_{sm}/m)(t_{ls}) \\ &= (0.365 \ m^{-0.244})(1,064.3 \ m^{0.20}) = 388 \ m^{-0.044}. \end{aligned} \tag{6-29}$$

Since this is the first report of liver SODase concentrations (in units per milligram of protein), the only baseline would be a linear regression of their values as a function of body mass:

$$\text{liver SODase} = 16.94 \ m^{0.04}. \tag{6-30}$$

This correlation is not significant ($r^2 = 0.19$, $p > 0.1$), so we can only use the 16.94 u/mg protein as a "constant" for comparison of specific SODase activities of the different species. MCC can be compared to the allometric prediction based on body mass for each species.

Now, to see if there is a correlation, we can run a linear regression of the ratios of observed (o) to predicted (p) values for MCC as a function of liver SODase, o/p. The correlation is highly significant, with $p < 0.001$, $r^2 = 0.846$:

$$\text{MCC}_o/\text{MCC}_p = -0.34 + 1.59(\text{SODase}_o/\text{SODase}_p). \tag{6-31}$$

Of course, we are still merely correlating, not proving cause and effect, but we have a very strong correlation using SODase activity levels by themselves, without "piggybacking" them onto another correlation.

Is this a trivial consequence of ignoring the 0.04 exponent in Eq. (6-30) because it was not a significant regression? The linear regression of Eq. (6-31) can be repeated using predictions from Eq. (6-30). The correlation of o/p ratios is not so tight, $r^2 = 0.47$, though it is still significant, $0.01 > p > 0.001$. Finally, we can go to the other extreme of handling the exponents -0.044 and 0.04, considering them as insignificant residual mass exponents, following Stahl (1962, 1967), and using the ratio of observed/coefficient (observed/388 for MCC, observed/16.94 for SODase), in which case r^2 would be 0.908, $p < 0.001$. From this it seems that the next step should be the determination of SODase activity from a sampling that is larger in number of data and phylogenetic variety, for further testing of the correlations between SODase and t_{ls} and between SODase and MCC.

Before we leave the subject of enzyme and longevity correlations, we should note that while many enzyme concentrations or activities are inversely correlated with body mass (and therefore with longevity), seeming to provide enhanced support for the high $\dot{E}_m/m$ of small mammals, a positive allometry is now known to occur for more than one enzyme. The SODase report of the Tolmasoff group (1980) was a first for correlating the specific activity level of an enzyme with maximum longevity, but catalytic activities of the glycolytic enzymes pyruvate kinase, lactate dehydrogenase, and phosphorylase are proportional to $m^{0.14}$, $m^{0.15}$, and $m^{0.09}$ respectively (Emmett and Hochachka, 1981; see also Chapter 5). This scaling indicates that the larger the animal (and therefore the longer its life span), the higher the glycolytic capacity of its skeletal muscles.

Life Span and Brain Mass. A regression of log life span vs. log brain mass can account for 79% of the variance. A similar relationship between log life span and log body mass accounts for only 60% of the variance (Sacher, 1959; Fischer, 1968; Mallouck, 1976; Western, 1979). The temptation to equate correlation with causation has proven irresistible in view of the tighter correlation between life span and brain mass. When one such causal hypothesis is discounted, the correlation is rediscovered independently and a new brain substance is hypothesized to enhance longevity.

The explanation of the differences in correlation coefficients is simple. Variation in body mass is due primarily to changes in fat content because of the balance between rate of food consumption and rate of metabolism. Unlike the rest of the body, the brain is protected by and confined within a rigid, bony cranium; it cannot expand to store a surplus of fat or water, so once adult size is attained the brain mass is not likely to change. This can be confirmed by the coefficient of variation (c.v. or standard deviation divided by mean value) for brain and body size, with the latter two or more times the former (man—body c.v. 0.24, brain c.v. 0.10; horse—body c.v. 0.33, brain c.v. 0.12; mammals in general—body c.v. 0.12 to 0.15, brain c.v. 0.06 to 0.07; from Yablokov, 1966; Lande, 1979; Lindstedt and Calder, 1981).

If we overlook the variance argument, the life span/brain correlation leads logically to the following reasoning. The brain controls physiological maintenance and homeostasis. The larger the brain, the greater the capacity to regulate; and the finer the regulation, the more favorable the cellular environment, the better the preservation of the body, and the longer it remains in good working order. Thus longevity comes to be associated with a high index of cephalization and reduced entropy generation (Sacher, 1978).

As an exploration of the implications of this argument, it is instructive to perform allometric cancellations on the allometric ratios of maximum life span (in years of captivity) to brain mass (in grams):

$$\text{eutherian mammals:} \quad \frac{t_{ls,max}}{m_{brain}} = \frac{11.6 \, M^{0.20}}{11.3 \, M^{0.67}} = 1.03 \, M^{-0.47}; \quad (6\text{-}32)$$

$$\text{birds:} \quad \frac{t_{ls,max}}{m_{brain}} = \frac{28.3 \, M^{0.19}}{8.2 \, M^{0.60}} = 3.45 \, M^{-0.41}. \quad (6\text{-}33)$$

From these derivations it appears that the ratio of years per gram of brain decreases significantly with size. A 500-kg mammal, for example, lives 20.3 days per gram of brain, while a 110-g rodent lives 2.9 years per gram of brain tissue. Second, birds tend to have smaller brains and longer lives, despite the fact that they experience wider variation in body temperature and plasma osmotic concentrations and plasma pH (Lindstedt and Calder, 1981). Of course, the brain is important in controlling the conditions necessary for life, and mechanisms which set longevity or program for aging may be discovered therein—but correlation, once again, is only correlation.

Physiological time is one of the most significant characteristics of living animals. There are too many time scales that are approximately proportional to body $M^{1/4}$ to deny its importance in repetitive maintenance processes and for stages within the life span devoted to development, growth, replacement, and decline. Nonetheless, we do not know why the exponent is $\frac{1}{4}$ rather than $\frac{1}{3}$. It would be fascinating to understand how natural selection or the genome control the physiological time scale—whether one function at a time, or as a single master function that pulls everything together so that a life span can be complete. Until biology advances sufficiently to be able to field such questions, much can be accomplished by empirical description. Because the $M^{1/4}$ scaling appears to be so extensive among the internal functions, we might expect that the animals' outward functioning in the ecosystem would show similar body size dependency. This possibility will be examined in Chapter 10. But first, we must explore the allometry of locomotion and environmental relationships, and then the reproduction and growth that produce properly scaled populations.

7 Locomotion

O for a horse with wings! . . .
He is at Milford-Haven: read, and tell me
How far 'tis thither. If one of mean affairs
May plod it in a week, why may not I
Glide thither in a day?

—WILLIAM SHAKESPEARE, *CYMBELINE* (1623)

EVEN IF the wish had been granted, the horse with wings could not have flown. The power required for horizontal flight scales as $M^{1.17}$, while the power available scales as the cross-sectional area of the flight muscles, proportional to $M^{0.67}$. The upper limit for flight can be predicted by the overtaking of power available by power required, which theory says would be at a body mass of about 12 kg, a reasonable approximation to the largest pelicans, bustards, turkeys, and swans (Pennycuick, 1969, 1972). The record body mass for the trumpeter swan (*Olor buccinator*), for example, is 12.5 kg (Banko, 1960).

Nevertheless, a horse can travel faster and its endurance will carry it much farther than a small running mammal. Among the runners, the migrants who range farthest are the larger ungulates, the caribou of the tundra, and the bigger antelopes of the African grasslands (Jarman, 1974).

Many effects of size upon locomotion are obvious, at least in a qualitative sense. The cadence stepping frequency, or wingbeat frequency, is inversely related to body size or limb length. The pace or stride length increases with size. The energy cost of travel increases with size, but the unit cost is cheaper by the truckload (cheaper yet by the trainload). Fliers migrate farther than runners. These are obvious facts that may not seem very interesting, but to demonstrate them quantitatively has posed a great challenge and stimulated real progress in comparative physiology over the past two decades. The proceedings of a 1975 international symposium on "Scale Effects in Animal Locomotion" held in England constitute a reference point of major significance for future work in many areas of the life sciences: biomechanics, bioenergetics, the applied physiology of exercise, comparative physiology, and allometric evolution (Pedley, 1977). The implications for ecology and life-history patterns are profound, although initial attempts by ecologists to incorporate allometric equations for the cost of locomotion into optimization models have not been without problems.

   The scaling of some basic components of locomotion has already been discussed: the proportions of body mass allocated to the muscular machinery and the skeletal framework (Chapters 2 and 3); the dimensions of the skeletal components, and limb proportions (Chapter 4); and the metabolic effort and cardiopulmonary delivery of oxygen and energy to this machinery (Chapter 5). Assuming that I am an average biologist with average proficiency in mechanics and mathematics, I would extrapolate to say that most biologists find the literature to be significant, interesting, and slow reading. It is likely that there are others who find it difficult to keep up, and who feel the need for a survey of the geometry and energetics of locomotion.

Running and Flying

EUTHERIAN MAMMALS

> "Nature," explained the philosopher, "always tries to make compensation. For instance, if one eye is lost, the sight of the other becomes stronger, and if a person grows deaf in one ear, the hearing of the other becomes more acute."
>
> "Faith," said Pat, "and I believe you're right; for I've noticed that when a man has one leg shorter, the other is always longer."

   To take the simple case of bilateral symmetry, both legs have presumably evolved to that same length adequate to reach from the body to the ground, with sufficient clearance to make locomotion practical. Leg length increases with body size, so we might expect stride length to be approximately the base of a triangle (actually, the arc described by the angle of excursion of the limb) from the farthest rearward extension to the farthest forward extension. This might be the case in walking, when there is no airborne leap through the air during the stride cycle. However, at a fast gait or in leaping, there is a gliding period when none of the feet are in contact with the ground. Thus, the stride length is projected to a longer distance than the arc of excursion of the legs.

The size dependency of limb length, angle of limb excursion, stride frequency, stride length, and running speed have been described in the many papers on running locomotion published by Taylor and his colleagues at Harvard University, to be cited below. In order to standardize for the size dependence of speed, the comparisons were made at the transition from a trot to a gallop, t-g. The stride frequency at this transition (f_{t-g}, in seconds^{-1}) was described by Heglund and colleagues (1974) for a series composed of mouse (30 g), rat (360 g), dog (9.2 kg), and horse (680 kg):

$$f_{t-g} = 4.48\ M^{-0.14} = 11.8\ m^{-0.14}; \qquad r = 0.99. \tag{7-1}$$

The stride length (l_{t-g}, in meters) for these animals was

$$l_{t-g} = 0.35\ M^{0.38} = 0.025\ m^{0.38}; \qquad r = 0.97. \tag{7-2}$$

The speed or velocity (in meters per second) attained at the onset of the gallop is the product of stride length and frequency:

$$u_{t-g} = (0.35\ M^{0.38})(4.48\ M^{-0.14}) = 1.53\ M^{0.24}. \tag{7-3}$$

This is in close agreement with what would be predicted from the elastic similarity model, with stride frequency scaling as $M^{-0.125}$ and speed as $M^{0.25}$, for a third mechanical geometry (three or more links—for example, flexing trunk + femur + fibia + metatarsals). However, of the animals used in the derivation of Eq. (7-1), all but the horse fail to conform to elastic similarity in the scaling of their linear dimensions (Alexander, 1982). Alexander's model of dynamic similarity, to be discussed shortly, predicts the same sort of scaling: $M^{-0.13}$ and $M^{-0.18}$ for the stride frequencies of the Bovidae and other mammals respectively.

A useful objective for future studies would be clarification of the scaling of stride frequency, based on a larger number of species and separating the Bovidae with their $M^{1/4}$ limb-length scaling from the other mammals with $M^{0.33}$ to $M^{0.37}$ leg-bone scaling (Chapter 4). It will be interesting to find out exactly what the true exponents are, but we can be sure that frequency will be inversely scaled to body size, and that the natural rhythmic patterns of locomotion are executed more rapidly in smaller animals.

There have been two hypotheses for the neural basis of regular, repetitive, or cyclic actions in running and walking. One is that sensory feedback from the propulsive action of one limb or set of muscles serves to stimulate the appropriate contractions for the next phase of the cycle. The other hypothesis attributes the rhythmic timing to a neural pacemaker or oscillator in the central nervous system. Experiments involving isolation, deafferentation of the sensory nerves as required by the first hypothesis, or paralysis have strongly supported the existence of a rhythmic capacity in the central nervous system (reviewed by Delcomyn, 1980). This means that the intrinsic central rhythmicity must also be scaled inversely to body size, just as the ionic leakage rate of the sinoatrial node culminating in cardiac contraction must be an inverse function of body size.

Does elastic similarity account for the $M^{0.38}$ scaling for stride length? Let us designate the angular excursion of the limb from full forward position (f) to full rearward extension (r) as angle θ. The length of the leg (l_{leg}) is approximately the hypotenuse for each of two triangles, the bases of which add up to the excursion distance l_{r-f} (see wapiti, Figure 7-1), while the angles between the leg and the perpendicular from the acetabular joint at maximum forward swing and maximum rearward swing add up to $\angle\theta$. In moving pictures of the same animal series used by Heglund and associates (1974) θ decreased with size, the larger animals tending to be relatively conservative in their stride excursions (McMahon, 1975b):

$$\angle\theta = 37.2\, M^{-0.10}. \tag{7-4}$$

If the muscle stress (force/cross-sectional area) is a size-independent constant and elastic similarity is preserved, $\angle\theta$ should be scaled as $M^{-0.125}$, which Eq. (7-4) approximates. The muscle fibers develop maximum force when the filaments overlap maximally, so the relative shortening should be independent of body size. The relative mechanical advantage of extensor muscles, or their moment:arm ratios, increase with body size ($\propto M^{0.22}$). To obtain maximum force with different moment:arm ratios, the angular stride excursion must decrease with size (see Figure 7-1 and Biewener, 1983).

The limb excursion distance l_{r-f} of the wapiti in Figure 7-1 is a geometric consequence of θ and limb length. McMahon (1975b) used the scaling of $\angle\theta$ (Eq. 7-4) and limb lengths of ungulates ($\propto M^{0.25}$) to predict the scaling of the excursion:

$$l_{r-f} \propto M^{0.15}. \tag{7-5}$$

As seen in Figure 7-1, the actual stride distance exceeds the excursion distance. McMahon expressed this as the "stride efficiency," or the ratio of stride length (Eq. 7-2) to l_{r-f}:

$$l_{t-g}/l_{r-f} \propto M^{0.38}/M^{0.15} \propto M^{0.23}. \tag{7-6}$$

Elastic similarity predicted $M^{0.25}$ for stride efficiency.

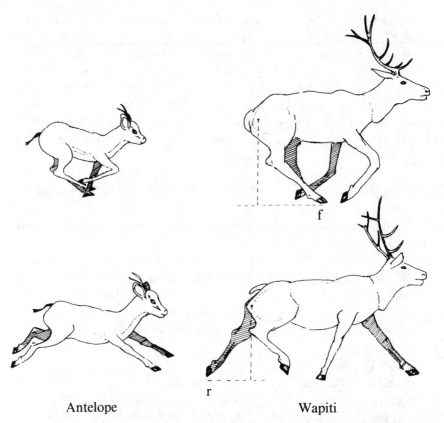

Antelope   Wapiti

7-1 Limb excursion angles during gallop decrease with increasing body size, as seen in this series of sketches of animals of different sizes (antelope, wapiti, and bison; after Muybridge, 1902, pp. 199, 201, 203). The excursion angle θ is the angle between the maximum forward extension (f) and maximum rearward extension (r) of the hind limb. The stride length is the distance between two successive footfalls of the same foot. It exceeds the length of the arc through which the foot moves, which is the sum of the bases of the two triangles bounded by the extended limb forward and the limb to the rear of the perpendicular from the acetabular joint and shown by the dotted lines in the wapiti sketch. (Drawing by Lorene Calder.)

The increase in stride efficiency with body size would not be as spectacular for the other mammals with limb lengths proportional to $M^{0.33}$ (see Table 4-2). The relationships for f_{t-g}, l_{t-g}, and $\angle\theta$ (Eqs. 7-1, 7-2, 7-4) were derived from these other-than-artiodactyls, and for them the excursion would be more like:

$$l_{r-f} \propto (l_{leg})(\sin \angle\theta) \propto (M^{0.33})(M^{-0.10}) \propto M^{0.23}, \qquad (7\text{-}7)$$

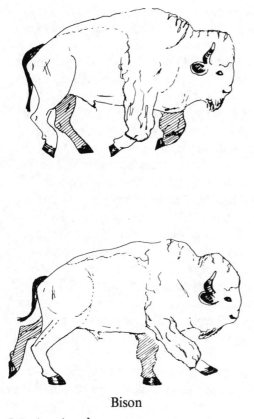

Bison

7-1 (*continued*)

which would result in a stride efficiency of

$$l_{t\text{-}g}/l_{r\text{-}f} \propto M^{0.38}/M^{0.23} \propto M^{0.15}. \tag{7-8}$$

This does not conform to either elastic ($\propto M^{0.25}$) or geometric ($\propto M^{0}$) similarity, but is somewhere between what the two models would predict.

Allometric studies of the limbs and limb bones of both mammals and running birds have revealed a number of differences between empirical descriptions and the corresponding predictions of scaling that would preserve elastic similarity. In addition to the fact that limb and bone lengths of mammals other than the Artiodactyla are scaled in proportion to greater body-mass exponents, even in the antelopes (Bovidae in the Artiodactyla) which generally fit the elastic model, some variables of limb allometry do not conform to an elastic similarity. These include some bone diameters and cantilever strengths, and cross-sectional areas of tendons (Alexander, 1977). While exponents for bone diameter and moment arms of muscles of running birds are closer to an elastic similarity scaling, the exponents for tendon cross-sectional areas and moment of inertia of the legs come closer to geometric similarity, and the vertical leg bones have lengthened beyond geometric similarity with greater body size. Maloiy and coworkers (1979) therefore expressed reservations about an explanation based upon elastic similarity. The gravitational forces that would cause sagging in the stationary animal are exceeded by the peak inertial forces during galloping locomotion. An individual metatarsal, tibia, or humerus must sustain peak vertical forces that were calculated to be 1.6 to 2.6 times the total body weight of a galloping 70-kg antelope (Alexander, 1977).

A stress similarity will be preserved in corresponding muscles if duty factors scale proportional to $M^{0.125}$. The duty factor for a foot is the fraction of the stride cycle that the foot is in contact with the ground. Each foot of a smaller mammal is thus "on duty" for a smaller portion of the period of the stride — the mouse, for instance, is airborne more of the time than the cat. This smaller duty factor means that the vertical forces on the four feet, which must average out to support the body weight for the entire cycle, are concentrated into a shorter fraction of time and attain peak values that are larger multiples of body weight.

Thus while the elastic similarity model appears to explain much of running locomotion, paradoxes remain and indicate the desirability of further research into, for example, the components of Eq. (7-3) for artiodactyls. If the components hold for artiodactyls and nonartiodactyls, why has natural selection in the latter accepted an apparently smaller emphasis on stride efficiency? What can be learned about scaling in the Perissodactyl progression from *Hyracotherium* to *Equus* (see Simpson, 1961)?

According to Alexander (1982), the fact that f_{t-g} scales as predicted does not confirm the existence of elastic similarity unless we deal with three problems. The first has been noted above, that three of the four animals used to determine the scaling of f_{t-g} have linear dimensions which do not conform to elastic similarity scaling. Second, the swinging of the legs in running is not maintained by elastic storage and transfer between forward and backward swings, because the kinetic energy of the backward swing considerably exceeds that of the forward swing. Finally, Alexander found that a dynamic similarity model anticipated a scaling essentially indistinguishable from that predicted by the elastic similarity model.

Dynamic similarity would be exhibited if the motions of a large and a small animal could be made identical by changing scales of length and time so as to have identical Froude numbers, $(\mathcal{F}\mathfrak{r})$:

$$\mathcal{F}\mathfrak{r} = u^2/gl, \tag{7-9}$$

where u is a characteristic speed, g is the gravitational acceleration constant, and l is a characteristic length. The Froude number is the ratio of inertial force to gravitational force, $m\, u^2 l^{-1}/mg$. Since m cancels out, this expression becomes the ratio of acceleration to gravity. In preserving the same $\mathcal{F}\mathfrak{r}$, the inertial acceleration is directly proportional to the gravitational acceleration. If animals of different dimensions l move with the same $\mathcal{F}\mathfrak{r}$, u^2/l is a constant, or $u \propto l^{1/2}$. With the same angular amplitude of limb excursions, the stride lengths are proportional to l_{leg}, thus:

$$f = u/l_{stride} \propto l^{1/2}/l \propto l^{-1/2}. \tag{7-10}$$

For the Bovidae with $l \propto M^{0.25}$, f scales as $M^{-0.125}$, the same as for elastic similarity. For other eutherian mammals having $l \propto M^{0.35}$, f scales as $M^{-0.175}$. This latter prediction is in accord with observed scaling of $M^{-0.14}$ to $M^{-0.18}$ (Alexander, 1982). The dynamic similarity model should not be regarded as incompatible with elastic similarity. As Alexander (1982) points out, if running were entirely dependent upon elastic storage and release, dynamic similarity would require elastic similarity as well.

The restrictions on the universal applicability of the various similarity models do not leave the dynamic model unscathed. There is an inverse relationship between size and Froude numbers when animals are running at maximum speed, a buffalo having a lower number than a gazelle (Alexander, 1982). Thus dynamic similarity does not apply to running at top speed. It would be simpler if one model could explain everything, but scaling is too complex for that. We return, as Alexander has done, to empirical description until the patterns can be discerned more clearly, instead of building an explanation on general principles that may not apply because of other unseen and overriding considerations.

The concept of symmorphosis, discussed earlier, stated that the cost of building and maintaining biological structures requires that there be no more of them than is necessary (Taylor and Weibel, 1981). This is related to present considerations. Alexander (1977b) concluded that antelopes were scaled so that, at the animal's maximum running speed, the bones, muscles, and tendons would perform near their stress limits. It would not make sense, symmorphologically, to be twice as strong as necessary or to refrain from use of full capacity when evasion or pursuit required it!

Does the $M^{0.24}$ scaling of speed at the trot-gallop transition characterize other gaits and levels of effort as well? Garland (1983a) has analyzed the relationships between maximal speeds and body size. For 106 species of mammals (eutherian and marsupial) ranging in size from 16 g to 6,000 kg, the maximal running speed (u_{max}) was approximately twice $u_{aero\,max}$ and scaled in agreement with dynamic similarity — $r^2 = 0.439$; 95% confidence interval (c.i.) of $b \pm 0.036$:

$$u_{max} = 23.6\, M^{0.17}. \tag{7-11}$$

However, in the large mammals there was a tendency for u_{max} to decrease with size after a maximum running speed of 56 km/hr, calculated to have been attained by an optimal animal of 119 kg body mass. Excluding the mammals of over 300 kg, those remaining ($n = 87$) yielded a tighter correlation with $r^2 = 0.570$, and with a scaling exponent indistinguishable (95% c.i. of $b \pm 0.042$) from that for speed at the trot-gallop transition, $u_{t\text{-}g}$:

$$u_{max} = 23.3\, M^{0.23}. \tag{7-12}$$

Converted to units of meters per second as in Eq. (7-3), this is over four times the speed at the onset of gallop!

Mechanical rather than cardiopulmonary limits are expressed in Eqs. (7-11) and (7-12), which were derived from observations in nature where anaerobiosis was undetected. Garland derived an equation for maximal aerobic speed ($u_{aero\,max}$, in kilometers per hour) from Taylor's equations for $\dot{V}_{O_2max}$ (Chapter 5) and for transport cost, which is speed dependent:

$$u_{areo\,max} = 11\, M^{0.15}. \tag{7-13}$$

Thus aerobic and mechanical limitations scale differently, a point to which we shall return.

The daily movement distance ($\dot{L}_{dm}$) for 76 species of mammals also has the dimension of speed (kilometers per day) and a scaling exponent indistinguishable from that for $u_{t\text{-}g}$ and u_{max} (95% c.i. of $b \pm 0.10$; $r^2 = 0.264$; $p < 0.001$; Garland, 1983b):

$$\dot{L}_{dm} = 1.038\, M^{0.25 \pm 0.10}. \tag{7-14}$$

Except for the aerobic limits, travel speeds seem to scale in a consistent pattern as approximately $M^{1/4}$. The reciprocal of speed is time per unit of distance, so Eqs. (7-3), (7-12), and (7-14) are also consistent with the general pattern for physiological time, as examined in Chapter 6.

BIRD FLIGHT

In most of the morphological and physiological measurements surveyed in previous chapters, the scaling exponents of birds and mammals have been similar. The scaling of brain size and some limb bones have been rather conspicuous exceptions. The geometry and mechanics of locomotion are also exceptions to this bird-mammal similarity.

Flight Velocity. Flight is not only faster than running, it appears to be slightly less size dependent, with scaling of $M^{0.12}$ to $M^{0.17}$ vs. the mammalian $M^{0.23}$ to $M^{0.25}$ (although smaller exponents were obtained when the largest mammals were included; see Eqs. 7-11 to 7-13).

There are two basic strategies for forward flight. One is important in migration or long-distance travel, and that is obtaining maximum range on the stored energy supply (per "fuel tank load"). The velocity to obtain maximum range is designated $u_{max\,r}$. The other strategy is used for staying airborne in a home area, as in a search for food, the object being to achieve the maximum endurance time at a minimum power requirement and thus settling for a lower velocity, $u_{min\,P}$. (Pilots of search-and-rescue aircraft use the same principles, flying from home base to the search area at $u_{max\,r}$, then throttling back to $u_{min\,P}$ while searching as the airplane becomes a platform for observation over a restricted area.)

The strategy is complicated by winds. If the desired direction of travel is downwind, the ground speed of the bird is the sum of its air speed and the speed of the tail wind, but if the bird must go directly into a head wind, ground speed is air speed minus head-wind speed. The effect of wind velocity on the total energy cost of travel is appreciated by considering the distance that the wind blows the bird forward or backward, the product of wind speed and duration of flight. With the benefit of a tail wind, the distance covered at no cost is increased by taking a longer time flying at a lower air speed. Even so, the total ground speed is greater than the no-wind air speed. The distance lost in a head wind is reduced by flying faster, getting the trip over with as soon as possible. This is what birds do (Tucker and Schmidt-Koenig, 1971; Schnell and Hellack, 1979), but the measurements are sophisticated and feasible only where the winds are steady, in that they involve determination of azimuth and altitude via radio-controlled, camera-recorded theodolites or Doppler radar, measurement of wind velocity and direction with balloons, and computer analysis. Clocking a bird's speed by driving down the highway and reading the speedometer is not sufficient. At the present time the data base is too small for meaningful allometric correlations. Variability due to the morphology of specialized flight types exceeds that due to body size. However, Greenewalt (1975a) tabulated average speeds from three reliable sources, and these do show a slight, significant correlation with $M^{0.07}$.

Information on the allometry of flight velocities therefore is heavily dependent upon theoretical derivation. Pennycuick plotted *calculated* speed ranges for $u_{min\,P}$ and $u_{max\,r}$ for flying insects, birds, and bats (1969, his figure 8). The best-fit line for these ranges is approximately

$$u = 13\,M^{0.17}. \tag{7-15}$$

(All speeds will be given in meters per second.)

The $u_{max\,r}$ that would minimize the cost of travel in a no-wind condition at sea level, according to Tucker (1974), would be

$$u_{max\,r} = 15.7\,M^{0.17}. \tag{7-16}$$

This assumes typical wingspan. If the wingspans were shorter than the general allometric predictions, as is the case for gallinaceous birds, for example, the coefficient would be greater than 15.7. Longer wingspans would decrease the coefficient. In either case the exponent would not change.

Rayner (1979) *calculated* from real morphometric data for "all birds" ($n = 68$ points):

$$u_{min\,P} = 5.70\,M^{0.16}, \tag{7-17}$$

$$u_{max\,r} = 10.89\,M^{0.19}. \tag{7-18}$$

He found these to agree fairly well with Greenewalt's (1962) ideal scaling:

$$u_{min\,P} = 5.72\,M^{0.20}, \tag{7-19}$$

$$u_{max\,r} = 10.04\,M^{0.24}. \tag{7-20}$$

Greenewalt (1975a) distinguished four types of flight morphometrics which he characterized as passeriform, shorebird, duck, and hummingbird types or models. These were not limited phylogenetically, but were based on wing loading (the lightest characteristic of passerines, the heaviest characteristic of ducks, with hummingbirds and shorebirds falling in between).

For three of these groups the calculated flight speeds were

passeriform model: $\quad u_{min\,P} = 9.5\ M^{0.12},$ $\tag{7-21}$

shorebird model: $\quad u_{min\,P} = 11.5\ M^{0.15},$ $\tag{7-22}$

duck model: $\quad u_{min\,P} = 14.6\ M^{0.16}.$ $\tag{7-23}$

The r^2 values were 0.764, 0.896, and 0.597, respectively. From real data for these groups Rayner (1979) gives (his table 6):

passeriformes $\quad u_{min\,P} = 5.04\,M^{0.13},$ $\tag{7-24}$
($n = 44$):

$$u_{max\,r} = 8.63\ M^{0.14}; \qquad (7\text{-}25)$$

shorebirds
$(n = 15)$:
$$u_{min\,P} = 5.85\ M^{0.10}, \qquad (7\text{-}26)$$

$$u_{max\,r} = 10.49\ M^{0.06}. \qquad (7\text{-}27)$$

Excluding ducks, for which the results were not statistically significant ($n = 9$), there is rough agreement among the various determinations.

Comparing $u_{max\,r}$ of birds (in general, Eqs. 7-16, 7-18, 7-20) to speeds of running mammals, we have:

$$u_{max\,r,\,birds}/u_{t\text{-}g} = \begin{cases} 15.7\ M^{0.17}/1.53\ M^{0.24} = 10.3\ M^{-0.07}, & (7\text{-}28) \\ 10.89\ M^{0.19}/1.53\ M^{0.24} = 7.1\ M^{-0.05}, & (7\text{-}29) \\ 10.04\ M^{0.24}/1.53\ M^{0.24} = 6.6\ M^{0.0}. & (7\text{-}30) \end{cases}$$

Even without the advantage of tail wind, which migrants wait for (Lowery and Newman, 1955), flight is obviously much faster—as well as cheaper—than running.

Flight Form and Function. Most lay persons older than age two know that birds lay eggs and that most birds can fly. Only recently have these attributes been approached analytically, but impressive efforts of the past two decades have made up for much lost time, with allometry a conspicuous and instructive part of the progress. There are two theories relative to the mechanics of generating the lift and thrust necessary for forward flight. The momentum jet theory, based on steady or continuous airflow, has been developed for the most part by Pennycuick (1969, 1975), Tucker (1973, 1974, 1977), and Greenewalt (1975a). The vortex theory, which takes account of unsteady lift and thrust generation by vortices in the bird's wake, was contributed by Rayner (1979). We speak fondly of being "free as a bird," a freedom conferred by the ability to fly—and a freedom that makes all the desired measurements almost impossible. Consequently, these studies have been based to a large extent upon morphological measurements, and mathematical derivatives of these measurements.

Measured: *Body mass*—M

Wing semispan—l_{span}, length of one wing, base to tip; or "wing length," the ornithologist's straight-line distance from bend or wrist, between metacarpals and phalanges, to the tip of the longest primary, omitting the airfoil provided by the brachium (Pettingill, 1970)

Wing area—A_{wing}, obtained by tracing the wing outline on graph paper and counting squares

Derived: *Wing loading*—$(M)(g)/A_{wing}$; usually the gravitational g is omitted and M/A_{wing} is expressed in grams per square centimeter

Disc loading—$(M)(g)/\pi(l_{span})^2$, the area of the disc described by the flapping wings

Aspect-ratio—l_{span}/l_{chord}, the chord being the average distance from the leading to the trailing edge of the wing and perpendicular to l_{wing}, calculated as $4(l_{span})^2/A_{wing}$

The significance of the derived characteristics can be appreciated by considering a simple flying machine, a propeller-driven airplane with wings that do not flap. In level, steady-state balanced flight, the thrust developed by the propeller must equal the drag forces tending to hold the plane back, and the forward motion sustained by that thrust produces the flow of air around the wings generating the force of lift that must equal the weight. Thus if lift is less than weight, the aircraft descends; if drag exceeds thrust, the aircraft loses air speed.

The amount of induced power required for flight is proportional to the disc loading, $(M)(g)/\pi(l_{span})^2$.

The lift, F_{lift}, which must equal $(M)(g)$, is given by

$$F_{lift} = (C_{lift}\rho u^2 A_{wing})/2, \tag{7-31}$$

in which C_{lift} = coefficient of lift and ρ = density of air. This can be solved for velocity:

$$u \propto [(M)(g)/A_{wing}]^{1/2}. \tag{7-32}$$

Thus the characteristic speeds such as stalling speed and maximum speed are proportional to the square root of wing loading. The cost of flight (power required per unit distance) is inversely proportional to the ratio of lift to drag being developed—or, put another way, the range that a bird or plane has per unit of fuel (fat) is proportional to the effective lift/drag ratio. The maximum lift/drag ratio is proportional to the square root of the aspect ratio $(l_{span}^2/A_{wing})^{1/2}$. Thus the morphometric measurements can give us a comparison of power requirements, speeds, and costs of flight in birds of different proportions (Greenewalt, 1975a; Pennycuick, 1975; Rayner, 1979; Vogel, 1981).

Two kinematic characteristics should also be considered, the period (t_{wing}) and the frequency (f_{wing}) of wingbeat. Treated allometrically, these together with the measured and derived characteristics can tell us a good bit about how the radiations of birds have adapted to various niches. The allometric descriptions are drawn together in Table 7-1.

Table 7-1 The allometry of flight morphometrics and kinetics.

Function	Birds	Passeriform	Shorebird	Duck	Hummingbird	Procellari-iformes	Divers[b]	Similarity model exponents[c] G	E
Wingstroke period (sec)	$\propto M^{0.29}$	$0.33\,M^{0.36}$	$0.25\,M^{0.19}$	$0.17\,M^{0.24}$	$0.012\,m^{0.61}$			0.33	0.13
Wingbeat frequency (sec^{-1})	$\propto M^{-0.29}$	$3.03\,M^{-0.36}$	$4.00\,M^{-0.19}$	$5.88\,M^{-0.24}$	$85.3\,m^{-0.61}$			-.33	-.13
Semispan (cm)		$66\,M^{0.42}$	$58\,M^{0.40}$	$47\,M^{0.41}$		$62\,M^{0.38}$		.33	.25
Wing length (cm)	(humerus: $11.8\,M^{0.48}$)	(corvids: $45\,M^{0.48}$)			$2.32\,m^{0.56}$			.33	.25
Wing area (cm^2)		$2{,}200\,M^{0.79}$	$1{,}300\,M^{0.71}$	$800\,M^{0.71}$	$3{,}376\,M^{0.96}$	$1{,}186\,M^{0.59}$	$444\,M^{0.41}$	.67	.75
Wing loading (kg/m^2)		$4.55\,M^{0.21}$	$7.69\,M^{0.29}$	$12.5\,M^{0.29}$	$2.96\,M^{0.04}$	$8.43\,M^{0.41}$	$22.5\,M^{0.59}$	.33	.25
Disc loading (N/m^2)		$7.15\,M^{0.16}$	$9.28\,M^{0.19}$	$14.4\,M^{0.19}$	$3.04\,M^{0.0}$ [d]			.33	.25
Velocity $\propto$ (wing load)$^{1/2}$	$\propto M^{0.07}$	$\propto M^{0.11}$	$\propto M^{0.15}$	$\propto M^{0.15}$	$\propto M^{0.02}$	$\propto M^{0.21}$	$\propto M^{0.30}$	.17	.13
Aspect ratio		$\propto M^{0.05}$	$\propto M^{0.09}$	$\propto M^{0.11}$	$\propto M^{0.16}$	$\propto M^{0.17}$		.00	-.25
Stride	$\propto M^{0.36}$	$\propto M^{0.47}$	$\propto M^{0.34}$	$\propto M^{0.39}$	$\propto M^{0.63}$			.50	.38
Chord		$\propto M^{0.37}$	$\propto M^{0.31}$	$\propto M^{0.30}$	$\propto M^{0.40}$	$\propto M^{0.21}$		.33	.50
$u_{wing\ tip}$		$\propto M^{0.06}$	$\propto M^{0.21}$	$\propto M^{0.17}$	$\propto M^{-0.05}$			.00	.13

Sources of relationships or data: Greenewalt, 1962, 1975; Warham, 1977; Prange et al., 1979; Rayner, 1979; Alexander, 1982; Brown and Bowers, MS.

a. Groups in parentheses are those of Greenewalt, 1975, based on wing loading, not necessarily phylogenic; others are phylogenic or as otherwise noted.

b. $n = 6$ spp.: 4 alcids, 1 diving petrel, 1 diving passerine (*Cinclus*).

c. G = geometric similarity; E = elastic similarity.

d. Regression: $2.8\,m^{0.10\pm0.08}$, exponent not significantly different from 0.00 at 95% confidence level; mean 3.403, $n = 26$ spp., 2.3 to 10 g.

The semispan and wing-length scalings are clearly neither geometric nor elastic similarity scalings (see Prange, 1977; Prange et al., 1979), nor is dynamic similarity maintained in flying animals (Alexander, 1982). Dynamic similarity would require equal Reynolds numbers, equal angles of attack, and equal lift coefficients throughout the size range of flying animals. This is not the case. The period of a pendulum is proportional to the square root of its length, and the scaling of a wingstroke period is approximately proportional to the square root of wingspan or length for the so-called shorebird and duck models (those having heavy and medium wing loading). For the passeriform model and the hummingbird observations, the period is scaled similarly to the scaling of wingspan or length. (Greenewalt, 1975a, found f_{wing} to be proportional to $l_{wing}^{-1.03}$.) The standard errors for the scaling exponent for t_{wing} are large, however, and there is some evidence that t_{wing} (or f_{wing}) is not fixed, but may vary with slow and fast forward flight (Rayner, 1979).

The vertical velocity of the tip of a beating wing should be proportional to the product of l_{wing} (in centimeters) and f_{wing} (in seconds^{-1}); if l_{wing} were increased without reducing f_{wing}, then in the same t_{wing}, the tip of the larger wing would travel farther. (The same proportional shortening, $\Delta l/l$, of the pectoral muscles would produce the same angle or proportional amplitude of the wingbeat.) There is in engineering a dimensionless constant, the Strouhal number, or ratio of wing-tip velocity (u_{wtip}), to $l_{wing}f_{wing}$, for which Günther and Guerra (1957) and Stahl (1962) calculated the scaling in insects and birds (in units of meters and seconds):

$$\frac{u_{wtip}}{l_{wing}f_{wing}} = \frac{5.1\ M^{0.01}}{(0.03\ M^{0.39})(48\ M^{-0.38})} = 3.5\ M^{0.00}. \qquad (7\text{-}33)$$

The scaling for u_{wtip} in the above, $5.1\ M^{0.01}$, can be compared to $M^{0.06}$ in the passeriform model and $M^{-0.05}$ for hummingbirds, exponents which probably are quite close to being size independent. However, the shorebird and duck models would appear to have $M^{0.21}$ and $M^{0.17}$ scaling for u_{wtip}, which would amount to more than a threefold difference in wing-tip velocity in the shorebird model (the exception being the albatross which uses dynamic soaring instead of continuously flapping flight), and a 1.7-fold range in the birds of the duck model (see Table 7-1).

Wing loading increases with body mass (Figure 7-2); in all but the hummingbirds and gliders, wing-surface area does not keep up with body-mass increases. This is the basis for calculating the scaling of flight speeds as indicated above. In other words, with weight gain the lift is amplified in two ways: by (inadequately) increasing wing surface and by increasing velocity

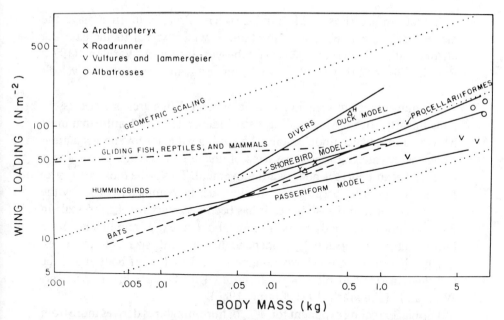

7-2 When the wing loading of flying vertebrates is scaled, bats, the duck model, and the shorebird model (from Greenewalt, 1975a) show approximate geometric similarity. The same could be said for the Procellariiformes (from Warham, 1977) which include the albatrosses (O) that Greenewalt pooled with his shorebird model. The passeriform model of Greenewalt has an $M^{0.21}$ scaling, including most of the small birds that feed on the ground and in close vegetation, where the necessity for short, quick flights predominates during most of the year (except in migration). The vultures (V) are also included in the passerine model. There have been two estimates of wing loading of *Archaeopteryx*, the earliest known bird fossil, Heptonstall's (ΔH; 1970) exceeding considerably the estimate of Yalden (ΔY; 1971), and the latter very similar to the present-day roadrunner (X, *Geococcyx californianus*, this study) which uses its wings primarily for short gliding escapes.

The divers that use their wings for underwater locomotion have wing loading that considerably exceeds any geometric scaling. Wing loading is size independent for both the hummingbirds and the nonflying gliders, both of which must derive lift without active forward flight. The hummingbirds (from Greenewalt, 1975a) must generate lift in direct proportion to body mass during hovering. The gliders (line approximated from fig. 2 of Rayner, 1981) are apparently optimizing the lift in proportion to their weight in order to stretch the glide as much as possible while not engaged in active forward flight; in either case, lift must scale with weight.

over that surface (Eqs. 7-24 and 7-25). The Procellariiformes have the highest scaling exponent for aspect ratio, which serves to increase the lift/drag ratio at high speeds. Based on the wing loading of Warham's (1977) Procellariiformes, they would have the highest scaling exponent for velocities.

The hummingbirds scale in a unique manner. Wing area is scaled essentially linearly, so that wing loading is size independent — a feature that must be related to reliance upon sustained hovering while feeding on nectar and while insect catching among swarms of gnats. In hovering, no airflow is induced because there is no forward movement. This is the most costly way to remain airborne, because the weight must be completely offset by the vertical, downward thrust; hence it has been necessary for A_{wing} to scale in direct proportion to body mass. The metabolic rate per gram of a hovering hummingbird is essentially independent of bird size, 41 to 52 ml $O_2/(g \cdot hr)$ for hummingbirds ranging from 3 to 10 g of body mass; the extremes in size had rates of 42.4 and 43.4 ml $O_2/(g \cdot hr)$, respectively (Wolf and Hainsworth, 1971; Epting, 1980).

The bigger scaling exponent for A_{wing} of hummingbirds derives more from increased length scaling than from mean chord scaling. The greater scaling of the aspect ratio confers additional benefit for economy in forward flight in progressively larger species, more than occurs in all models or groups except the Procellariiformes. When hummingbirds move from flower patch to flower patch or migrate, the size-independent speed prediction means that there is no enhancement of lift by higher speed for heavier hummingbirds.

The density of water is one thousand times that of air, so a pair of wings flapped underwater for locomotion in diving must work against much more inertia than in air. The viscosity is also increased about a hundredfold. Wing area is sacrificed. In a "vertical allometry" (reflecting adaptive radiation, more of which is discussed in Chapter 12) for a series of 500-g birds, there is a progressive increase in wing loading from lightly built, usually airborne, harriers to herons to crows to shorebirds to puddle ducks to diving ducks to alcids; the alcids have over seven times the wing loading of the harriers (Greenewalt, 1975a). There is also convergent evolution of heavy wing loading in the diving petrel *Pelecanoides urinatrix* (Warham, 1977). When the diving passerine *Cinclus mexicanus* is added to the regression for wing area, the change in scaling exponent is not significant, nor is the change in coefficient. *Cinclus* (water ouzel or dipper) has only 51% of the surface predicted for a 49.5-g passerine; the diving petrel has 51% of the surface area predicted for a 119-g procellariiform. Neither has gone as far as the alcids in reducing wing area. Alcid wing areas are about one-third the areas predicted from the shorebird model.

The hypergeometric increase of avian wing lengths not only helps to increase wing area but reduces power requirements for maintaining the greater flight speeds required to keep a larger body mass airborne. This is seen in the approximations of power input requirements for several wing lengths (Table 7-2). As wing length increases relative to the typical or body-mass–predicted wing length, the power requirement decreases, whether the comparison is made at sea level or in the thinner air at 6,000 m altitude. The velocity at minimum cost of transport is also inversely related to relative wing length; while the power requirement is less, the trip takes longer. The velocity decrease is not as great as the decrease in power required, so that the minimum cost of travel drops as wing span increases.

It is easy to appreciate why there would be a selective advantage for greater wing lengths if limitations on elastic or cantilever strength could be out-maneuvered. Still, there must be trade-offs with other considerations that counterbalance the selection for greater wing length, such as the limits on the cross-sectional area of muscle required to generate sufficient force to flap longer wings with greater moment arms, and the need for maneuverability in close vegetation or when facing risks of predation.

Many factors have had to be weighed in the evolution of avian flight morphometrics and kinetics. Neither geometric nor elastic similarity offer satisfactory descriptions of what is observed, as can be seen by comparison of these exponents with the model exponents at the far right in Table 7-1. Other possibilities have been reviewed by Alexander (1982). Perhaps the adaptive significances of intraspecific trends in wing length are due for reexamination, now that the allometry of locomotion has been studied intensively for more than two decades after the original approaches to the topic (Snow, 1958; Hamilton, 1961; James, 1970).

There is unquestioned need for more careful measurements of flight speed (with data on wind speed and direction), of wing dimensions, and of wingbeat frequency in various modes of flight — the last is especially desirable in conjunction with measurements of flight speed. There is already good reason for the use of allometry to give the "adaptive radiation of birds" a quantitative description (Storer, 1971). The case for pursuing this further can be seen in the lift equation, (7-31). The product of $u^2 A_{wing}$ should scale linearly, providing lift that scales as $g \cdot M^{1.0}$. That is the way birds appear to be designed, using expressions for u and A_{wing} from Table 7-1. However, my reasoning here is circuitous, because the velocity scalings were calculated from Eq. (7-32), which was derived from (7-31). The only escape is to obtain enough real measurements of velocity over a significant size range, and with the sophistication used by Tucker and Schmidt-Koenig (1971) and by Schnell and Hellack (1979).

Energy Costs of Locomotion

I have already stressed the importance of appropriate dimensions and units as a way of reducing confusion and promoting communication. There are more rewards than these, however, for the yield of insight may be enhanced whenever data can be recombined to calculate properties beyond the initial tabulation of numbers. The slope of a graph correlating the data for a dependent variable with the independent or experimentally manipulated variable should have biological meaning. For example, the slope of a regression of metabolic rate as a function of environmental temperature has dimensions of energy requirement rate per degree of temperature difference and thus tells how much the metabolism must be increased to maintain body temperature as the environment chills. This in turn must be inversely proportional to the amount and quality of insulation and vasomotor control, as will be discussed in Chapter 8.

The data on metabolic power input have been similarly exploited to determine the minimum cost of transport ($) from the metabolic rate and speed or velocity:

$$\$ = \frac{energy}{time} \Big/ \frac{distance}{time} = energy/distance, \tag{7-34}$$

as has been done in the last column of Table 7-2. In fact, it is this manipulation by Tucker (1970) and by Taylor and associates (1970) which has put us on the track, conceptually, to an understanding of locomotory energetics. This will prove to be a major breakthrough for ecology and evolutionary biology as well, for as long as reproduction can exceed the energy supply, minimization of the energy costs of moving about will be a consideration in animal design and animal behavior. Without a quantitative basis for expressing this cost, any discussion of strategies in those fields would amount to little more than medieval theological arguments.

TERRESTRIAL LOCOMOTION

Let us start with the energetics of terrestrial locomotion, then make comparisons with the energetics of flight. The terrestrial vertebrates include both bipedal and quadrupedal runners. Earlier, bipedal locomotion was thought to be more costly than quadrupedal locomotion, but the considerable progress in measurement of the locomotion metabolism of a diversity of animals has led to a reexamination of the costs. The conclusion is that bipedal is not significantly more expensive than quadrupedal running (Paladino and King, 1979 — $n = 69$; Fedak and Seehermann, 1979). In their

Table 7-2 Calculated flight velocity and requirements for power input at minimum cost of transport for various wingspan proportions, assuming no winds and using equations from Tucker (1974).

Relative wing length (typical = 1.0)	Velocity (m/s)	Power input (watts)	Minimum cost (J/m)
0.8	$(954.7 \times 10^{-6}\,h + 17.66)\,m^{0.169}$	$(8.60 \times 10^{-3}\,h + 122.9\ )\,m^{0.974}$	
1.0	$(853.3 \times 10^{-6}\,h + 15.73)\,m^{0.169}$	$(6.43 \times 10^{-3}\,h + 94.15)\,m^{0.974}$	
1.2	$(760.7 \times 10^{-6}\,h + 14.28)\,m^{0.169}$	$(5.09 \times 10^{-3}\,h + 76.21)\,m^{0.974}$	
At sea level:			
0.8	$17.66\,m^{0.169}$	$122.9\,m^{0.974}$	$6.96\,m^{0.805}$
1.0	$15.73\,m^{0.169}$	$94.15\,m^{0.974}$	$5.99\,m^{0.805}$
1.2	$14.28\,m^{0.169}$	$76.21\,m^{0.974}$	$5.34\,m^{0.805}$
At 6,000 m:			
0.8	$23.39\,m^{0.169}$	$174.5\,m^{0.974}$	$7.46\,m^{0.805}$
1.0	$20.85\,m^{0.169}$	$132.7\,m^{0.974}$	$6.37\,m^{0.805}$
1.2	$18.84\,m^{0.169}$	$106.8\,m^{0.974}$	$5.67\,m^{0.805}$

studies these authors included eutherian, marsupial, and monotreme mammals, ratite and carinate birds, and lizards, there being no significant phylogenetic differences, although Paladino and King excluded data from "hopping and waddling animals."

Studies by C. R. Taylor and his colleagues have refined and extended our understanding of the energetics and mechanics of terrestrial locomotion. Four recent papers parallel the brilliant series on respiratory design and the support of the oxygen demands of the locomotory efforts (see Chapter 5). Because many of the same animals and techniques were used, the two series and the data and relationships derived therein are uniquely compatible for further analysis.

When the oxygen consumption or equivalent energy requirements are plotted as a function of speed, an essentially straight line is obtained from an intersect with the vertical axis at zero velocity (Figure 7-3). The intersect, an extrapolation from the running metabolism and considerably greater than zero, represents the resting metabolism and cost of postural maintenance. From that zero-speed point, the metabolic rate increases in linear proportion to speed. Combining the data from 62 species of mammals and birds, the metabolic rate while running on the ground is summarized allometrically:

$$\dot{V}_{O_2}/M = 0.300 \, M^{-0.303} + 0.533 \, M^{-0.316} \, (u_g), \qquad (7\text{-}35)$$

where $\dot{V}_{O_2}/M$ is in milliliters of oxygen per second per kilogram, M is body mass in kilograms, and u_g is speed in meters per second. This is converted into units of metabolic power, $\dot{E}_{met}$ (in watts per kilogram, at 20.1 joules per milliliter of oxygen) as

$$\dot{E}_{met}/M = 6.03 \, M^{-0.303} + 10.7 \, M^{-0.316} \, (u_g). \qquad (7\text{-}35a)$$

At the midspeed for each animal 90% of the values calculated from these equations came within $\pm 25\%$ of the observed values. When the data were analyzed separately for all mammals, all birds, all wild animals, all domesticated animals, Marsupialia, Insectivora, Artiodactyla, Carnivora, Rodentia, and Primata, none of the resulting allometric equations differed from the pooled one at the 95% level of confidence. Thus we have a reliable estimating equation for the homeothermic vertebrate runners. From Eq. (7-34) the net cost for transporting 1 kg of body mass ($\$/M$, kJ m^{-1}kg^{-1}) is then

$$\$/M = 10.7 \, M^{-0.316}. \qquad (7\text{-}36)$$

This is similar to the net cost of locomotion in reptiles (Bennett, 1982):

$$\$/M = 13.5 \, M^{-0.25}. \qquad (7\text{-}37)$$

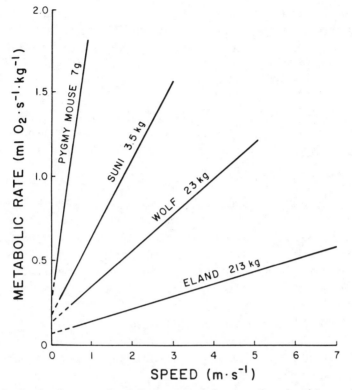

7-3 The metabolic rates of running mammals increase linearly with speed. The increase per kilogram of body mass is inversely related to size, as seen in this series of wild mammals from the data of Taylor and colleagues (1982). The slope of the line is the mass-specific cost of transport.

The mammal and reptile curves cross at a body mass of 31 g, below which the mammalian net cost is predicted to be greater; at larger sizes, the predicted reptilian net cost would increase above the mammalian costs. The scaling factor of the reptilian $/M falls outside the 95% confidence interval for the mammalian scaling factor, but the statistical and theoretical aspects of the difference need further study.

It has been traditional in comparative physiology, and useful in the analysis of the energetics and mechanics of locomotion, to express metabolic rates per gram or kilogram of body mass, and to express cost of transport as energy/(kilogram × kilometer). These mass-specific expressions are inversely related to body size; thus the transport cost *per kilogram* becomes less expensive as body size increases. For beasts of burden and cargo vessels the cost per (kilogram × kilometer) is meaningful, but in the natural state the biological unit is the whole animal (McNab, 1971). The biologically relevant expression thus is the travel cost per animal — obtained quite simply, by multiplying the traditional "per kilogram" by $M^{1.0}$.

The total metabolic rates for terrestrial locomotion of entire animals are therefore predicted by

$$\dot{E}_{met} = 6.03 \ M^{0.697} + 10.7 \ M^{0.684} \ u, \qquad (7\text{-}38)$$

and the incremental cost of locomotion in kilojoules per kilometer (or joules per meter) is

$$\$ = 10.7 \ M^{0.68}. \qquad (7\text{-}39)$$

The exponents in Eqs. (7-38) and (7-39) are close enough to the scaling for resting metabolic rates that the ratio of running to resting metabolism appears to be a size-independent constant. The larger the animal, the greater the cost of locomotion, although it is cheaper per kilogram, and this would apply in terms of transporting energy (fat) or water reserves. The larger animal can, and apparently does (Figure 2-1), carry proportionately more fat.

Heglund and colleagues (1974) and Taylor and associates (1982) have used the speed at which the gait is in transition from a trot to a gallop as a "physiologically equivalent speed," given in Eq. (7-1), which they have substituted into Eq. (7-38):

$$\dot{E}_{met,t\text{-}g} = 6.03 \ M^{0.70} + (10.7 \ M^{0.68})(1.53 \ M^{0.24})$$
$$= 6.03 \ M^{0.70} + 16.4 \ M^{0.92}. \qquad (7\text{-}40)$$

Zero-speed (extrapolated intercept) and gallop-speed metabolic rates are plotted in Figure 7-4. For the 10-g mammal these two costs, which are additive, are equal. This means that the difference between standing still and going places at minimum gallop speed would only be a doubling of the metabolic rate, exclusive of the acceleration. In contrast, a 100-kg mammal's metabolic rate while galloping is 8.5 times the zero-speed rate, and a 450-kg (horse-sized) mammal would be metabolizing at 11.8 times the zero-speed rate while galloping. Therefore it is relatively simpler for a mouse to gallop than it is for a horse. This may explain why, in a cabin at night, we often see deer mice running around, never just walking, even though they seem to be oblivious to our presence. In fact, Heglund and coworkers (1982a) observed:"It was extremely difficult to obtain good walking records from small mammals, and we were never able to obtain them from the chipmunks and ground squirrels. At slow speeds, small animals normally moved in a series of short bursts, alternating with stops, rather than at a

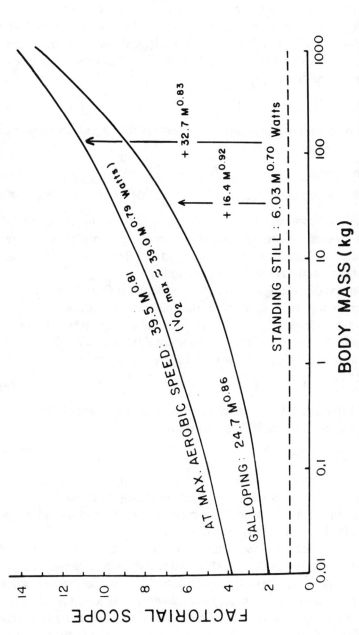

7-4 Differences in the scaling for the zero-speed (standing-still) component and the incremental costs of galloping at transition or maximum aerobic speeds mean that the larger the mammal, the greater the factorial increase from standing still to running. See text for sources of equations.

constant speed." This greater multiple of the metabolic increment for galloping relative to the resting rate may be pertinent to Garland's observation that u_{max} decreases with body sizes above 119 kg (see Eqs. 7-12 and 7-13). Garland (1983a) has calculated the ecological cost of transport over observed daily movement distances, and expressed this as a percentage of daily energy expenditures:

$$\dot{E}_{transport}/\dot{E}_{total} = 1.39 \; M^{0.24}. \tag{7-41}$$

Separating his data into two groups to account for the larger home ranges of predators, he found:

carnivores: $\qquad \dot{E}_{transport}/\dot{E}_{total} = 5.17 \; M^{0.21}$; $\qquad\qquad$ (7-42)

other mammals: $\quad \dot{E}_{transport}/\dot{E}_{total} = 1.17 \; M^{0.21}$. $\qquad\qquad$ (7-43)

These equations confirm our conclusion that the cost of transportation becomes a progressively more significant part of an animal's daily energy budget as size increases, thereby explaining any inhibitions we humans may have about continuously scampering around.

Note also that the scaling exponent for the increase in metabolic rate from resting to galloping, $M^{0.92}$, approaches the "paradoxical" scaling for oxygen diffusing capacity (Eqs. 5-57 and 5-57a). The total $\dot{E}_{met,t-g}$ can be obtained by solving Eq. (7-40) for hypothetical animals of 0.001 kg to 1,000 kg and is represented by the "galloping" curve in Figure 7-4, which is

$$\dot{E}_{met,t-g} = 24.7 \; M^{0.86}. \tag{7-44}$$

For the metabolic rate while traveling at the maximum aerobic speed, Eq. (5-58a) is converted from ml O_2 s^{-1} to watts:

$$\dot{E}_{aero\;max} = 39.0 \; M^{0.79}. \tag{7-45}$$

Because of rounding and derivation from a slightly different version of (7-38), Eq. (7-11) comes back to an insignificantly different $39.5 \; M^{0.81}$, while inclusion of 7 species of domestic mammals with the 14 species of wild mammals (Taylor et al., 1981) yields $38.5 \; M^{0.81}$.

Now compare the t-g and "aero max" metabolic rates with the zero-speed rate that is the first term in Eq. (7-38). This is shown as the factorial scope, or ratio of activity to zero-speed rates (Figure 7-4). While the trot-gallop transition represents a mechanically equivalent speed for comparison, it is certainly not a physiologically equivalent demand for mammals of different sizes; for a 1-kg mammal's metabolic rate is increased by a factor of four by going from a standstill to the t-g transition, while a 450-kg mammal would increase by a factor of almost eleven to attain this mechanical equivalent.

The increase from zero speed to aero max is also much more dramatic for the large mammals. Per gram of tissue, a 10-g mammal standing still would be burning oxygen at about twice the rate of one gram of a 450-kg mammal running at its aerobic maximum. Even limited to its factorial scope of four, the 10-g mammal's gram can metabolize aerobically at nine times the rate of the horse-sized mammal with a factorial scope of eleven, when both are running.

Finally, these equations show the limitations imposed on running by extremely large size. For an elephant-sized (4,500-kg) mammal the predicted metabolic rate at the transition to a gallop exceeds $E_{aero\,max}$ by 14%; thus a mammal of this size could not aerobically sustain a gallop. Furthermore, if maximum-speed scaling (such as Eq. 7-11) could be maintained in sizes up to 4,500 kg, the net metabolic rate would be 64 times the zero-speed metabolism, vastly exceeding the known metabolic scope for any mammal. Hence, for metabolic as well as for mechanical reasons, speed was sacrificed in the evolution of large body size.

Suppose that muscle efficiency works out to be the same, with or without incurring an oxygen debt, once the debt has been repaid. The scaling of energy expenditure at u_{max} could then, from Eq. (7-12) converted to meters per second and Eq. (7-38) for mammals under 300-kg body mass, be predicted as:

$$\dot{E}_{u_{max}} = 6.03\,M^{0.70} + (10.7\,M^{0.68})(6.47\,M^{0.23}). \tag{7-46}$$

Simplified by solutions for hypothetical sizes and regression of totals, this becomes:

$$\dot{E}_{u_{max}} = 78.3\,M^{0.89}. \tag{7-46a}$$

This scaling does not exceed the $M^{0.95\ to\ 0.99}$ for pulmonary diffusing capacity (Eqs. 5-57 and 5-57a) as determined by Weibel and colleagues (1981). Thus the $D_{L_{O_2}}$ appears to be scaled adequately, yet reliance upon an oxygen debt increases progressively with size when mammals run at their mechanical maximum speed. One possibility is that the limits are reached first in the cardiovascular stage of oxygen transport. Yet this would still not explain the paradox of a greater scaling for pulmonary diffusion in the larger mammals that could deliver oxygen faster than the circulation could subsequently handle.

From Eqs. (7-1), (7-3), and (7-35a) Taylor and associates (1982) calculated that the metabolic energy expended per stride and unit of mass was size independent, a relatively constant 5 J/(stride × kg). This is further argument for the existence of a mechanically equivalent basis for adequate design of the oxygen supply. The fact that it takes less energy to move one unit of mass over one unit of distance for the larger animal (see Eq. 7-49) suggests that although there is something less efficient in the oxygen delivery, there is something more efficient in the locomotion of the large beast. Efficiency, as it applies here, is the ratio of work performance rate to energy consumption rate. The mechanical work required to make the animal go is the sum of the kinetic energy ($\dot{E}_{tot}/M$) required to maintain the oscillations of limb and body movement and the potential energy of the animal's body lifted in the aerial phase against gravity.

The rates at which running animals perform mechanical work are size independent and do not explain the linear increase in metabolic rates with velocity and the size dependence of locomotory costs. Instead of scaling allometrically, they are functions of velocity:

$$\dot{E}_{tot}/M = \dot{E}_{ke} + \dot{E}_{cm} = 0.478\,u_{t\text{-}g} + 0.685\,u_{t\text{-}g} + 0.072, \qquad (7\text{-}47)$$

where $\dot{E}_{ke}$ is the total kinetic power requirement for movement of the limbs and $\dot{E}_{cm}$ is the total power requirement for lifting and reacceleration of the center of body mass, both in watts per kilogram, and velocity u is in meters per second.

The metabolic energy consumption is represented by the approximately 5 J/(stride × kg) mentioned above. At the trot-gallop transition, $\dot{E}_{tot}/M$ is 0.07 J/(stride × kg) for a 10-g animal, 0.46 for a 1-kg animal, and 0.35 for a 100-kg animal, representing efficiencies of less than 2%, 9%, and 63% in this size progression. There is a loss of about two-thirds of the energy in metabolic fuels such as carbohydrates and fats, leaving much less than the 63% to be accounted for in the mechanical performance, unless some of the energy is being recovered in elastic recoil after storage during deceleration of moving limbs or gravitational fall of the body mass. The combination of studies of both the energetics and the mechanics of locomotion thus have shown the importance of elastic storage and recovery of energy in galloping mammals, but as yet have neither confirmed nor ruled out the existence of elastic storage for small runners (Fedak et al., 1982; Heglund et al., 1982a,b).

The apparent size dependency of efficiency may result from a number of factors, one of which could be the high rates of contraction required for the high stride frequencies of small mammals. There are qualitative differences in the strides of small animals: they decelerate with the forelimbs and reaccelerate with the hind legs, and do not appear to transfer between gravitational potential and kinetic energies. Kangaroo rats' tendons are relatively thick, and perhaps too stiff to store much energy (Fedak et al., 1982), but that only raises the question of why kangaroo rats have thick tendons.

FLIGHT

The cost of transportation in flight was related to body size by Tucker (1970) as follows, in kilocalories per kilogram per kilometer:

$$\$ = 1.25 \, M^{-0.227}, \tag{7-48}$$

which we convert for comparison to an expression for whole animals, in S.I. units of kilojoules per kilometer, to:

$$\$ = 5.23 \, M^{0.77}. \tag{7-49}$$

Equations (7-39) and (7-49) may now be used to compare the travel costs of running and flying:

$$\frac{\text{trot-gallop running}}{\text{flying}} = \frac{10.7 \, M^{0.68}}{5.23 \, M^{0.77}} = 2.05 \, M^{-0.09}. \tag{7-50}$$

While the exponents differ somewhat, perhaps not significantly, at the 1-kg size, birds can travel for half fare, even without the tail winds that they seem to await before migrating. Flying also makes it possible to travel up to ten times as fast as by running (Eq. 7-28) over geographic barriers, and to travel at higher altitudes where the weather patterns provide those tail winds!

Several allometric relationships for the terms of Eq. (7-34) are brought together in Table 7-3. On the basis of their dimensions, the products of speed and cost should show the same mass exponents as the metabolic rates, but they appear to be different in both flying and running at the trot-gallop transition. If these exponential differences are statistically significant and obtained by valid combination, several possibilities suggest themselves:

(1) The expansion of metabolic rate and oxygen consumption in loco-motion is not size independent, and the structural design in alveolar surface area that is adequate for peak demands should have a mass exponent greater than ¾ (see Table 7-3). In both flying and running locomotion, the body mass is lifted against gravity; ignoring aerody-namic lift, the work done by raising the body is proportional to $M^{1.0}$. Since locomotion metabolism is considerably greater than standard metabolism, the $M^{3/4}$ scaling for the latter could be easily masked.

(2) The flight metabolism equation (Table 7-3) comes from a wide variety of methods of determination including body mass and fat loss, hovering instead of forward flight, carbon dioxide production, oxygen consumption, sometimes for very short periods. With addi-tional data obtained under standardized procedures, this relation-ship should be verified.

Table 7-3 The allometry of power input, speed, and travel costs for flying and running.

Group and function	Speed, m/s (with reference)	×	Cost, J/m (with reference)	=	Metabolic rate increase, W (with reference)
Birds					
Flight	$13.0\ M^{0.17}$ (estimated from Pennycuick, 1969)	=	$5.23\ M^{0.77}$ (Tucker, 1970)	=	$68\ M^{0.94}$
	$15.7\ M^{0.17}$ (Tucker, 1974)		$5.99\ M^{0.81}$ (Tucker, 1974)	=	$94\ M^{0.98}$
Flight metabolism					$53\ M^{0.73}$ (Hart and Berger, 1972)
Mammals					
Running at trot-gallop transition	$1.53\ M^{0.24}$ (Heglund et al., 1974)	=	$10.71\ M^{0.68}$ (Taylor et al., 1982)	=	$17\ M^{0.92}$
					$23\ M^{0.82}$ (calculated from table 1, Taylor, 1977)
At max $\dot{V}_{O_2}$					$37\ M^{0.77}$ (Taylor et al., 1981)

(3) By whatever determination, the metabolic demands of flight loco-motion in birds exceed markedly the demands of running locomo-tion. It may be faster and cheaper to fly, but the rate of oxygen supply needed is dramatic.

(4) It may be that the body-mass exponents for travel cost (third column of Table 7-3) have been exaggerated by pooling of data from different types or morphologies of fliers or runners, and that future studies will refine these.

(5) The trot-gallop transition offers a definable standard for compari-son. However, $\dot{V}_{O_2}$ is higher for horses at the trot-gallop transition than it is when they are either trotting at a lower speed or galloping at a higher speed (Hoyt and Taylor, 1981).

Clearly, it must be acknowledged that at present there is not perfect agree-ment in mass exponents.

Migration

As indicated earlier in this chapter, every aspect of locomotion is size dependent: speed, endurance, range, stride length, stride or wingbeat fre-quency, and energy costs. It would seem impossible to analyze migration without giving considerable attention to size. However, a recent treatise on animal migration, including chapters on migration strategies and bioener-getics, makes only incidental and qualitative mention of this all-important factor. Greater body size gives migratory fish the benefit of greater fecundity (assuming that maturation times are similar, although actually they vary). As Gauthreaux (1980) says, "Large ungulates may migrate long distances." How far southward migrant whales will penetrate into colder waters "is correlated with body size." For migrant birds flight speed increases while the ratio of power input to weight decreases with size increase, but larger birds deposit proportionately less lipid reserves. The net result is that "flight duration and range of flight increase with body weight and percentage lipid composition." Perhaps allometry can be a little more specific!

The speeds of birds and mammals were compared in Eqs. (7-28) and (7-30), but the difference may have been exaggerated by using $u_{t\text{-}g}$ for mammals. What if we compare $u_{aero\,max}$ of mammals with $u_{max\,r}$ of birds? Converting Eq. (7-16) to units of kilometers per hour, we have:

$$\text{birds/mammals:} \quad u_{max\,r}/u_{aero\,max} = 56.52\,M^{0.17}/11\,M^{0.15}$$
$$= 5.14\,M^{0.02}. \qquad (7\text{-}51)$$

Not only do birds have a clear advantage in speed when compared on an equal-size basis, but as seen in Eq. (7-50), flying is half as costly (in kilojoules per kilometer) for a 1-kg bird as for a 1-kg runner, but if the small negative r.m.e. is significant, running is only 63% more costly for a 12.5-kg runner than for the largest flier. Obviously, if the animal can fly it is a better candidate for migration than if it cannot (see Table 7-4).

How long can a bird remain airborne and how far can it travel? An equation for approximate potential flight duration (t, in hours) was given by Berger and Hart (1974):

$$t = 36 \ F \ m^{0.28} = 249 \ F \ M^{0.28}, \tag{7-52}$$

where $F =$ initial fractional fat content.

Equations (7-16) and (7-52) can be combined to give the maximum range, with no wind:

$$\text{km} = u_{max \, r} \, t = (56.52 \ M^{0.17})(249 \ F \ M^{0.28}) = 14,073 \ F \ M^{0.45}. \tag{7-53}$$

The rufous hummingbird *(Selasphorus rufus)* winters in Mexico, and some populations breed as far north as Cordova, Alaska. The return flight to Mexico is thought to be along the crest of the Rocky Mountains. The body mass of males captured in southwestern Colorado ranges from 3.03 g to 4.60 g. Assuming that the minimum body mass is that of an exhausted arrival, and that the maximum is a bird with fat reserves fully restored, we have an initial fractional fat content of 34%. Equation (7-53) predicts a range of 426 km. This would be one of perhaps 11 flights necessary to go from coastal Alaska to Mexico, a total of more than 5,000 km. The Pacific golden plover *(Pluvialis dominica fulva)* migrates 3,900 km from the Aleutian Islands to Wake Island on 20% to 25% lipid deposited in its 130-g body (Johnston and McFarlane, 1967), whereas Eq. (7-53) predicts a range of only 1,405 km. Either the equation underestimates the bird's abilities, or the golden plover is very careful to take advantage of strong tail winds.

The migrations of land mammals are rather short by comparison. The longest is thought to be about 600 km for the North American caribou *(Rangifer tarandus),* body mass about 148 kg. Migrating ungulates can feed as they go, so endurance may not be crucial. At the daily movement speed predicted from Eq. (7-20), they would travel only 3.62 km/day, taking 165 days for the complete trip. If they could maintain trot-gallop transition all day (Eq. 7-3), they could cover 219 km and the trip would take less than 3 days. The truth must lie somewhere in between.

The point of a regular annual migration is to take advantage of better

Table 7-4 Size and effect on potential for migration and dispersal.

Variable	Birds				Mammals				
	10 g	100 g	1 kg, eq.	References	10 g	100 g	1 kg, eq.	100 kg	References
Speed of travel km/hr	21.7	32.3	$47.4\,M^{0.17}$	Calculated from Pennycuick, 1969	1.8	3.2	$5.5\,M^{0.24}$	16.6	Heglund et al., 1974
Travel time hr/1,000 km	46	31	21		556	316	182	60	
da/1,000 km (@ 12 hr/day)	3.8	2.6	1.8		46	26	15	4	
Energy cost kJ/km	0.17	1.0	$5.23\,M^{0.77}$	Converted from Tucker, 1970	0.68	2.7	$10.7\,M^{0.60}$	170	Converted from Taylor, 1977
Distance on fat km(@0.2 M_b)	463	788	1,337		118	295	741	4,685	
da/1,000 km (@0.2 M_b/day)	2.15	1.27	0.75		8.50	3.33	1.35	0.22	

feeding conditions in a seasonally changing environment. This involves going from one situation that is favorable in certain months only, to a wintering ground capable of sustaining the animals until food and space are once again available on the breeding grounds. If the travel is overland, a considerable distance might have to be covered before a pronounced seasonal improvement would be realized. Very small mammals, therefore, would be unable to better their situations because of their short travel distances. Travel speeds, and the potential distances covered per day, scale as $M^{-1/4}$ (Eqs. 7-3, 7-16, and 7-20). We could extend a relationship of $M^{1/4}$ scaling from the 148-kg caribou that can migrate 600 km:

$$\text{practical migratory distance} = 172\ M^{1/4}. \qquad (7\text{-}54)$$

On this basis, a 50-g vole could migrate 81 km and a 1.5-kg hare 190 km; there must be other considerations, for to my knowledge such trips are not attempted, perhaps because they would take too long or because snow cover provides enough protection to make them unnecessary.

The same advantages of being able to fly or, barring that, of being a large terrestrial animal, that apply to seasonal and repetitive migrations apply also to emigration via temporary openings or bridgings of biogeographic barriers. It would seem then that species of larger size, such as *Rangifer tarandus* and *Cervis elaphus,* should tend to have wider, even circumpolar, distributions. However, several genera of microtine rodents have not been inhibited by allometric doubt, being as widely dispersed as those genera of large mammals.

Uphill and Downhill

Whereas birds can fly level once they have gained sufficient altitude to clear the topography, the running locomotion of mammals is not all level. Small mammals can travel just about as fast uphill as on the level, but this ability is gradually lost with increase in body size.

The mechanical work of lifting, whether for mouse or horse, is the product of mass times gravitational acceleration of free-fall times increase in height, $(M)(g)(\Delta h)$. If the efficiency is the same, the energy expended in raising 1 kg up 1 m should be the same, 9.81 J. However, the specific cost of level transport, in joules/(kilogram × kilometer), is greater for the smaller animal, as is the mass-specific metabolic rate at rest. Thus the proportional increase above level running is a lower demand for energy and oxygen on the part of the mouse.

The costs of transport per kilogram meter measured by Taylor and colleagues (1972) in a comparison of 30-g mice and 17.5-kg chimpanzees are set forth in Table 7-5. The increment of energy expenditure for the vertical component of uphill travel is similar for the two sizes, 16.3 and 14.8 J/(kg · m). Thus the mechanical efficiencies (work done/energy expended) are similar. Nevertheless, while the vertical cost for the mice is only 36% of the cost of running along a treadmill inclined 15 degrees, the cost per rise for the chimpanzee is 168% of the travel cost along the treadmill plane.

Rates of oxygen consumption of both resting and running animals are equivalent to rates of energy expenditure when multiplied by the factor of 20.1 J per ml O_2 (assuming a respiratory quotient of 0.8; see Chapter 3). The cost of running, in J/(kg · m), has been expressed as the ratio of the rate of energy expenditure per kilogram of mass [J/(kg · hr)] to running speed (m/hr), in which time cancels out. The net energy cost of the vertical component is then

$$\text{cost} = \frac{\text{climbing} - \text{level}}{\sin \theta} \tag{7-55}$$

in which sin θ is the fraction of a meter climbed vertically per meter of running when the treadmill is inclined at the angle θ.

Table 7-5 summarizes the calculations from treadmill "climbing" measurements. Note that the expenditure and efficiency values are not totals but partial efficiencies for the vertical component only. The expenditure per kilogram of mass appears to increase with body size (20.0 $M^{0.04}$), but this trend is not significant ($r^2 = 0.19$; $p > 0.1$). Since the efficiency values are 9.8 J/(kg · m) divided by these expenditure values, the trend in efficiency likewise is not significant. Brockway and Gessaman (1977) suggested that the 60% and 66% efficiency values for the mouse and chimpanzee may be impossibly high and artifacts of the calculations. For the present, then, we must stay with the mean partial efficiency value of 4% and assume that this is size independent.

While the partial expenditure for the vertical component of the running cannot be considered to be size dependent, its significance to the animal is; for in level running, the mouse uses over eight times as much energy per unit of body mass as does the much heavier chimpanzee. Thus the incremental increase for climbing is proportionately much greater for the chimp. If both were running at a speed of 2 km/hr, going from a level run to a 15-degree incline would increase the mouse's oxygen consumption and energy requirements 23.5% and the chimp's requirements 189%. Extrapolated to a 1,000-kg mammal, the increase would be 630% (Taylor et al., 1972).

Table 7-5. Energy requirements and mechanical efficiency of lifting body mass (vertical component) of mammals running on a treadmill inclined 14 to 18 degrees.

Species	Body mass (kg)	Vectoral energy expense [J/(kg · m)]		Efficiency[b] (%)	Energy recovery [J/(kg · m)]	Efficiency (%)
		Treadmill plane	Vertical[a]			
Mouse	0.030	45.8	16.3	60	8.53[c]	87
Red squirrel	0.252	—	22.9	41	—	—
Chimpanzee	17.5	17.5	14.8	66	9.32[c]	95
Sheep	29.0	—	26.9	36	—	—
Red deer	68.3	—	21.5	45	—	—
Man	70.0	—	28.7	34	3.35[c]	34
Mean ± 1 s.d.			21.9 ± 5.6	47.0 ± 13.1	7.07 ± 3.24	72 ± 33.2

Sources: From Taylor et al., 1972; Wunder and Morrison, 1974 (red squirrel, Tamiasciurus hudsonicus); and Brockway and Gessaman, 1977 (red deer, Cervus elaphus).

a. $\frac{(\$_{up} - \$_0)/(kg)}{\sin \theta}$, where $\$_{up} - \$_0$ = difference in cost (metabolic rate ÷ speed) for uphill and level travel, and sin θ = fraction of meter climbed per meter run; a negative expenditure signifies energy recovery (use of gravity to accelerate legs while running downhill).

b. Mechanical efficiency calculated as mechanical work ÷ energy expended, $M \cdot g \cdot \Delta h$.

c. Approximations only, based on indistinguishable differences among costs for uphill, downhill, and level running.

The extent to which the relative energetic advantage of smaller size in hill climbing is modified by the recovery in returning downhill is apparently equivocal, from the three efficiency values shown in Table 7-5. The energetics of travel on natural slopes is the first of a series of environmental couplings that we shall consider.

Traveling the Terrain

A Colorado lawyer once told me that if his state were totally flattened, it would be larger than Texas. I do not know how that was quantified, but ten summers of biological research in the West Elk Mountains, a part of the Colorado Rockies, have not caused me to doubt it. My feelings have been a mixture of awe and envy whenever a deer, wapiti, bear, snowshoe hare, or mouse has run uphill, startled by my slow, hyperventilated approach; they all seem to travel so easily. Without measurements, it appears easier for a large deer than for a mouse, probably because of the size-dependent speeds involved.

However, natural history has progressed beyond the days when "it seems to me" was adequate (Aristotle's conclusion, for instance, that the migratory disappearance of swallows was instead a metamorphosis into frogs, which then hibernated in the mud). Laboratory studies of animals running on inclined treadmills have yielded real insight, which has in turn been confirmed by field observations.

If uphill travel is relatively more demanding for larger animals, one might expect that when going up a mountain, the trail of a larger mammal would be less steep; a mouse might go straight up, while a wapiti might make more switchbacks to reach a point directly upslope. To observe whether or not this is true would require years of direct observation. A more ingenious approach was devised by Reichman and Aitchison (1981), who identified the tracks of mammals and measured the angles of the trails and the slopes, selecting only fresh trails in which the purpose appeared solely to go somewhere without foraging or breeding distractions en route. The results confirmed the expectation that smaller animals would pick steeper routes. From 130 trails these authors found a highly significant inverse relationship between trail angle ($\angle_{tr}$) and body mass. However, the trail angle also varied with the slope angle ($\angle_{sl}$); the steeper the slope, the steeper the angle taken by the animal:

$$\angle_{tr} = 7.21 + (-1.37) \log M + 0.53 \angle_{sl}, \tag{7-56}$$

where angles are in degrees and mass is in kilograms; $r = 0.57$; $p < 0.001$. Expecting that there might be a practical upper limit to the steepness of the climb, Reichman and Aitchison repeated the analysis using the maximum trail angle for each body size and found a tighter correlation ($r = 0.72$, $n = 22$, $p < 0.001$):

$$\angle_{tr} = 24.1 \; + (-4.65) \log M + 0.304 \angle_{sl}. \tag{7-57}$$

The clear relationship between body mass and angle of trail chosen demonstrates the relevance of the laboratory pattern to behavior in the wild. Less obvious is the reason for picking a steeper trail on a steeper hill. If the objective is to gain elevation as efficiently as possible and the trail angle is too shallow, an excessive amount of travel and energy goes into switchbacking. If the angle is too steep, the animal is overstressed, wastefully exceeding the optimal loading per contraction cycle. The authors attempted to explain the positive correlation between trail and slope angles on the basis of the net energy cost of the vertical component (Eq. 7-55), but the relevant cost may be *total* cost; factoring out the cost of unnecessary forward travel does not explain the correlation. The basis for the decision may be not the minimization of net cost (which would not be minimized in the authors' explanation) but the minimization of time of movement (Hainsworth, MS), which would leave more time for foraging or other pursuits.

Environmental Coupling

The smaller the animal the . . . [larger] the convection coefficient
. . . and the animal surface temperature is coupled tightly to the air
temperature . . . Because of the larger body size . . . the sheep temperature is decoupled from the air temperature.

—W. P. PORTER AND D. M. GATES (1969)

UP TO THIS POINT, we have considered the magnitude of internal
proportions, internal functions, and internal schedules. All of these manifest
the consequences of body size and can be analyzed with the aid of allometry.
The available information has been acquired in the laboratory environment, where conditions may or may not be representative of the real world.
While laboratory data are an indispensable foundation for basic biology and
for biology as applied to medical and agricultural practice, any extrapolation
must be done with caution.

So, from our background on the anatomy and basic physiology of body
size, we should now attempt to see how size affects the coupling or interfacing of animal and environment—in heat exchange, in energy cost of life in
the wild, in endurance of displacements from normal energy and water
balance, in sensory capacity, and in meshing of physiological time with
absolute "geotime." In such an effort the biologist faces great complexity,
for once an animal has been cornered or surrounded with a metabolic
chamber, it is no longer something in the wild being measured, but a
creature in the microenvironment (and perhaps terror) of artificial boundaries. Telemetry can provide wireless data, from presumably natural behav-

ior, on body temperature, location, and cyclic events such as heartbeat, respiration, or wingbeat. While telemetry cannot follow metabolic rate (though heart rate has been correlated with metabolic rates of captive subjects), turnover of doubly labeled water over periods of time between successive recaptures can give a value for total metabolism, an integration of activity and rest over a day or more. Careful observation from a perspective of the understanding of basic physiology and anatomy can yield time budgets — or, with added assumptions, time and energy budgets. No doubt the state of the art will advance, but for the present the account must be a preliminary one.

Temperature and Heat Exchange

The impact of climate and other environmental factors is very much size dependent. One need only recall how quickly one's fingers numb in the cold, or think of photographs of cattle in a snowstorm or a rodent in its burrow, to realize that the small animal is more vulnerable, and therefore more likely to seek a moderate microenvironment (into which a cow could not fit). It is well known that a small animal has relatively more surface area facing the environmental challenges and relatively less tissue volume generating heat or regulating its *milieu intérieur*.

What we need to do is to put this perhaps vague awareness in a clear quantitative framework. Among other lessons, we shall learn that the "magic" $M^{3/4}$ scaling of metabolism is peculiar to thermoneutral temperature ranges, the width of which is a function of body mass. At lower or higher temperatures metabolic rate may scale quite differently. A brief outline of comparative thermoregulatory physiology is perhaps necessary as background before we scale it.

Metabolic reaction rates are temperature dependent. The tissue temperature represents the net effect of the various factors in total heat balance: metabolic heat production (+), evaporative heat loss (−), conduction (±), convection (±), and radiation (±). The last three processes are proportional to the difference between the temperatures of the animal's surface and the surrounding environment (see Campbell, 1976, for details). Because these fluctuate almost continuously, an animal's temperature responses are studied in a simplified environment (such as a snap-lid can painted flat black inside and located in a temperature-controlled chamber). The metabolic rate (power) is calculated from measurements of oxygen consumption rate (20.1 J/ml O_2 for postabsorptive, resting animals), and metabolic rate can be plotted as a function of the controlled ambient temperature. Concurrently,

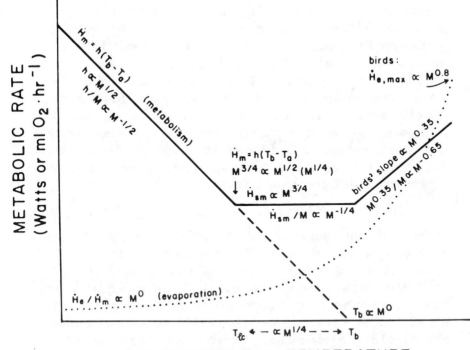

ENVIRONMENTAL TEMPERATURE

8-1 Metabolic heat production *(solid lines)* and evaporative heat loss *(dotted curve)* for birds and mammals are typically related to environmental temperature during the inactivity as shown. The *dashed line* is the theoretical extrapolation of Eq. (8-1) to body temperature, T_b (see text). Except for T_b and the ratio of evaporative heat loss to metabolic heat production ($\dot{H}_e/\dot{H}_m$), all expressions of thermal qualities and responses are strongly size dependent. Heat transfer coefficients (h) are proportional to $M^{1/2}$ and $\dot{H}_{sm}$ to $M^{3/4}$, so the difference between T_b and the lower critical temperature (T_{lc}, which marks the lowest temperature at which $\dot{H}_{sm}$ is sufficient to maintain T_b) must scale as $M^{1/4}$. At temperatures below T_{lc}, $\dot{H}_m$ must scale like $H \propto M^{1/2}$. The allometry of heat stress has not been described in mammals, and quantitative explanation is not obvious in birds.

the water that the animal has evaporated into the air is collected and used to calculate heat loss by evaporation. The latent heat absorbed in evaporation from a 30°C surface amounts to 2.427 J/mg H_2O. It is insignificantly temperature dependent, 1% greater at 20°C, 1% less at 40°C.

For homeothermic animals, typical curves are shown in Figure 8-1. Each segment of these curves can be analyzed and expressed allometrically, because metabolic rates, surface areas, thickness of pelage or plumage, and so on, are all size dependent. There is a flat region in the middle of the metabolic curve, at which the minimal (basal or standard) metabolic rate is adequate to maintain constant body temperature (T_b). This is accomplished by modifying body position (the amount of surface exposed is less when the animal is curled into a ball or has its head tucked under its wing) by fluffing the feathers or erecting the fur, and by control of blood flow distribution.

At the lowest temperature for this thermoneutrality (the lower critical temperature, T_{lc}) piloerection and cutaneous vasoconstriction have done about as much as possible to reduce heat loss and cost. Any further decline in ambient temperature (T_a) must be offset by a higher metabolic rate $\dot{H}_m$ so that production equals loss, which is an approximately linear proportion of the temperature difference being maintained:

$$\dot{H}_m = \dot{H}_{loss} = h(T_b - T_a). \tag{8-1}$$

The proportionality constant (h) is the heat transfer coefficient (often called the thermal conductance). The thicker the insulation, the more difficult it is for heat to flow, and the smaller the value of h (joules/$^\circ$C · hr, or W/$^\circ$C), and the shallower the slope of the curve below T_{lc}. Extrapolated in the other direction, this line might be expected to intercept the T_a-axis at T_b, since when $T_b = T_a$ there is no heat loss by conduction, convection, or radiation and no need for a metabolic rate for the sake of generating heat. Quite often the line does not intercept T_b, which means that the straight line is only a crude approximation; there may be even more adjustment of heat-conserving mechanisms below T_{lc}.

With this brief background we can now examine the scaling of the components of Eq. (8-1) and Figure 8-1: body temperature, insulation, thermoneutral zone, and rates of heat production during thermal stress.

BODY TEMPERATURE

Body temperatures of calm, resting birds and mammals show no size dependency. For birds, T_b averages $40^\circ \pm 1.5^\circ$ C (Calder and King, 1974), with no passerine/nonpasserine distinction. Monotreme mammals show no size trend in the adult size range of 1 g to 16.5 kg, with body temperatures of $31.7^\circ \pm 0.6^\circ$ C (Dawson et al., 1979). Marsupial mammals have basal body temperatures that are higher, but size independent:

$$^\circ C = 35.6 \, M^{0.01} \tag{8-2}$$

($r^2 = 0.44$; calculated from Dawson and Hulbert, 1970). In the Eutheria body temperature is also size independent and still higher, 36° to 38° C (Schmidt-Nielsen, 1979; Bartholomew, 1982).

There is, however, a circadian cycle in resting body temperature. The amplitude of this temperature change ($\Delta T_{b,\alpha-\rho}$/day) is size dependent (Aschoff, 1981a):

primates: $\Delta T_{b,\alpha-\rho}$/day $= 16.33\ m^{-0.287};$ (8-3)
$2.25\ M^{-0.287};$

Eutheria except primates: $\Delta T_{b,\alpha-\rho}$/day $= 4.76\ m^{-0.197};$ (8-4)
$1.23\ M^{-0.197};$

passerine birds: $\Delta T_{b,\alpha-\rho}$/day $= 4.18\ m^{-0.125};$ (8-5)
$1.76\ M^{-0.125};$

nonpasserines: $\Delta T_{b,\alpha-\rho}$/day $= 10.86\ m^{-0.396};$ (8-6)
$0.70\ M^{-0.396}.$

These relationships could be considered to express the mean *rates* of T_b change (per day), so their reciprocals are expressions of time per degree Celsius change, times proportional to $M^{0.13}$ to $M^{0.40}$, mean $M^{0.25}$—interesting, but perhaps only coincidental.

INSULATION

The lower critical temperature for the average 10-g bird (nonpasserine) or mammal (eutherian) is about 30° C; if the size were increased to 100 g, this T_{lc} would only go down to 23° to 25° C (Calder and King, 1974; Bartholomew, 1982). These temperatures are warmer than the animals often must face, so the importance of insulation is readily apparent.

Having more insulation is better for heat conservation, but for an animal that must move around, insulation adds to the mass of what the animal must carry. If it were especially thick it could also impede movement (picture the problems of a mouse with fur longer than its legs). Thus, there is an energetic compromise between thermoregulation and locomotion.

The mass of the birds' plumage (fx = feathers) scales as

$$m_{fx} = 0.09\ m^{0.95},$$ (8-7)

essentially in linear proportion to body mass (Turček, 1966), while the body-surface area scales (Walsberg and King, 1978) as

$$A_b = 8.11\ m^{0.667}\ (s_{\log y \log x} = 0.0397;\ r = 0.998).$$ (8-8)

Thus, per unit area, there is less insulation on the smallest birds (Calder and King, 1974). If the density of plumage is size independent, then volume of insulation would be proportional to mass, so that

$$V_{fx} \div A_b \propto m^{0.95}/m^{0.67} \propto m^{0.28} \propto l_{fx},$$ (8-9)

where l = feather thickness. Feather thickness is, of course, adjustable by means of the ptilomotor muscles, which fluff the feathers by erecting them to a greater angle from the skin, thereby trapping a thicker air layer. The angular change from "sleeked" to "fluffed" is probably the same in birds of all sizes, so that the allometry of feather thickness, all fluffed or all sleeked, would have about the same $m^{-0.28}$ exponent; in other words, it would be another characteristic body linear dimension, scaled like other lengths, as the ¼ to ⅓ power of body mass.

A similar pattern occurs for the fur of mammals. The mass of the fur (see Table 3-4) is

$$M_{fur} = 0.032 \ M^{0.98}. \tag{8-10}$$

The surface area of the pelt (Tables 5-1 and 5-2) is

$$A = 1,100 \ M^{0.65}. \tag{8-11}$$

The thickness (l_{fur}), again assuming a size-independent density, would then be

$$l_{fur} \propto M^{0.98}/M^{0.65} \propto M^{0.33}. \tag{8-12}$$

More meaningful than geometric measurements is the direct evaluation of insulative quality, which can be done in vivo by determining heat transfer or thermal conductance in steady-state oxygen consumption (determining h in Eq. 8-1). This has also been measured in the past by running cooling curves on warmed carcasses (Morrison and Tietz, 1950; Herreid and Kessell, 1967; Bakken, 1976); heat loss is estimated from body mass, body-temperature drop, and a literature value for specific heat. Although the carcass method lacks normal internal convection (circulation) and piloerectional control, it gives similar results (mW/° C). From a combination of data from metabolic and cooling curve determinations, we derive the following relationships (Hart, 1971; Calder, 1974):

eutherian mammals: $h = 5.30 \ m^{0.50}$,   (8-13)

birds (Passeriformes): $h = 5.29 \ m^{0.46}$,   (8-14)

birds (other): $h = 4.72 \ m^{0.46}$.   (8-15)

If h is expressed per unit area of surface, it scales as $m^{-0.21}$ for birds, $m^{-0.17}$ for eutherian mammals, and $m^{-0.13}$ for marsupials (Dawson and Dawson, 1982). Insulation (resistance of heat flow) is the inverse of heat transference or conductance, with units of ° C differential/unit flow rate of heat = ° C/mW or ° C(mW · cm²)⁻¹, and would accordingly scale as $m^{0.17}$ to $m^{0.21}$.

These are somewhat smaller exponents than the dimension of fur or feather thickness or path length for heat flux, as might be expected — as if quality of insulation were more than a matter of mere thickness.

Thermal conductivity is thermal conductance per unit of surface area per unit of thickness. Birkebak (1966) collected the available data on thermal conductivity of mammal fur and showed that the insulative quality actually decreases with thicker pelage. Although a heavy coat is warmer than a thin one, a doubling of thickness will not double the insulation (or cut the conductance in half), probably because the thicker fur is made up of coarser hairs, which are better conductors (Figure 8-2).

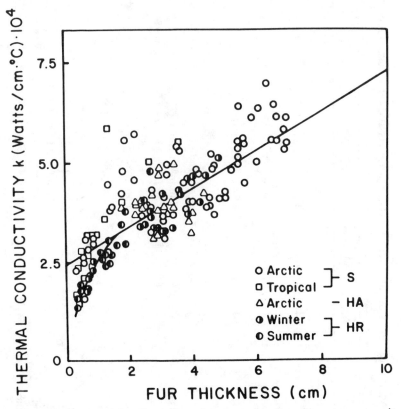

8-2 Thermal conductivity (heat flux rate per unit of surface area per unit of thickness) actually increases with thickness. While a thicker coat is warmer than a thin one, the increase in insulation is not directly proportional to thickness, for thicker fur is coarser fur. S, HA, and HR indicate different data sets used by Birkebak. (Redrawn from Birkebak, 1966, and converted to S.I. units. Copyright Academic Press.)

Table 8-1 The scaling of heat transfer coefficients of mammals and birds. Units have been converted from the original reports.

Animals	Size range (g)	mW/°C	r	References
Diurnal Phases Pooled				
Mammals	4–150,000	$4.24\,m^{0.57}$; $224\,M^{0.57}$	0.94[a]	Bradley and Deavers, 1980
Dasyuridae marsupials	7–5,050	$5.10\,m^{0.54}$; $208\,M^{0.54}$		MacMillen and Nelson, 1969
Eutheria	4–6,660	$4.94\,m^{0.54}$; $203\,M^{0.54}$		Aschoff, 1981b
Eutheria, "small"		$5.71\,m^{0.50}$; $174\,M^{0.50}$		Bartholomew, 1982
Chiroptera	10–598	$8.60\,m^{0.46}$; $206\,M^{0.46}$	0.97[a]	Bradley and Deavers, 1980
Heteromyidae	8–103	$3.46\,m^{0.56}$; $166\,M^{0.56}$	0.78[a]	Bradley and Deavers, 1980
Cricetidae	7–1,100	$5.75\,m^{0.46}$; $138\,M^{0.46}$	0.97[a]	Bradley and Deavers, 1980
Muridae	6–390	$4.69\,m^{0.53}$; $182\,M^{0.53}$	0.95[a]	Bradley and Deavers, 1980
Sciuridae	47–998	$3.29\,m^{0.62}$; $239\,M^{0.62}$	0.82[a]	Bradley and Deavers, 1980
Mean exponent		$m^{0.53\pm0.05}$		
Birds	11–2,755	$5.33\,m^{0.46}$; $128\,M^{0.46}$		Herreid and Kessel, 1967
Passeriformes	6–1,130	$5.29\,m^{0.46}$; $126\,M^{0.46b}$		Calder and King, 1974
Nonpasserine	3–2,755	$4.72\,m^{0.46}$; $113\,M^{0.46b}$		Calder and King, 1974
Passerines and nonpasserines	3–2,430	$4.74\,m^{0.49}$; $140\,M^{0.49}$		Lasiewski et al., 1967
Mean exponent		$m^{0.47\pm0.02}$		

Active (α) and Rest (ρ) Phases Separated

Mammals			
Eutheria, α	79–6,660	$8.59\ m^{0.48}$; $242\ M^{0.48}$	Aschoff, 1981b
Eutheria, ρ	4–4,400	$5.71\ m^{0.48}$; $158\ M^{0.48}$	Aschoff, 1981b
Birds			
Nonpasserine, α	2.7–2,430	$5.28\ m^{0.52}$; $187\ M^{0.52}$	Aschoff, 1981b
Nonpasserine, ρ	40–2,020	$5.29\ m^{0.42}$; $94\ M^{0.42}$	Aschoff, 1981b
Passeriformes, α	6–1,130	$4.79\ m^{0.54}$; $195\ M^{0.54}$	Aschoff, 1981b
Passeriformes, ρ	10.3–360	$3.22\ m^{0.54}$; $133\ M^{0.54}$	Aschoff, 1981b

Mean exponent $\qquad m^{0.51\pm0.06}$

a. It appears that all of the regressions have been for mass-specific conductance [ml O_2/(g · hr · °C) or W/(g · °C)] as a function of body mass. Since the dimension of body mass is common to both x and y variables, there is a degree of inevitability to the correlation; the correlation coefficients therefore are exaggerated.

b. Difference in coefficients not significant.

Since Eqs. (8-13) to (8-15) were published, the size dependence of h (the so-called thermal conductance) has been examined in further detail. Table 8-1 provides a summary of recent as well as earlier allometries of this variable. The statistical parameters have either been ignored or exaggerated, so it is really from the overall consistency of the exponents that we feel we can trust them. Note that conductance is usually expressed in mass-specific terms, ml O_2 increase/(g · hr · °C) or calculated as heat loss/(g · hr · °C), in the traditional way of physiologists. Less often it is calculated per unit of surface area as estimated from body mass. In either case, the y-axis or variable is given a dimension in common with the x-axis or variable. This leads to an inevitable correlation (Riggs, 1963). Indeed, one can take random numbers, divide them by body mass and get a fairly impressive correlation (Calder and King, 1972). Such a practice exaggerates the r^2 values.

When a body of data is divided into phylogenetic or ecological subsets, the range in body sizes decreases, along with the decrease in sample size; the precision of the regression equations to be derived is thereby reduced, as is the potential for generalization in subsequent comparisons. It appears (Table 8-1) that the passeriform birds have less insulation than the others, but the 95% confidence intervals for the original equations overlap, and the two are not statistically distinct (Calder and King, 1974). The bats appear to have higher h, perhaps due to the poor insulation of wing membranes. However, y-intercepts cannot be compared simply, because of differences in exponents which lead all of the expressions to cross between body masses of 1 g and 1 kg, a range in which most of the families summarized are fairly well represented. The mean values for the exponents, $m^{0.53(\pm 0.05 \text{ s.d.})}$, do not appear to differ significantly from the previous derivations for eutherian mammals. Exponents for birds are similar.

For the most part, the conductance values have been calculated from data on metabolic rates and temperatures, both of which undergo a circadian cycle (Aschoff and Pohl, 1970a,b; Aschoff, 1981a, 1982). While the insulative material of the feathers or hair probably cannot change in the period of a regular daily cycle, the conductance is relatively higher in the active (α) part of the daily cycle (Table 8-1), by about 40% (Aschoff, 1981b). The one exception is in the ρ phase for nonpasserine birds, but Aschoff states that this was the least reliable equation.

Technically, the conductance values used in these analyses represent more than conducted heat loss. The slopes of the metabolic regressions below T_{lc} (Figure 8-1) represent the increased heat production needed to offset heat losses by conduction, convection, radiation, and evaporation.

Before convection and radiation take the heat, it must be conducted across the skin and plumage or feathers. However, much of the evaporation takes place via the respiratory tract and as vapor leaving the skin surface. A true conductance ("dry conductance") is calculated by subtracting the rate of heat loss by evaporation from the total heat loss (the latter assumed to equal total metabolic heat production) before dividing by the temperature differential.

The expression for h in Table 8-1 represents the so-called wet conductance, a bizarre internal contradiction in terms that attempts to say that the calculation has not excluded the nonconducted evaporative heat loss. Aschoff (1981b) attributes the differences between α and ρ conductance equations to circadian differences in evaporation rates. A next step would be to review the allometry of evaporative heat loss, separated into α and ρ determinations, to see if there is a circadian cycle. The set relationship between temperature and vapor content of saturated air makes it unlikely that this will occur.

THERMONEUTRALITY

Let us now return to the scaling of the thermoneutral range, over which the animal maintains body temperature without increased costs of energy or water. Intuitively we would expect that the larger and woolier the animal, the lower the temperature it could handle without increasing the metabolic rate. Regression analysis of this variable, however, is confused by many other variables. If, for example, the animal has an elevated $\dot{H}_{smr}$, then the intercept (T_{lc}) of that baseline with the regression based on h (refer to Figure 8-1) will be estimated to be a lower T_a. In other words, elevated metabolism would appear to affect the T_{lc} just as would arctic adaptation or large body size.

Even if the standard metabolic rate determination is not quite accurate, the slope of the subneutral regression is fairly reliable; consequently it has a tighter correlation with body mass. We can then go back to Eq. (8-1) and solve it for $(T_b - T_a)$ at $T_a = T_{lc}$:

mammals: $\quad (T_b - T_{lc}) = \dot{H}_m / h \propto m^{0.73} / m^{0.50} \propto m^{0.23}$;   (8-16)

birds: $\quad\quad (T_b - T_{lc}) = \dot{H}_m / h \propto m^{0.72} / m^{0.46} \propto m^{0.26}$.   (8-17)

Thus, the difference between body temperature (which appears to be size independent) and the lower critical temperature scales as $M^{-1/4}(M^{0.23}$ to $M^{0.26})$ in birds and mammals (Morrison, 1960; Calder and King, 1974), so T_{lc} is inversely related to size, as expected.

In the case of birds, it appears that this temperature difference is directly proportional to plumage thickness (see Eqs. 8-8 and 8-17), or, put another way, that the gradient ($^\circ$C/cm) is size independent:

$$(T_b - T_{lc})/l_{fx} \propto M^{0.26}/M^{0.28} \propto M^{-0.02}. \qquad (8\text{-}18)$$

For mammals, we could consider Eqs. (8-12) and (8-16) to obtain

$$(T_b - T_{lc})/l_{fur} \propto M^{0.23}/M^{0.33} \propto M^{-0.10}. \qquad (8\text{-}19)$$

This is not a size-independent, direct relationship between insulative thickness and thermoneutral range as in birds, but does not take the size dependence of insulative quality into account (see Figure 8-1).

COLD STRESS

If we lower the temperature until few animals or none are in their thermoneutral range, and expose all to the same T_a (which unlike $T_a = T_{lc}$ does not include a size-dependent scaling) we would find:

$$\dot{H}_m = h(T_b - 0^\circ C) \propto m^{1/2}(m^0) \propto m^{1/2}. \qquad (8\text{-}20)$$

It might come as a shock, after pages of trying to relate biology to $m^{3/4}$ scaling for metabolism — and having found a surprising degree of conformity — to learn that the metabolic rate below thermoneutrality is not scaled to $m^{3/4}$ (Table 8-2), nor need its oxygen and food delivery rates, or the resources upon which it depends, scale to $m^{3/4}$. This could mean one of several things:

(1) We are adrift in a world of imaginary coincidences or biased conclusions.

(2) For the most part, animals are either in milder microclimates or are absorbing radiation or generating extra heat in activity, so that they are in a thermoneutral environment.

(3) The animal and its requirements, whether internal (physiological) or external (ecological, behavioral), are scaled to active metabolism, such as in sustained locomotion, which does in fact scale approximately to $M^{3/4}$.

Analytical biology needs patterns, and that need may have an important subjective effect on what we see. We should not, therefore, rule out any of the above possibilities. However, number (2) appears capable of resolving matters. There have been a number of studies on the metabolic energy-sparing effect of artificial and natural radiation on birds, in which the thermo-

Table 8-2   The effect of environmental temperatures on the allometry of resting metabolic rates.

Animals	Conditions	Metabolic rate (watts)[a]	Source of equations or data
Passeriform birds	0°C, night, summer	$0.167\ m^{0.53}$	Kendeigh et al., 1977
	0°C, night, winter	$0.150\ m^{0.53}$	Kendeigh et al., 1977
Other birds	0°C, night, summer	$0.127\ m^{0.57}$	Kendeigh et al., 1977
	0°C, night, winter	$0.088\ m^{0.59}$	Kendeigh et al., 1977
Eutherian mammals	0°C	$0.270\ m^{0.63}$	Tracy, 1977
Eutherian mammals (mostly rodents)	0°C[b]	$0.211\ m^{0.50}$; $0.198\ m^{0.51}$	Herreid and Kessell, 1967 Hart, 1971 (table 2)
Eutherian mammals (6 rodents, 1 rabbit)	5°C, max NST[c]	$0.167\ m^{0.55}$	Heldmaier, 1971
Arctic mammals (4 rodents and 4 carnivores, 0.32–12.9 kg)	0°C	$0.394\ m^{0.40}$	Withers et al., 1979
	10°C	$0.260\ m^{0.39}$	Withers et al., 1979
	20°C	$0.090\ m^{0.56}$	Withers et al., 1979
	30°C	$0.061\ m^{0.62}$	Withers et al., 1979
Eutherian mammals	Thermoneutral	$0.020\ m^{0.76}$	Bartholomew, 1982

a. Units have been converted (Calder, 1981) from the original reports.
b. Calculated from h as $\dot{H}_m = h(T_b - 0°C)$, assuming $T_b = 37°C$ if not otherwise stated.
c. Nonshivering thermogenesis, measured as $\dot{V}_{O_2}$ after noradrenaline injection.

neutral range for the adequacy of standard metabolism extends to lower temperatures than without the radiation (Calder and King, 1974).

Kendeigh (1969, 1970) has summarized the "existence" metabolic rates (daily energy requirements) of caged birds. The allometric equations indicate that heat production in even the restricted activity of caged existence is about 25% above the standard or basal level. For a 30-g passerine bird, this in itself would lower the lower critical temperature an additional 4°C before the $M^{3/4}$ standard metabolic rate would be inadequate. Metabolic rates in sustained flight or running are about ten times those at the resting level, and still approximate $M^{3/4}$ scaling. Although the forced convection of moving through the air would increase the rate of heat loss, it is easy to envision $M^{3/4}$ thermoregulation at 0°C during much of the daily cycle.

HEAT STRESS

As the temperature differential is reduced by progressively higher environmental temperatures, adequate dissipation of metabolic heat depends on enhancement of heat transfer by "dry" and evaporative means. Conductance is increased by relaxing piloerection, reducing the thickness of the trapped air and insulation, and posturally exposing areas with less insulation (such as axillary regions). If $T_a > T_b$, heat flow is in a disadvantageous direction, so the insulation should be increased, as has been documented, for example, in ostriches, jackrabbits, and passerine birds (Crawford and Schmidt-Nielsen, 1967; Dawson and Schmidt-Nielsen, 1966; Hinds and Calder, 1973).

The allometry of response to heat stress had received little attention before Weathers (1981) analyzed data from birds exposed to high tempera-

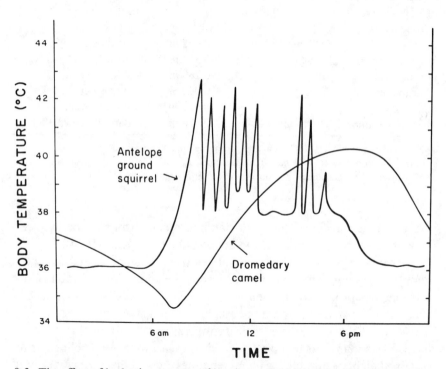

8-3 The effect of body size on rates of heating and cooling of desert animals. The camel, with a body mass of about 500 kg, stores heat during the day and dumps it into the cooler nocturnal environment, saving the 5 l of water that otherwise would be needed to dissipate the heat by evaporation (Schmidt-Nielsen, 1964). The ground squirrel has 1/5,000 of the thermal inertia of a camel, heats rapidly, and soon must go underground to a cool burrow to recover from the overheating; it repeats this cycle several times per day. (From G. A. Bartholomew, 1964, *Symposium of the Society for Experimental Biology* 18:13; with permission of Cambridge University Press.)

tures. He defined a "coefficient of heat strain," which has the same dimensions as the common expression of conductance, and derived the following relationship:

$$mW/(g \cdot {}^\circ C) = 12.5 \, m^{-0.65}, \tag{8-21}$$

for which $n = 26$, $r^2 = 0.86$, $S_{yx} = 1.49$, $S_b = 0.174$. Converting to a whole-animal basis for comparison with scaling in Table 8-1, this becomes

$$mW/{}^\circ C = 12.5 \, m^{0.35} = 140 \, M^{0.35}. \tag{8-21a}$$

This curve intersects with the Lasiewski et al. (1967) equation at 1 kg and thus appears to be of the same order of magnitude as h below thermoneutrality. The puzzle lies in the smaller exponent, which does not even fit as a coincidence with other allometries.

One response to heat stress that conserves water is to store heat by accepting a tolerable degree of hyperthermia during midday in the desert heat, and then get rid of this stored heat to a cooler environment, either in the open at night or by returning to the cooler underground burrow in the daytime. The time course of such oscillations in body temperature is size dependent, as depicted in Figure 8-3, and once again the theme is in evidence that life's needs must fit into a physiological time frame.

Endurance

In Chapter 5 I attempted to assemble an allometric account of the steady-state flow of oxygen and energy from the environment to the working tissues, via the respiratory exchanges and the internal convection of the circulatory system. There are periods when the organism may be temporarily cut off from its sources of supply, such as during the underwater dive of an air breather, the fast imposed when a snow or ice storm covers up the food supply, or the trek between water holes over a hot desert.

In any of these cases the animal enters a period of deprivation with some amount of the necessary substance stored: oxygen, fat, crop contents, or fully hydrated tissues. The rate at which such reserves are used is roughly proportional to the metabolic rate. The endurance time will be the amount of reserves divided by the rate of their depletion. If the amount of reserves scales with a higher exponent of body mass than does the rate of depletion, the larger animal will last longer. This is yet another manifestation of the physiological time scale, that of the smaller animal being much compacted, with a shorter period to spare before the steady state must be restored or the equilibrium of nonlife will set in. It is thus a physiological time scale that becomes an ecological time scale as the animal interacts with environmental caprice.

If the impact of an environmental fluctuation is size dependent, natural selection can lead to body-size changes and trends. Such may be the basis for ecological or natural history "rules" such as Bergmann's Rule and Cope's Rule for size increase (see Lindsey, 1966; Rosenzweig, 1968; McNab, 1971; Calder, 1974). Obviously, this could only be one side of natural selection, for the persistence of hummingbirds and shrews makes it obvious that there are counterbalancing factors, or some compromise between considerations. Hence it seems worthwhile to survey some size aspects of endurance.

FASTING ENDURANCE

This is one quotient of total body energy reserves divided by the metabolic rate of their utilization. The energy reserves consist of gut contents, blood sugar, liver and muscle glycogen, and adipose fat. The allometric relationships of some of these are hard to find.

Adipose fat contains more reserve energy per gram than undigested gut content or carbohydrate storage forms, so we would expect that the amount of fat would have the greatest effect on fasting endurance. Unfortunately, I have found no direct analysis of the allometry of body fat in "normal" mammals or birds. I noted in Chapter 2 that addition of all of the allometric predictions for skeletal muscle, skeleton, blood, integument, and major organs accounted for 97% of a body mass of 7 g, down consistently to 81% of a 6.6-metric-ton mammal (elephant size). It is likely that stored fat makes up much of the unaccounted portion, to judge from the scaling of the two, $M_? = 0.09\, M^{1.13}$ and $M_{fat} = 0.075\, M^{1.19}$ (Eqs. 3-31 and 3-32), and the general agreement between predictions from these equations and observed amounts of fat. Averaged over the year, *Peromyscus* had 5.7% lipids, compared to 5.6% predicted by Eq. (3-31). A normal (70.6-kg) man had 13.6% lipids, compared to 15.8% predicted. On the other hand, two species of shrews had seven to eight times as much fat as the prediction for $M_?$.

In a study of the potential of African wildlife as a source of food for humans, Ledger and associates (1967) reported live weights, and carcass as a percentage of live weight, with and without fat. From the change in *percentage* when the fat was removed, the mean calculated amount of fat for each of 17 species of ungulates ranging in mean body mass from 20 kg for Thomson's gazelle to 1,384 kg for the hippopotamus gave an essentially linear relationship of carcass fat (M_{cfat}, in kg) to live body mass (M_b):

$$M_{cfat} = 0.11 \, M_b^{1.02}; \qquad r^2 = 0.975. \tag{8-22}$$

This, however, includes only carcass fat, while much of the body fat is cutaneous fat. Pace and coworkers (1979), for example, found that the body fat of laboratory rodents (mice, rats, hamsters, guinea pigs) was almost equally divided between skin, viscera, and carcass, while the carcasses of laboratory rabbits had over one-half of the total body fat, the viscera almost one-third, and the skin only one-eighth. These were not wild animals, and it would be difficult to identify criteria for "normal" body mass and fat content of caged animals with food ad libitum and little exercise. Lab mice and rats tend to weigh considerably more than feral representatives of their species.

Thus, a good allometric equation for normal fat content (as contrasted with obesity and scrawniness) would be a useful standard for defining normalcy, as well as for analyzing fasting endurance times. Alas, our predicament is circular. To decide which are normal animals from which to take data, one needs to have a standard that can only be derived from data on normal animals.

The situation is about as equivocal for small birds (Calder, 1974). Maximum values for mass of body fat (M_{fat}, in g) scale as:

passerines: $$M_{fat} = 0.03 \, M_b^{1.57} \tag{8-23}$$

$$n = 19 \text{ species}, M_b = 12.4 \text{ to } 41.3 \text{ g}, r^2 = 0.50;$$

nonpasserines: $$M_{fat} = 0.06 \, M_b^{0.83} \tag{8-24}$$

$$n = 3 \text{ species}, M_b = 5 \text{ to } 72 \text{ g}, r^2 = 0.98.$$

In the decade since these were derived an increment of data scattered in the literature should have accrued for refinement of the above, but so far there is no clear basis for rejecting the usual assumption that the amount of fat stored is a linear function of body mass. If this assumption is in fact true, the endurance time on fat reserves would be

$$\text{time} = \text{amount of fat/rate of metabolizing} \propto M^{1.0}/M^{3/4} \propto M^{1/4}. \tag{8-25}$$

This would be true at rest in a thermoneutral environment of microclimate, or in activity which also scales as $M^{3/4}$ (Hemmingsen, 1960; Taylor et al., 1978). The relationship was first articulated by Morrison (1960) in the context of the improvement of fasting endurance by hibernation.

In progressively colder (subneutral) temperatures, endurance time would scale more like $M^{1/2}$, since the exponent for resting metabolism approaches that for heat transfer coefficients at the reciprocal of total insulation, $M^{1/2}$, giving the larger animal proportionately more time than the smaller one. Likewise, if the $M_?$ component of body mass (Eq. 3-31) turned out to be a good indicator of normal fat reserves, the large animal would also have a better chance of surviving a fast:

$$t \propto M^{1.13}/M^{0.75} \propto M^{0.38}. \tag{8-25a}$$

Some animals carry reserves of energy stored as undigested food in cheek pouches or in the crop. In order to evaluate the significance of such reserves, the cheek pouches of heteromyid rodents were filled to maximum capacity with millet seeds. Cheek pouch volumes (V_{cp}, in cubic centimeters) were correlated with body mass (9 g to 116 g) as follows:

$$V_{cp} = 0.065 \ m^{0.89} \tag{8-26}$$

($r^2 = 0.728$; $n = 62$; Morton et al., 1980). Filled with seeds of the storksbill *(Erodium cicutarium),* an exotic now used commonly by heteromyid rodents in the deserts of California and Arizona, the energy contained (in joules) was

$$E_{cp} = 8.08 \ m^{0.89}. \tag{8-27}$$

The daily energy expense of heteromyid rodents, estimated from rates of oxygen consumption (at 2.6 times $\dot{E}_{sm}$) scales as

$$\dot{E}_{tot} = 5.18 \ m^{0.70}. \tag{8-28}$$

The endurance time on one crop load (in days) should therefore be

$$t = 8.08 \ m^{0.89}/5.18 \ m^{0.70} = 1.56 \ m^{0.19}. \tag{8-29}$$

Crop volumes of hummingbirds appear to be linearly related to a small range in body mass (2.7 g to 10.1 g):

$$V_{cp} = 0.092 \ m + 0.224 \tag{8-30}$$

($n = 75$, $s_{yx} = 0.152$; Hainsworth and Wolf, 1972). However, because the linear regression extrapolates to a volume of 0.22 ml at a body mass of 0 g, it is likely that the relationship above is a small segment of a power function rather than a linear scaling. Converting to such a function by using Eq. (8-30) to predict crop volume for body masses of 3 g, 6 g, and 10 g and regressing these values, we get

$$V_{cp} = 0.23 \ m^{0.68}. \tag{8-31}$$

However, the small size range of the original data and the crudity of my secondary treatment are such that Eq. (8-31) cannot be used in an allometric cancellation with any confidence.

DESICCATION ENDURANCE

The time an animal could withstand desiccation would be equal to body-water content divided by rate of loss. The body-water content during normal hydration seems to be a standard fraction of body mass, independent of size (that is, body water is a linear function of body mass, Table 5-12). Rates of loss by evaporation (in grams per day) at moderate temperatures are roughly proportional to metabolic rate:

mammals: $\quad \dot{M}_{H_2O} = 38.8\ M^{0.88}$; $\qquad\qquad\qquad$ (8-32)

bats: $\qquad \dot{M}_{H_2O} = 40.7\ M^{0.67}$; $\qquad\qquad\qquad$ (8-33)

birds: $\qquad \dot{M}_{H_2O} = 24.2\ M^{0.61}$. $\qquad\qquad\qquad$ (8-34)

During heat stress birds evaporate at much higher rates (milligrams per minute):

$$\dot{M}_{H_2O,max} = 259\ M^{0.80}. \qquad\qquad\qquad (8\text{-}35)$$

(The exponents in Eqs. 8-32 and 8-34 are not statistically distinguishable.) In mammals the rate of water loss (in milliliters per day) via the urine is

$$\dot{V}_{H_2O} = 60.9\ M^{0.75}. \qquad\qquad\qquad (8\text{-}36)$$

It would appear, from these fragments, that desiccation endurance times can also be assumed to roughly parallel physiological time scales.

DIVING ENDURANCE TIME

This time could be limited either by depletion of oxygen stores or by chilling in a highly conductive medium. Let us consider first the possible limitation due to oxygen supply. At the start of a dive, the fully oxygen-saturated animal has between 30 ml and 52 ml of oxygen per kilogram of body mass. The largest portion is in the blood, the volume of which is essentially linearly proportional to body mass, and the hemoglobin content of which is size independent. The air in the lungs is similarly scaled, volume $\propto M^{-1.0}$ and composition $\propto M^0$. The muscle mass is proportional to $M^{1.0}$, but the myoglobin concentration does have a positive allometry, $M^{0.22}$. The tissue water also has some dissolved oxygen, which would be directly proportional to size but small in total amount. Thus, with the exception of the muscle store of oxygen, about one-tenth of the total (Scholander, 1940), all compartments are approximately linearly proportional to body mass. Owing to its higher oxygen affinity, the myoglobin does not release oxygen to the blood, so it is not available to the brain and cardiac muscle tissues which only function aerobically and thus limit endurance in terms of oxygen. Diving time constraints should therefore be as seen in Figure 8-4:

$$\text{amount of } O_2/\text{rate of consumption} \propto M^{1.0}/M^{3/4} \propto M^{1/4}. \quad (8\text{-}37)$$

The amount of oxygen an animal can take up before the dive varies with the extent of evolutionary adaptation for diving (Andersen, 1966). The rate of oxygen consumption, of course, depends on how actively the diver is exerting itself; a vigorous chase cannot last as long as stationary underwater hiding.

Now consider the limiting effects that may be imposed by chilling. Kanwisher pointed out that porpoises are the equivalent of marine shrews. They can remain in the water permanently only because of basal metabolic rates that are elevated above the Kleiber mouse-to-elephant prediction, and because of a disproportionate dedication of body mass to an outer blubber layer (Kanwisher and Sundnes, 1966; Kanwisher, 1977). Smaller divers cool during diving in cold water and must therefore alternate between swimming and periods of drying and warming, out of the water.

The smallest diving homeotherms are water shrews, the North American species, *Sorex palustris,* being the smallest with body mass ranging from 8 g to 18 g. During a dive of 30 seconds' active swimming, about two-thirds of the apparent maximum endurance in water of 10° to 12°C, the water shrews' body temperatures dropped 1.43 ± 0.46°C from a predive 39.7 ± 0.4°C. Although this cooling was considerably more rapid than that of larger animals, in dives of this duration the shrews were not hypothermic by mammalian standards at 38.3°C; two individuals that cooled to 34° and 32°C appeared to be slowed and less coordinated, but this amount of cooling was atypical (Calder, 1969). The calculated time ($t_{\Delta 1°}$, in seconds) for cooling 1°C by a series of swimmers (water shrew, rat, rabbit, and dog) can be summarized as

$$t_{\Delta 1°} = 8.4 \, m^{0.35}; \quad r^2 = 0.957. \quad (8\text{-}38)$$

This scaling is somewhat steeper than that expected by considering oxygen the limiting factor, but not so steep as would be predicted from scaling of heat transfer coefficients. The small sample size makes it a crude approximation at best, but does qualitatively confirm the time constraints experi-

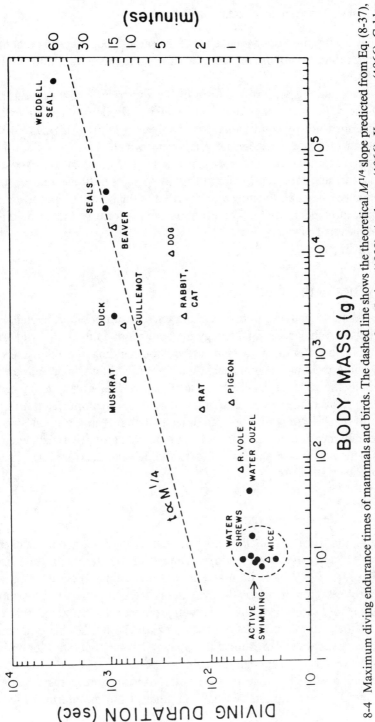

8-4 Maximum diving endurance times of mammals and birds. The dashed line shows the theoretical $M^{1/4}$ slope predicted from Eq. (8-37), extended from the bladdernose seal (Scholander, 1940). Other data are from Irving (1939), Andersen (1966), Kooyman (1966), Calder (1969), and Murrish (1970). ● = survival of forced or natural dives; △ = time to last movement in drowning.

enced as a consequence of small size, at the same time suggesting that the apparent exhaustion of oxygen stores occurs before excessive chilling in the water shrew.

Diving is, of course, just one form of the approach to asphyxia during breath holding. Bird songs that consist of sustained trills appear to be accomplished not by continuous expiration, as in human singing, but by oscillation of tracheal dead-space air, during which the thorax does not deflate by expiration until the end of the song. A 21-g canary, *Serinus canaris,* can sing continuously for at least 27 seconds until the lung p_{O_2} has presumably fallen and the lung p_{CO_2} has risen to the point that the air must be replaced (Calder, 1970). The duration of this form of breath holding approaches that of small divers.

Energetics in the Wild

A considerable gap exists between oxygen consumption in metabolic chambers and the actual total energy metabolism of a day in the wild—a day that includes temperature changes, activity, standing or perching, feeding, sleeping, evading predators or chasing prey, caring for young, and so on. Eventually this gap must be filled if we are to know what is really going on in nature in terms of the energetics. For if we describe something qualitatively, we think we understand it; if we can draw up a quantitative account based on accurate measurements, chances are not only that we will understand it, but that the previous qualitative "understanding" was in error. Quantification produces surprises!

EXISTENCE METABOLISM

Realizing that basal metabolism does not tell us anything about the energetics of existence in the real world, we can decide to measure the total cost of living in a cage. There normal activities add, above the physiological baseline of basal maintenance, the increments of feeding, preening, thermoregulation, and some movements. This environment probably comes fairly close to the natural scenario for small mammals, if they have nesting materials and space to run around; it is perhaps less representative for birds, unless the cage is large enough for flight.

The relationships in Table 8-3 were obtained from continuous monitoring of oxygen consumption by small mammals, from monitoring of food

Table 8-3 Average daily metabolism in captivity and estimated in the wild. Units have been converted from the original reports.

Group	Ambient temperature	Metabolic rate (kJ/day)	Predicted metabolic rates (kJ/day)		References
			For 20 g	For 50 g	
Captivity (caged, existence "energy")					
Rodents	20° C	$8.57\ m^{0.54}$	43.2	70.8	Grodzinski and Wunder, 1975
Insectivores	20° C	$14.30\ m^{0.43}$	51.9	76.9	Grodzinski and Wunder, 1975
Passerine birds	30° C	$6.58\ m^{0.62}$	42.2	74.4	Kendeigh, 1969, 1970
Nonpasserines	30° C	$2.26\ m^{0.75}$	21.4	42.5	Kendeigh, 1969, 1970
Passerines and nonpasserines	0° C	$18.15\ m^{0.53}$	88.8	144.3	Kendeigh, 1969, 1970
Estimated from field observations					
Rodents	—	$7.42\ m^{0.67}$	55.2	102.0	King, 1974
Birds	—	$13.05\ m^{0.61}$	81.1	141.9	Walsberg, 1982
Doubly labeled water determinations					
Mammals	—	$8.40\ m^{0.66\pm0.08}$	60.7	111.0	Garland, 1983b
Iguanid lizards	—	$0.224\ m^{0.80\pm0.023}$	2.5	5.1	Nagy, 1982

consumption and waste production, and by using bomb calorimetry of food and waste materials to obtain calculated energy use.

Because of a relatively small size range, the 1-g intercepts and exponents for the metabolic expressions in Table 8-3 make a poor basis for comparison. If we project them to a prediction for an average 20-g small homeotherm, we see the sort of relative differences that were found for resting metabolism of mammals and birds: insectivores higher than rodents, passerine birds higher than nonpasserines, and an inverse relationship between metabolic rate and ambient temperature. These predictions are, of course, lower than those estimated from time and activity observations in the field, so have limitations that leave us a bit short of our goal of understanding what is really happening in nature.

TOTAL ENERGY REQUIREMENTS

Unfortunately the technology of field metabolic determinations is sufficiently complicated that there is not enough unassailable information for a full-scale allometric summary. Most of the field techniques (reviewed in Walsberg, 1982, 1983) utilize a composite estimate from observations of daily time budgets. The amount of time spent in a given activity or inactivity is then multiplied by metabolic rates measured in the laboratory at comparable environmental temperatures and activity levels, with intermediate assumptions such as "perching is 1½ times basal." It is hoped then that the total of the estimates for various categories at various times approximates what the animal needs for a day in its natural environment.

The diurnal and conspicuous nature of bird behavior has resulted in there being more estimates for birds than for mammals. The total estimated daily energy expenditure ($\dot{E}_{tot}$ in kJ/day) of 42 species of free-living birds, studied by almost that many different investigators (not all using the same techniques), shows a good correlation ($r^2 = 0.981$, $p < 0.001$, $S_{yx} = 0.415$, $S_b = 0.0199$; Walsberg, 1982):

$$\dot{E}_{tot} = 13.05 \ m^{0.61} = 882 \ M^{0.61}. \tag{8-39}$$

Walsberg then separates birds that forage in flight from those that feed from a perch or the ground, obtaining separate expressions with higher and lower intercepts but the same scaling exponents (Table 8-4, Figure 8-5). For 18 species of rodents, a comparable correlation ($r^2 = 0.903$; $p < 0.001$; $S_{(\log y)(\log x)} = 0.104$; $S_b = 0.213$; King, 1974) is

$$\dot{E}_{tot} = 7.40 \ m^{0.67} = 757 \ M^{0.67}. \tag{8-40}$$

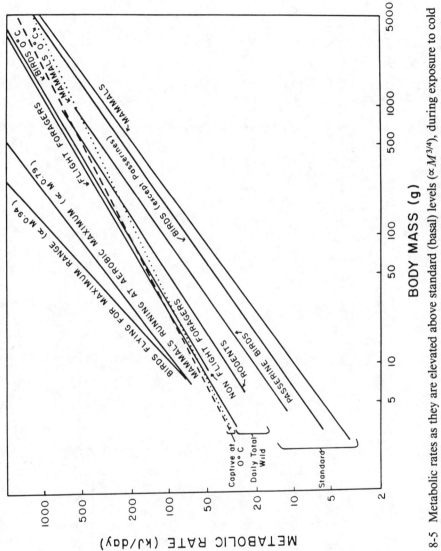
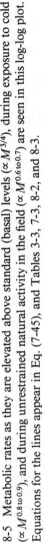

8-5 Metabolic rates as they are elevated above standard (basal) levels ($\propto M^{3/4}$), during exposure to cold ($\propto M^{0.8 \text{ to } 0.9}$), and during unrestrained natural activity in the field ($\propto M^{0.6 \text{ to } 0.7}$) are seen in this log-log plot. Equations for the lines appear in Eq. (7-45), and Tables 3-3, 7-3, 8-2, and 8-3.

Table 8-4 Allometric equations for avian energetics, where m = body mass in grams. (From Calder, 1983d.)

	Milliwatts	Kilojoules/day	Relative to standard power[a]	%	References
Flight metabolic rate, passerines and nonpasserines	$341.4\,m^{0.73}$	—	$7.32\,m^{0.01}$ p		Calder, 1974[b]
			$11.91\,m^{0.00}$ np		Walsberg, 1982
Portion of active day spent flying				$44.3\,m^{-0.603}$	Walsberg, 1982
Estimated total energy per day, "all birds"		$13.05\,m^{0.605}$	$3.24\,m^{-0.10}$ p		Walsberg, 1982
			$5.26\,m^{-0.11}$ np		
Flight foragers		$14.17\,m^{0.607}$	$3.52\,m^{-0.10}$ p		Walsberg, 1982
			$5.71\,m^{-0.11}$ np		
Nonflight foragers		$12.84\,m^{0.610}$	$3.19\,m^{-0.10}$ p		Walsberg, 1982
			$5.18\,m^{-0.11}$ np		
Resting metabolic power at 0°C					
Passerines	$230.9\,m^{0.42}$	$19.95\,m^{0.42}$	$4.95\,m^{-0.30}$		Calder, 1974[b]
Nonpasserines	$161.8\,m^{0.53}$	$13.98\,m^{0.53}$	$5.64\,m^{-0.20}$		Calder, 1974[b]
Standard (basal) metabolic power					
Passerines	$46.63\,m^{0.72}$	$4.03\,m^{0.72}$	$1.0\,m^{0}$		Calder, 1974[b]
Nonpasserines	$28.67\,m^{0.73}$	$2.48\,m^{0.73}$	$1.0\,m^{0}$		Calder, 1974[b]

a. p = passerine, np = nonpasserine.
b. Units have been converted from the original papers.

Garland (1983b) gives a similar equation based on doubly labeled water turnover:

$$\dot{E}_{tot} = 802 \ M^{0.66 \pm 0.08}. \tag{8-40a}$$

The exponents for the two classes do not differ significantly, but the coefficients do, with the $\dot{E}_{tot}$ of birds being higher than that of mammals of the same size.

The mass exponent for $\dot{E}_{tot}$ of birds is significantly lower than that for the standard metabolic rate, but the same distinction cannot be made for mammals at this time. It is no surprise that life in the wild appears from the foregoing to require more energy to support a combination of higher activity levels and the need to thermoregulate. We saw earlier that the allometry for resting metabolic rates during exposure to $0°C$, for birds and mammals as well, had exponents consistently smaller than ¾, which tells us that the increase above the standard level is greatest for small animals and not so significant for larger animals. Thus the scaling of $\dot{E}_{tot} \propto M^{-0.6}$ could reflect a mixture of time spent at thermoneutrality ($M^{0.75}$) and in cold stress ($M^{-0.5}$).

The intermediate scaling of $\dot{E}_{tot}$ for birds can also be approximated by a combination of times spent at basal metabolism and at flight levels, even though these may both carry essentially the same scaling ($M^{0.73}$; see Table 7-3). This occurs because the proportion of the active day spent flying is size dependent ($\propto M^{-0.60}$; see Table 8-4). To appreciate the combined effect, an oversimplified budget is postulated for seven hypothetical birds, which are assumed to fly for the size-dependent fraction of a day and drop to the standard metabolic level when not flying, with no thermoregulatory demands. A regression on the values for their total energy estimates as a function of the sizes used (hummingbird to goose) was (Calder, 1983c):

$$\dot{E}_{tot} = 5.87 \ m^{0.67}. \tag{8-41}$$

While both the coefficient and the exponent are less than what would be predicted from Eq. (8-39), it is evident that an inverse scaling of percentage of time spent in flight can "pull down" the exponent from the 0.73 for both resting and flying.

The general picture from the allometry of metabolic support (minute ventilation, cardiac output, glomerular filtration rate) was of an approximately $M^{3/4}$ scaling—not, as we see, for metabolic allometry when exposed to cold. The maximum recorded metabolic rates (summit metabolic rate, J/hr) during cold stress were summarized (Calder, 1974) as

$$\dot{E}_{max} = 797 \ m^{0.65}, \tag{8-42}$$

This is 1.47 $m^{0.04}$ times the average hourly rate estimated for wild birds (Eq. 8-39) and 0.65 $m^{-0.08}$ times the rate in flight. However, Eq. (8-42) for summit metabolism underestimates the metabolism observed in the careful experiments of Dawson and Carey (1976), who recorded sustained oxygen consumption rates during $-70°C$ exposure that were 62% above the prediction from Eq. (8-42) for a 12.7-g goldfinch and 36% above the prediction for a 13.8-g pine siskin (*Spinus tristus* and *S. pinus,* respectively). At least for this size, the $\dot{V}_{O_2max}$ in shivering is about the same as would be predicted for flight.

The Dimension of Time

Considerations such as feeding conditions, heat balance, and avoidance of enemies have led to specialization for either diurnal or nocturnal habits. From nightfall, the diurnal animal must wait a half-day before there is sufficient light to feed again, at which time the nocturnal animal must hide and wait many hours until it is dark again. These long waits are about the same whether the animal is large or small. However, the smallest land mammal is living about 32 times as fast as the largest. Thus the earth's rotation may seem interminably long to a shrew and insignificantly short to a busy elephant. As A. V. Hill (1950) stated the problem, "The physiological time-scale of an animal has to compromise with the constant time-scale of the external world." Animals must have physiological mechanisms or behavioral patterns that help the physiological time, $\propto M^{1/4}$, to mesh with the indifferent environment.

This interfacing of physiological and environmental time scales seems to be the role of the "biological clock" or circadian rhythm, a physiological anticipation of the environmental cycles that the animal must exploit to survive. This was concluded independently by Lindstedt and Calder (1981) and by Heusner (1982b). There is now a fascinating and extensive literature on chronobiology, the study of biological rhythms observed in a wide variety of animals and plants. In considering *The Clocks That Time Us,* Moore-Ede and his colleagues (1982, p. 4) mention the frequency spectrum of endogenous rhythmic processes of animals, from electroencephalographic and cardiac to circannual periodicities. They do not distinguish between the cycles that are scaled allometrically and those that must synchronize with the environment. Endogenous physiological rhythms are related to the exogenous cycles which they approximate and to which they are entrained. Ultimately the clock must have a cellular basis. It would be a simple matter to hypothesize a class-wide cellular oscillation that fits a 24-hr cycle if all mammals had the same biochemical times and rates incorporated in their physiology. What is different about the cellular mechanism of man, who takes roughly 7.3 times as long as a mouse to turn over the same proportion of his body fluids or chemical substrates? The physiological time of a 3,500-kg elephant is 31.6 times that of a 3.5-g shrew or hummingbird; what gears their biochemistry to the same 24-hr cycle of daylight and darkness?

In hard times a small business is more likely to fail than a large corporation with huge reserves. In the same way, the smallest animals would seem to have the greatest problem in extending their resources until opportunities reappear. The solutions are of two types. One is to conform to the daily cycle of the environment; the other is to subdivide the 24-hr day into several of the organism's own cycles. Opting for one solution or the other is related to many considerations such as cyclicity of food supply, which of the senses is relied upon for feeding, and what mechanisms there are for storing and conserving energy during the fasting half of a diel cycle. Those who enjoy pondering the "strategies" of Nature will find it fortunate that Nature took one alternative for the smallest birds and the opposite for the smallest mammals. It seems appropriate at this point to diverge on a "narrow allometry" (in the sense of Smith, 1980), or a "vertical allometry" (Chapter 12) to compare the two groups.

The most highly developed sense in birds is vision, and foraging is guided largely by this sense, whereas the sense of smell is of greater significance for most mammals. The smallest birds, hummingbirds, must therefore wait for light intensities of about 4 lux to begin feeding in the morning (Calder, 1975). Apparently the energetic demands for these tiny birds are strong enough that this daily "first-light" search is almost independent of length of day. Except in the extremely long days experienced in Alaska by the rufous hummingbird *(Selasphorus rufus)*, hummingbirds take just about all the daytime they can get (Figure 8-6). In fact, if heavy rainfall forces a female hummingbird to remain on the nest for extended periods, thereby reducing her feeding time for the day, it is often necessary for her to go torpid in the middle of the night, dropping her body temperature in an effort to save enough energy to resume feeding flights at dawn (Figure 8-7). Both the entry into torpor and the arousal from it appear to place heavy reliance upon a sense of time. How long must the available energy supply (in fat and crop contents) last? Once torpid, when should arousal begin in order to be ready to fly at first light? In fact, the "decision" to enter must involve not only a biological clock but a biological fuel gauge (Calder and Booser, 1973).

One reason the larger homeotherms (> 1 kg) do not find hypothermic

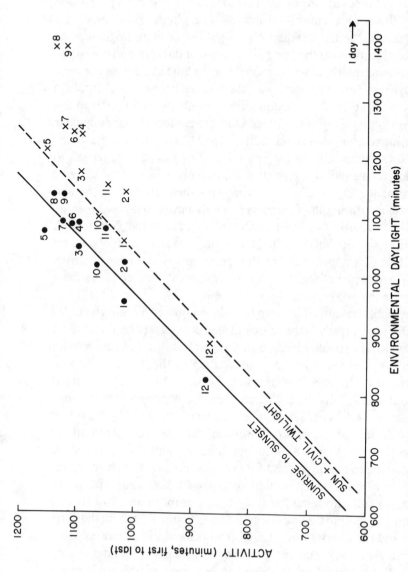

8-6 The active day of North American hummingbirds extends as late as daylight is available for feeding, except at the highest latitudes colonized by these tiny nectar burners. The solid and dashed lines are least-squares regressions for data from Anna's, black-chinned, calliope, broad-tailed, and ruby-throated hummingbirds. The numbered points are from rufous hummingbirds at different breeding and migratory locations; two points appear for each, ● relative to sunrise and sunset and ✕ relative to solar daylength plus periods of civil twilight. (From Calder, 1976b, p. 32; with permission of the *International Journal of Biometeorology*.)

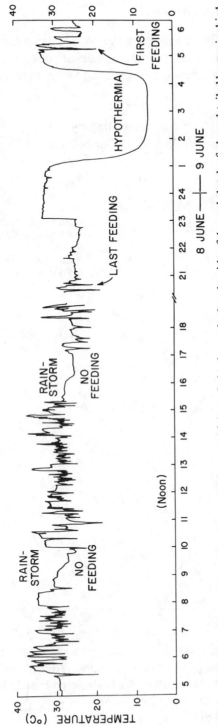

8-7 Nest temperatures recorded from a synthetic egg, which tended to be pushed to the side of the real clutch of a broad-tailed hummingbird at 2,910 m in the Colorado Rockies. At about 0520 the female began feeding and the "egg" cooled in her absence. Between feedings she rewarmed the nest. Rainstorms from 0830 to 0945 and from 1520 to 1710 kept the hen on the nest, thereby depriving her of feeding, so that at the day's end she had missed about one-fifth of her normal number of feedings. This deficit apparently necessitated entry into hypothermic torpor at 0100 to conserve her remaining energy. Arousal from this state began soon after 0400, so that the bird was at a normal temperature for flying to resume at daybreak. The abrupt temperature rise at 2300 was probably due to movement of the sensor egg to a more favorable position. (Data from Calder and Booser, 1973.)

torpor useful is that it takes them a long time to cycle in and out of normal and hypothermic states. Torpor entry in three species of hummingbirds and a poorwill (Figure 8-8) approximated a relationship for mean cooling time:

$$\text{min/}^\circ\text{C} = 2.08 \ m^{0.51}; \qquad r^2 = 0.85 \tag{8-43}$$

(Lasiewski and Lasiewski, 1967). For rewarming by metabolic heat in a sample of birds and mammals, Heinrich and Bartholomew (1971) showed

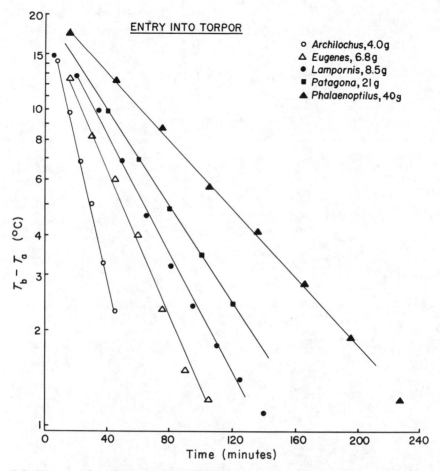

8-8 The rate of cooling is inversely related to body mass. This means that in a 12-hr night a 4-g hummingbrd could spend 10.7 hr at the energy-saving torpor temperature, whereas a 40-g poorwill *(Phalaenoptilus)* would have only 6 hr at the lowest T_b. Arousal times (rewarming) are scaled similarly. (From Lasiewski and Lasiewski, 1967, p. 41; with permission of *Auk*.)

that the rate of warm-up was

$$°C/min = 2.03 \ m^{-0.40} \tag{8-44}$$

(see Figure 8-9). When insect data were included with bird and mammal data, this became

$$°C/min = 3.22 \ m^{-0.51}. \tag{8-45}$$

Thus it is advantageous to be small if the energy conservation of hypothermic torpor is to be utilized: the smaller the animal, the faster it gets down to a lower temperature and reduced metabolic rate, and the more rapidly it can return to normal life.* Size, timing, and the nature of the food supply are all tied together in Figure 8-10. The consequence of failure to integrate these factors successfully is seen in Figure 8-11.

Now consider shrews. They live largely out of sight, in and under the litter of the forest floor, in crevices or old root channels, under banks or logs, where their predominantly insect diet can be secured. They have a keen sense of smell and also use echolocation, so are not dependent upon ambient light (Lorenz, 1953; Crowcroft, 1954; Gould et al., 1964; Lindstedt, 1980a). Some species tend to be more active before dawn and less active in late morning. Recordings from six species of Soricinae show that a 24-hr day is divided into many periods of activity amounting to cycles of sleep and wake (Table 8-5). The number of activity periods varies widely, even in one individual, from day to day (column 4), but appears to be inversely related to body size. From what is known of physiological time, we might expect the reciprocal of activity frequency (mean time from each activity period to the next one), to scale as $m^{1/4}$. In fact, it does — $m^{0.27}$ — but the correlation is not significant over this small fivefold body-size range.

One exception to this subdivision of the 24-hr day into a series of activity periods extending through both day and night is the gray shrew, *Notiosorex crawfordi,* which has restricted most of its activities to the nighttime, in apparent response to the daytime heat of its desert scrub environment (Lindstedt, 1980a,b). This pattern of reliance primarily on nocturnal activity is characteristic of the other shrew subfamily, the Crocidurinae. It is interesting that the shrews that conform to a day-night cycle as hummingbirds do (except with the opposite phase) are also capable of hypothermic torpor as an energy-saving mechanism. The Crocidurinae have lower stan-

---

* Avian T_b during hypothermia has been correlated with previous weight loss (Cheke, 1971), characteristic minimum T_a (Hainsworth and Wolf, 1978), and body size, smaller birds regulating at lower T_a (Calder and King, 1974).

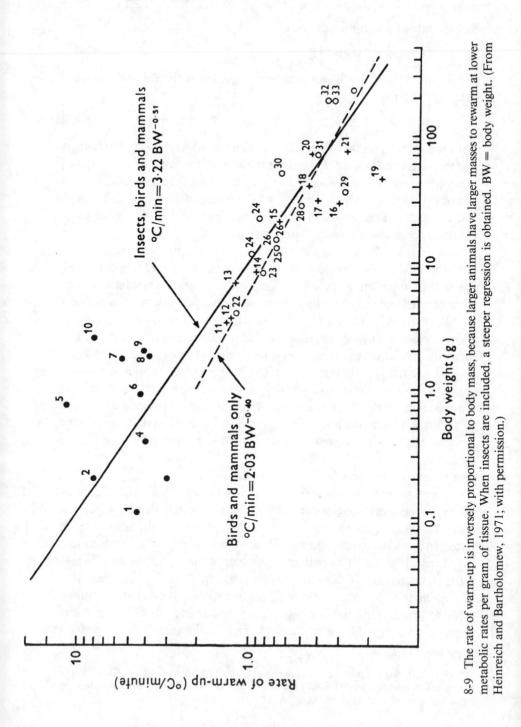

8-9 The rate of warm-up is inversely proportional to body mass, because larger animals have larger masses to rewarm at lower metabolic rates per gram of tissue. When insects are included, a steeper regression is obtained. BW = body weight. (From Heinreich and Bartholomew, 1971; with permission.)

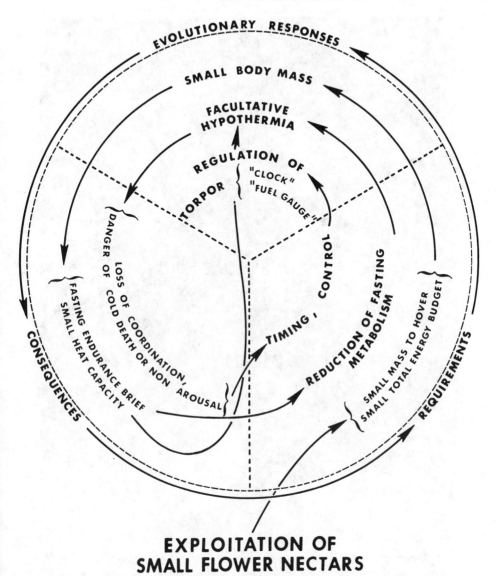

EXPLOITATION OF
SMALL FLOWER NECTARS

8-10 The requirements, the evolutionary responses, and their consequences in the adaptation to nectar feeding by hummingbirds, an integration (or adaptive suite) of flower nectar droplet yields, hummingbird body size, and energy conservation. (From Calder, 1974; with permission of the Nuttall Ornithological Club.)

8-11 When the rate of energy or water expenditure is not balanced by the rate of replacement, the hummingbird enters a final bout of torpor, proceeding to equilibrium and the end of fitness. This mummified carcass was found by William Calder IV, M.D., in Cañon del Oro, Santa Catalina Mountains, Arizona.

dard metabolic rates than the Soricinae in general, and they have longer life spans, at least in captivity. How many of these characteristics evolve as an "adaptive suite"? The only soricine shrew known to have a nocturnal-active diurnality, *Notiosorex* is also capable of hypothermic torpor (Walker, 1975; Lindstedt, 1980a,c; Vogel, 1980). Thus in the vertical allometry, or *y*-diver-

Table 8-5   Activity periods of shrews (Soricinae).

Species	Body mass (g)	Activity periods per day Average	Range	References
Sorex cinereus	3.6	16	13–19	Pearson, 1947; Buckner, 1964
S. minatus	4.9	24.6	19–31	Crowcroft, 1954
S. arcticus	5.4	14	—	Buckner, 1964
S. araneus	7.1	13.7	12–17	Crowcroft, 1954
Neomys fodiens	13.9	14.9	12–19	Crowcroft, 1954
Blarina brevicauda	20.1	10.5	10–11	Pearson, 1947; Buckner, 1964
Allometry summary		$26.0\ m^{-0.27}$		($r^2 = 0.41$, not significant)

sity, of the smallest homeotherms we see that A. V. Hill's "compromise with the constant time-scale of the external world" can be made in two different ways, but apparently when the "choice" is made, it is not for just one characteristic, but for several interrelated ones in this adaptive suite.

Rather than being a continuous period of hypothermia, mammalian hibernation is interrupted by short episodes of warming to normal (euthermic) body temperatures. The frequency of such midwinter arousals depends upon metabolic rate during the hypothermic torpor, and appears to be related to the need to restore chemical balances. The euthermia is more costly energetically than hypothermia, so it would be desirable to return to the latter state as soon as possible. French (MS "a") showed that the duration of the euthermic episode (in hr) that interrupts the midwinter hibernation of rodents was size dependent for 71 observations from four species, ranging in size from 24 g to 4,000 g:

$$t = 1.44\ m^{0.32}; \qquad r^2 = 0.92. \tag{8-46}$$

This scaling is within the range of other physiological times, a scaling no doubt selected with regard to metabolic regulatory requirements on the one hand, and energy reserves on the other (Morrison, 1960). If the scaling of stored fat in hibernators is similar (parallel) to that of mammals in general (see Eqs. 3-31 and 3-32), and if the metabolic rates in hibernation scale in parallel to the Kleiber equation for homeothermy (Eq. 3-13), the calculated endurance time would be

$$t \propto M^{1.13}/M^{0.76} \propto M^{0.37}, \tag{8-47}$$

or $t \propto M^{1.19}/M^{0.76} \propto M^{0.43}. \tag{8-48}$

Assuming for the moment that the exponents are precise and significant, these equations put the euthermic arousal time between a simple mass-specific metabolic scaling and a fat-endurance scaling. As French pointed out, this scaling is also the approximate inverse of that for mass-specific metabolic rates. While the hypothermic torpor is a physiological departure from the normal active state, it is a quantitatively proportional departure for different sizes of mammals, as well as a departure that is approximately proportional to the regular homeothermic timetable.

Finally, springtime emergence from hibernation appears also to be size dependent. Larger hibernating mammals can afford to gamble on early emergence, having relatively more energy reserves left. This earlier emergence is desirable because the pace of growth of the new season's offspring is slower (see Chapter 10). Reproduction early in the season is necessary if the young are to attain adequate growth and fat storage for their first hibernation (French, MS "b").

Sensory Design: Hearing

The sense of vision is limited to a relatively small part of the electromagnetic spectrum in which the photons of light have sufficient energy to activate molecular reactions (photons in the infrared do *not*) but not so much that they will damage molecular and cellular structures (photons in the ultraviolet and beyond *do* cause ionized damage). The wavelengths of visible light are so short, 400 to 700 nm, that body-size dimensions are irrelevant; that is, body size is not a factor limiting how far into the long (near infrared) or short (ultraviolet) end of the spectrum the animal can see.

Size, however, is a significant factor for the sense of hearing. This is apparent from Figure 8-12, taken from a fascinating study of hearing in the Asiatic elephant *(Elephus maximus)* by Heffner and Heffner (1980). If an audiologist tests your ears, the job is simplified by the fact that both of you speak the same language. The elephant had to be trained to "communicate" her ability to hear different frequencies and intensities of sound by pushing a button after a tone was presented, in order to be rewarded with 30 ml of fruit-flavored drink. The resulting audiogram had a typical mammalian form, but the elephant could hear nearly an octave lower than man at an intensity level of 60 dB. Her upper limit was also nearly an octave lower.

Thus the largest terrestrial animal tested has a sense of hearing that functions at lower frequencies than in other mammals (see also Figure 8-13).

Heffner and Masterton (1980) and Heffner and Heffner (1980) found a strong inverse relationship between the high-frequency cutoff (at 60 dB) and the time difference between arrival of sounds at the two ears in a series of mammals, bat and mouse to cow and elephant (Figure 8-12).

Between the time that your left ear hears a sound that originated directly to your left and the time your right ear hears it, there is a brief delay (Δt) which is equal to the distance between your ears divided by the speed of sound. This time difference is one of the means by which you have stereophonic hearing (the others being differences in intensity and differences in sound spectra caused by obstructions such as your head or an animal's pinna).

It turns out that there is a strong inverse correlation ($r^2 = 0.787$; $p < 0.001$) between the Δt of sound arrival and f_{max}, the high-frequency cutoff at 60 dB (Heffner and Masterton, 1980; Heffner and Heffner, 1980). This has the form

$$f_{max} = 502 \, \Delta t^{-0.44}. \tag{8-49}$$

Plotted on log-log coordinates, as in Figure 8-12, this formula gives a straight line. Since this Δt is a linear dimension divided by a constant, the speed of sound, and since the frequency or pitch of a sound is inversely related to the length of a harp string or a drum diameter, the inverse quality of this relationship is what we would expect.

H. E. Heffner kindly supplied me with the original data from which Eq. (8-49) was derived, so that Δt (in microseconds) and f_{max} could be expressed as separate functions of body size:

$$\Delta t = 195 \, M^{0.295} \tag{8-50}$$

($n = 31$; $r^2 = 0.923$; $p < 0.001$, $S_{yx} = 0.0232$; $S_b = 0.0158$). The upper limit, f_{max} (in kilohertz) at 60 dB, scales as

$$f_{max} = 50.2 \, M^{-0.13} \tag{8-51}$$

($n = 31$; $r^2 = 0.743$; $p < 0.001$; $S_{yx} = 0.0209$; $S_b = 0.0142$).

So mice squeak and elephants bellow! Another property of sound related to the difference in frequencies at which animals communicate is the attenuation of sound with distance, as the result of internal friction and heat conductance of gases. If I_o was the intensity at the source of the sound, then I_x, the intensity at distance x, has attenuated exponentially:

$$I_x = I_o \, e^{-\alpha x}. \tag{8-52}$$

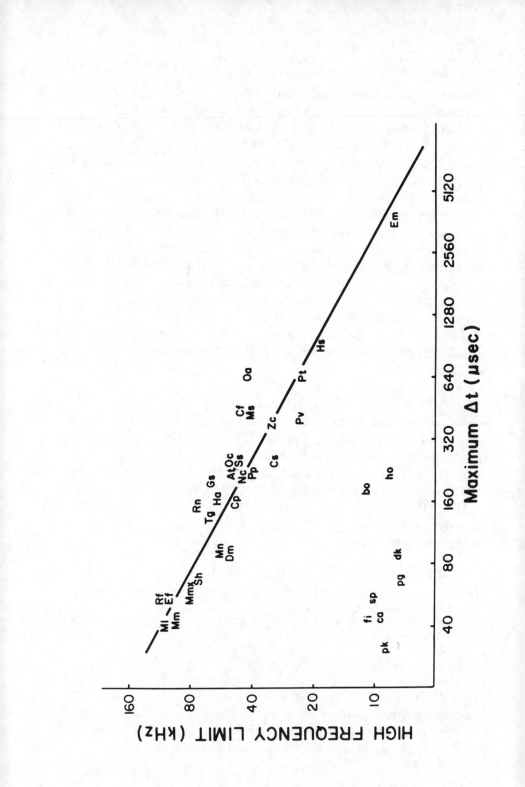

MAMMALS:

Ml = *Myotis lucifugus* (bat)
Mm = *Mus musculus* (feral mouse)
Rf = *Rhinolophus ferrumequinum* (bat)
Ef = *Eptesicus fuscus* (bat)
Mmx = *Mus musculus* (laboratory mouse)
Sh = *Sigmodon hispidus* (rat)
Mn = *Meriones unguiculatis* (gerbil)
Dm = *Dipodomys merriami* (kangaroo rat)
Tg = *Tupaia glis* (tree shrew)
Rn = *Rattus norvegicus* (laboratory rat)
Cp = *Cavia porcellus* (guinea pig)
Ha = *Hemiechinus auritus* (hedgehog)
Gs = *Galago senegalensis* (bush-baby)
At = *Aotus trivirgatus* (monkey)
Oc = *Oryctolagus cuniculus* (rabbit)
Nc = *Nycticebus coucang* (loris)
Ss = *Saimiri sciureus* (monkey)
Pp = *Perodicticus potto* (potto)
Cs = *Chinchilla sp.*

Zc = *Zalophus californianus* (sea lion, in air)
Pv = *Phoco vitulina* (seal, in air)
Ms = *Macaca sp.* (macaque)
Cf = *Canis familiaris* (dog)
Oa = *Ovis aries* (sheep)
Pt = *Pan troglodytes* (chimpanzee)
Hs = *Homo sapiens* (man)
Em = *Elephas maximus*

BIRDS:

pa = parakeet (*Melopsitticus undulatus*)
fi = finch (*Carpodacus mexicanus*)
ca = canary (*Serinus canaria*)
sp = sparrow (*Metospiza melodia*)
pg = pigeon (*Columba livia*)
dk = duck (*Anas platyrhynchos*)
bo = barn owl (*Tyto alba*)
ho = horned owl (*Bubo virginianus*)

8-12   The highest frequency that mammals can hear (standardized at an intensity of 60 dB) is inversely related to the time difference in arrival of a sound at the maximum interaural distance (longest dimension between the ears). The mammals range in size from a 6-g bat and a 14-g mouse (Ml and Mm) to the Asiatic elephant (Em). No such correlation seems to occur in birds (lower-case identifiers), which vary in size from a 14-g sparrow to a 2-kg duck. From left to right the species are distributed in approximate order of body size (see Eq. 8-50). (Redrawn after Heffner and Heffner, 1980, with the addition of bird data from Dooling, 1980, after Knudsen, 1980. Original figure copyright 1980 by the American Association for the Advancement of Science; with permission.)

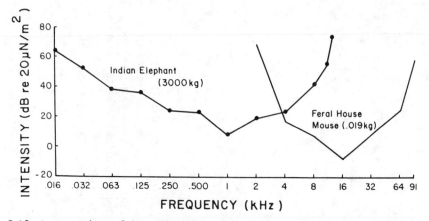

8-13 A comparison of the audiograms of the elephant and the mouse. (Redrawn from Heffner and Masterton, 1980, and Heffner and Heffner, 1980; with permission. Elephant curve copyright 1980 by the American Association for the Advancement of Science; with permission.)

When the loss in intensity is due only to internal friction and heat conduction, the attenuation (α) depends on the frequency:

$$\alpha = kf^2. \tag{8-53}$$

Thus higher frequencies attenuate more rapidly (Meyer and Neuumann, 1972). The territory and home-range size of animals are approximately linearly related to body mass, so the fact that the mouse's squeaks only travel a short distance is all right, at least qualitatively.

Since the difference in intensity detected between the two ears is another cue for directionality, we might examine the regressions of the Heffners' data further. According to the Weber-Fechner Law, the minimum detectable difference between I_x and I_o is a constant ratio. Substituting Eq. (8-53) into a rearrangement of (8-52) we have:

$$I_x/I_o = \text{constant} = e^{-kf^2x}. \tag{8-54}$$

Considering together the relationships of attenuation to frequency and of frequency to body size (Eqs. 8-51 to 8-54) it appears that the barely perceptible difference in intensity, the attenuation from I_o to I_x, occurs over progressively shorter distances as frequency increases or as body size decreases. The implications of this scaling for communicating or advertising claims to a home range or territory will be examined in Chapter 11.

The body-size-dependent shift in audiograms for upper cutoff frequency appears to be approximately paralleled by the scaling of the frequency of maximum sensitivity (f_{best}, in kilohertz), although the correlation is not quite so good ($r^2 = 0.272$; $0.001 < p < 0.01$; $n = 28$; data from H. E. Heffner, personal communication):

$$f_{best} = 6.9 \, M^{-0.14 \pm 0.046}. \tag{8-55}$$

The agreement in exponents between f_{max} and f_{best} is excellent; despite the fact that audiograms obtained in octave steps can only provide a crude approximation to the best frequency, the standard errors of the exponents overlap.

If these characteristic frequencies were related to the natural frequency (f_{nat}) of a simple first geometry (see Chapter 7) wherein $f_{nat} \propto l^{-1}$, we would expect a body-mass scaling of $M^{-1/4}$ if elastic similarity were maintained, or $M^{-1/3}$ if geometric similarity prevailed. Neither seems to be the case. The linkage between the eardrum and the oval window of the cochlea consists of three articulating ossicles. These serve functions of impedance matching (Prosser, 1973, p. 526) and amplification (Eckert and Randall, 1983, p. 253), and may in part determine the frequency response characteristics of the ear (Heffner and Masterton, 1980). The last of these suggested functions may be related to the fact that the articulations of the ossicles would be categorized by McMahon's (1975) description of second or third geometry, with one or more joints in the linkage. The natural frequency of second- or third-geometry systems is proportional to diameter/l^2. In the case of geometric similarity, the frequency would be scaled to $M^{-1/3}$, as in the case of a first geometry. However, if an elastic or dynamic similarity has been preserved:

$$f_{nat} \propto dl^{-2} \propto (M^{3/8})(M^{1/4})^{-2} \propto M^{-0.125}. \tag{8-56}$$

This is indistinguishable from the best frequency (Eq. 8-55) and from high-frequency cutoff scaling (Eq. 8-51).

The sounds important to an animal are those of intraspecific communication and environmental sounds that may provide information important for survival. The problems of size dependency in intraspecific communication were commented upon by Moog (1948) in a comparison of normal-sized humans (such as Gulliver), the tiny Lilliputians, and the huge Brobdingnagians. If Gulliver's vocal cords were of normal length with their vibrations centered at 256 Hz (middle C), a first-geometry scaling of $f \propto l^{-1}$, Lilliputians scaled down linearly by a factor of 12 would have voices at 37 kHz, seven octaves above Gulliver's voice, while the Brobdingnagians would center their voices at 3 Hz. Either extreme would be beyond Gulliver's usual hearing range.

In addition to intraspecific communication, in which size-dependent isolation has a neutral to beneficial effect, there is the problem of coupling to the environment. Is the roar of an approaching thunderstorm, forest fire, or herd of elephants significant information for a mouse or shrew? Can a large herbivore benefit from the calls of birds which have taken note of a foraging lion? Would it be better to reduce the scaling from $M^{-0.25}$ of a first geometry to $M^{-0.13}$ of a third geometry, to provide greater overlap in the audible frequency ranges of large and small mammals? An allometric examination of Heffner's data may yield insight into the evolution of hearing in mammals.

First, note that the body-mass exponents are essentially the same, which means that in general the evolution of a higher high-cutoff frequency tends to be coupled with evolution of a higher "best" frequency, or that the audible range is selected as a "package" or adaptive suite, with a ratio of f_{high}/f_{best} of $50.2/6.9 = 7.3$. Comparison of high/best ratios for particular species show quite a range, from 2.6 (potto) and 2.8 (slow loris) to 24.5 (rabbit) and 52 (kangaroo rat); perhaps these extremes would be a fruitful point of departure for further research. Some of the variation is no doubt caused by difficulties in the precise determination of the best frequency. Thus it would be useful to compare the departures of observed frequencies from the allometric prediction for the species size.

The extremely high ratio for the kangaroo rat is not due to specialization at the high limit, but rather at the best frequency, 1 kHz, versus the predicted 11 kHz. Kangaroo rats have very large auditory bullae which, according to Webster (1962, 1975), serve as resonating chambers that enhance sensitivity to low frequencies that may be emitted in the attacking flight of an owl or the strike of a rattlesnake. The best frequency of 1 kHz observed by Heffner seems to confirm this. The gerbil and the rabbit may have similar needs for hearing low frequencies of their predators in open country at night.

Bats have good use for a short-wavelength, high-frequency sonar system to detect small insects in their flight path. All three species have higher high-frequency cutoffs than predicted from size alone. Thus while kangaroo rats "stretched" for a lower best frequency and sustained a proportionately much smaller loss at the high end, bats "stretched" for the higher high cutoff but lost even more of the lower part of the audible range, perhaps in the absence of selective pressure to "retain" an ability to hear owls, which are not likely to outmaneuver them.

If this explanation of the scaling of high-frequency cutoff and best frequency to $M^{0.13 \text{ to } 0.14}$ is worth pursuing, one must consider that there does not appear to be the same kind of inverse relationship between high-frequency cutoff and interaural time interval in birds that there is in mammals (Knudsen, 1980; see also Figure 8-12). Knudsen used the data of Dooling (1980) to show no size-dependent trend in birds from the size of the field sparrow *(Spizella pusilla)* and canary to the great horned owl, a hundredfold range that should reveal a size dependence. It would be helpful to confirm this with birds ranging from the 3-g Costa's hummingbird *(Calypte costae),* which sings up to 13 kHz and would, I presume, be able to hear that frequency, and the ostrich—if high-frequency data were available from these birds. Knudsen concludes that either birds do not use the interaural intensity difference (ΔI) in sound localization or they are more sensitive to small ΔI, perhaps by virtue of mechanisms not possessed by mammals. However, a regression of high-frequency cutoff on body mass in birds is significant at the 5% level:

$$f_{max} = 7.0 \, M^{-0.08} \tag{8-57}$$

(data from Dooling, 1980; $n = 16$; $r^2 = 0.315$, $0.02 < p < 0.05$), so there is a possible size dependence that deserves further consideration.

9 Reproduction

The size of the organism in the cycle takes on a particular significance. Size is correlated with time, for in general large organisms have long cycles and furthermore the different parts of the cycle can be readily classified on the basis of their size characteristics.

—J. T. BONNER, *SIZE AND CYCLE,* (1965)

REDUCED TO the simplest terms, the life history of an animal comprises the receipt of a set of genes and an allotted lifetime in which to use that genetic information to survive and successfully reproduce, thereby passing the genetic information to the replacement generation. The animal's body is the physical vehicle that bridges two successive genetic transmissions. This has been accurately described by the statement that "a hen is only an egg's way of making another egg" (Samuel Butler, *Life and Habit,* 1877).

Correlation with Size

The life span can be subdivided into a period of growth and learning, a plateau of life's prime, and an eventual decline in ability to reproduce or even maintain the body functions. Reproduction, growth, and homeostatic maintenance all require energy. Since the homeostatic energy requirements at levels ranging from basal to maximal exertion seem to be proportional to $M^{-3/4}$ (Chapters 3, 5, and 7) and the pace of homeostatic processes is scaled to $M^{-1/4}$ (Chapter 6), we might expect to find a continuity or consistency in the scaling of aspects of reproduction and growth. After all, the mouse replaces itself with another mouse, not with an elephant, and a hummingbird need not lay an ostrich egg just to get another hummingbird.

Regardless of body size, a particular animal must cope with environmental factors such as day length, season length, and climatic ranges that are not synchronized to its $M^{3/4}$ metabolic rate or its $M^{1/4}$ physiological time. Studies of natural reproduction and the ecology of reproduction are difficult because so many different factors are involved; there is no simple relationship between reproduction and, say, temperature. For example, our efforts to understand the breeding biology of broad-tailed hummingbirds have only scratched the surface, partly because no two years have been climatically, chronologically, or biologically the same (Calder et al., 1983).

Despite these complexities, to begin unraveling the mysteries we need an allometry of reproduction and growth. Field studies usually do not include body-mass records, so the data do not provide as neat an allometric pattern as those generated in the laboratory. However, it is in real life and not in the laboratory that biology has meaning, and for this reason I shall grope for some first approximations to principles of reproductive allometry.

The evolutionary ecologist's objective in studies of reproduction is the measurement of fitness, the differential success in reproduction of genotypes. The animal that leaves the most offspring has the greatest "fitness." This fitness has several components: the general health and vigor of the animal, its ability to select the best mate, and the timing and reproductive behavior—the reproductive strategy. Comparative physiologists study functions and mechanisms as if they were engineering problems and speak of *design.* Evolutionary ecologists may be uncomfortable with this term because it smacks of teleology; they prefer instead to talk in terms of *strategy,* which in turn sounds anthropomorphic to the physiologist. My goal is to elucidate a set of constraints imposed by the principal variant, body size, on reproductive "strategy" or "design." To speculate on reproductive strategy in a comparative context before evaluating what body size alone would predict is to put the cart before the horse. For example, Stearns (1976) proposed that "the eggs should be laid at a size which yields the maximum growth on the parental investment." As we have already seen, egg size is primarily a function of hen size, at least within a phylogenetic group. A bird probably does not just decide to increase its egg size, independently of other scaling changes (Calder, 1979).

It seems reasonable to assume that reproductive fitness is size independent; for if by virtue of body size alone one animal could outreproduce another, the species of the size leaving the most offspring would eliminate species of other sizes using the same resources. The world would be populated with only one size—that one at the physical limits, maximum or minimum, depending on the sign of the scaling factor.

The problem of size-independent fitness can be approached in several different ways. One is to account for number of offspring as an index of reproductive fitness:

$$\Sigma n_{f_1} = (t_{ls}) \left(\frac{n}{\text{clutch or litter}} \right) \left(\frac{\text{clutches or litters}}{\text{year}} \right) \qquad (9\text{-}1)$$

where Σn_{f_1} is the total number of offspring in the f_1 generation per lifetime, t_{ls}. (Note that Σn_{f_1} of animal a divided by Σn_{f_1} of animal b would be an expression of relative reproductive fitness.) This relationship has made for interesting philosophical discussion (Haukioja and Hakala, 1979), but qualitative reflection does not seem to settle anything. A baseline for quantitative evaluation might result if allometric expressions could be found for each term in Eq. (9-1). It would then be possible to distinguish between adaptive enhancement of fitness and the consequences of size and ancestry. Table 9-1 attempts to accomplish this.

Static counts of eggs, chicks, pups, calves, cubs, fawns, kittens, and uterine scars are relatively easy to come upon in natural history studies, so there are records from a large variety of mammals and birds. Both litter and clutch sizes are inversely correlated with body size. The relatively small scaling exponents are statistically significant, despite the relatively large degree of adaptive variation around the general size trends, such as the tendency for larger clutches to exist at higher latitudes (Pianka, 1983).

It is more difficult to go from static counts to measurement of annual birth rates, because of the necessity for continuous records of known individuals. Since this is nearly impossible for various wild mammals, especially the smaller ones, much of the information has come from records of zoological parks; a treasury of such data has been collected by Eisenberg (1981). Again, with over half the variability resulting from adaptive and phylogenetic factors rather than size, there is a significant inverse correlation with size for the number of litters per year. Of the hundred species of terrestrial eutherian species for which body mass is given in Eisenberg's appendix 2, 45 had only one litter per year, animals ranging in size from 8 g to 1,277 kg—although the equation (Table 9-1) predicts this specifically for a 14.1-kg mammal. The mean interbirth interval (t_{ib}) scales similarly to other physiological times:

$$t_{ib} = 6.95 \, M^{0.19 \pm 0.023 s_b}; \qquad r^2 = 0.511, \, n = 67. \qquad (9\text{-}2)$$

Table 9-1 A preliminary allometry of reproduction, expressed as functions of adult body mass. Parentheses indicate an assumption that the general relationship for the class holds for the smaller taxa within.

Taxon	Litter or clutch size		Litters or clutches/yr[a]		Births/yr		Maximum life span[b]		Potential offspring/lifetime
Eutherian mammals	$3.43\,M^{-0.16}$[g]		—	×	$1.26\,M^{-0.33}$[c]	×	$11.6\,M^{0.20}$	=	$14.6\,M^{-0.13}$
Terrestrial eutherians[d]	$2.4\,M^{-0.12}$	×	$1.42\,M^{-0.13}$	=	$3.44\,M^{-0.25}$	×	$11.6\,M^{0.20}$	=	—
	—		—		$3.90\,M^{-0.26}$				$45.2\,M^{-0.06}$
Birds	$3.6\,M^{-0.05}$[e]	×	$\propto M^{-0.17}$[f]		$M^{-0.22}$	∝	$17.6\,M^{0.20}$	∝	$M^{-0.02}$
Anatidae[f]	$7.0\,M^{-0.16}$	×	$(\propto M^{-0.17})$		$M^{-0.33}$	∝	$(17.6\,M^{0.20})$	∝	$M^{-0.13}$
					$M^{-0.33}$		$16.9\,M^{0.14}$		$M^{-0.19}$
Phasianidae[f]	$5.8\,M^{-0.14}$	×	$(\propto M^{-0.17})$		$M^{-0.31}$	∝	$9.4\,M^{0.22}$	∝	$M^{-0.07}$

a. Direct determination unknown; scaling exponent for litter or clutch frequency assumed to be the inverse of gestation or incubation period. It is also assumed that the period between broods scales the same way as gestation or incubation period. Subsequently confirmed by May and Rubenstein (MS), whose graph shows litter $f \propto M^{-1/8}$.

b. Maximum life spans from Sacher, 1959, for mammals ($n = 63$; $r^2 = 0.60$) and Lindstedt and Calder, 1976, for birds ($n = 152$; $r^2 = 0.49$); Anatidae: $n = 13$; Phasianidae: $n = 2$.

c. Ungulates, carnivores, small mammals, but no rodents; from Western, 1979. Births/yr: $n = 23$ spp., $r^2 = 0.98$.

d. Excludes Chiroptera, Pinnipedia, and Cetacea; data from appendixes 2 and 4 in Eisenberg, 1981. Litter size: $n = 281$; $r^2 = 0.286$; litters/yr: $n = 100$; $r^2 = 0.403$; births/yr: $n = 99$; $r^2 = 0.500$.

e. Calculated as clutch weight/egg weight from Western and Ssemakula, 1982.

f. From equations of Rahn et al., 1974, 1975.

g. Millar and Zammuto, 1983.

Of these hundred species, 63% had intervals of 12 months, predicted by Eq. (9-2) for a size of 17.2 kg only. Litter frequencies of one per year, and periods of 12 months, suggest that there is a compromise between the underlying physiological pace and the reality of adaptation to environmental seasons. If those cases are excluded, the correlations are improved considerably:

$$\text{litters/yr} = 1.59 \, M^{-0.19 \pm 0.023}; \qquad r^2 = 0.547, \, n = 55; \qquad (9\text{-}3)$$

$$t_{ib} = 5.78 \, M^{0.27 \pm 0.026}; \qquad r^2 = 0.788, \, n = 25. \qquad (9\text{-}4)$$

Equation (9-3) predicts one litter a year for a 12.2-kg mammal; Eq. (9-4) predicts an interbirth interval of a year for a 14.4-kg mammal, so the exclusions have not made a major qualitative change. The exponents are somewhat more steep in each case, reflecting the deletion of many smaller species that were adaptively stretching their physiological schedules for the relatively long seasonal calendar. When individual values for litter size and litter frequency, which have been adaptively traded in emphasis to get the same annual birth rate, are multiplied before running the allometric regression, the variability is smoothed out further ($r^2 = 0.500$).

In any case, derivations from Eisenberg's tables and Western's equation for annual birth rate show clearly that small mammals produce more young per year than do large ones. They must do so, because they have shorter life spans in which to replace themselves and are forced into a faster reproductive pace. When maximum life span is inserted into the equation, the apparent inverse relationship between reproductive potential and size is reduced considerably. The mass dependency of total offspring per maximum potential life span does not appear to support my initial assumption that fitness should be size independent. If smaller mammals produce more young, and their populations oscillate around a stable average, they must be more vulnerable to predation and other mortality than is the case for the large mammals. Column 5 in Table 9-1 overstates the reproduction of the smaller mammals, because not all of the maximum life span is available for this function. A more appropriate calculation would involve life expectancy, or life expectancy minus time required to attain sexual maturity. As will be discussed in Chapter 11, life expectancy may, in fact, scale with a larger exponent than maximum longevity.

Despite the long history of field ornithology, I have not found enough ingredients for a complete avian reproduction allometry. In the absence of concrete information on the scaling of the number of clutches per year, one can perhaps assume that it is inversely related to the duration of incubation, although a parallel was not found for mammals (Eqs. 9-3 and 9-4).

Allometric regressions can give only a general picture of the effects of size, and crude (but useful) predictions for specific sizes because they include data from a size range of several orders of magnitude. By comparing generalizations with specific cases, we can begin to appreciate the limitations, as well as the usefulness, of allometry. Smith's (1978) study of the comparative demography of pikas (*Ochotona*) provides data on litter sizes for pikas in California (in Bodie and Sierra), where mean adult body mass is 132.5 g, and from similar latitudes in Colorado (in the Elk Mountains and Front Range), where mean body mass is 172 g, 30% greater.

The litter size predicted from Table 9-1 for a 132.5-g mammal is 3.06, 5% less than the observed mean of 3.21 for California pikas. The predicted litter size for a 172-g mammal is 2.96, 3% less than the observed mean of 3.06 for Colorado pikas. Alternatively, consider that the ratio of predicted litter sizes for these two body sizes is 1.03, while the observed ratio is 1.05. The accuracy of the prediction is of limited significance because standard deviations around the means for observed litter sizes are 0.93 and 0.92 respectively. Thus the allometric predictions are close, but the observed variability in actual observations could easily mask a size dependence where both the range in body sizes and the body-mass exponent are small. Furthermore, a fifth population from a much higher latitude (in Alberta) has a mean body mass similar to the California pikas, but a much smaller litter size (mean 2.33).

Reproduction Energetics

"Although reproductive effort is conceptually quite useful, it has yet to be adequately quantified" (Pianka, 1983). The "effort" is the relative investment of energy, energy divided between self-maintenance and reproduction. Since small animals such as rodents have short life spans and larger numbers of offspring, it is sometimes thought that smaller animals invest a higher proportion of energy in reproduction, whereas larger animals must invest relatively more in self-preservation. Is there quantitative evidence to support this assumption?

We have noted from Table 9-1 that clutch or litter sizes seem to be inversely related to body mass, although the exponent is small. Smaller size is also correlated with a shorter gestation period, which means that there could be more clutches in sequence per year or per breeding season. Taken together, these factors result in higher annual fecundity. However, "annual" does not take into account the fact that for a small animal living at a faster physiological pace, a year is a much greater portion of its lifetime. The energetics of reproduction are a portion of the lifetime energy budget. The allocations of both time and energy must be considered if meaningful statements are to be made about the relation of size to relative costs of reproduction and self-maintenance.

The life span can be divided into prereproductive and mature stages; the prereproductive in turn can be subdivided into premating and embryonic phases. Let us now examine the scaling of energy costs for each. Rahn (1982) has derived allometric equations for the total cost of producing a litter (E_{tot}, converted to kilojoules) for eutherian mammals and precocial birds:

$$\text{mammals:}\quad E_{tot/litter} = 17{,}113\ M^{1.24}; \tag{9-5}$$

$$\text{birds:}\quad E_{tot/clutch} = 15{,}983\ M^{1.00}. \tag{9-6}$$

Litter and clutch sizes are inversely related to size (Table 9-1), so the costs per neonate would be

$$\text{mammals:}\quad E_{tot/litter}/n_{litter} = 17{,}113\ M^{1.12}/2.9\ M^{-0.10}$$
$$= 5{,}900\ M^{1.34}; \tag{9-7}$$

$$\text{birds:}\quad E_{tot/clutch}/n_{clutch} = 15{,}983\ M^{1.00}/6.2\ M^{-0.05}$$
$$= 2{,}578\ M^{1.05}. \tag{9-8}$$

Thus the cost until birth actually appears to increase in proportion to size for mammals and is in linear proportion, or slightly greater, in birds.

The maintenance cost from birth to reproductive maturity is the product of the metabolic rate and the duration of that period. Heusner (1982a) calculated intraspecific allometric equations for seven species of mammals and found that the average scaling was not the $M^{3/4}$ of interspecific shrew-to-elephant regressions, but $M^{0.67\pm0.03}$. Although he considered mature animals, the scaling may be more appropriate for the period up to sexual maturity than the interspecific Kleiber scaling. If we use this and the mean coefficient a for Heusner's seven species with the scaling of periods until mating begins (t_r, in days, from Table 6-1), the maintenance cost until reproduction will be

$$\text{mammals:}\quad (t_r)(\dot{E}_{sm}) = (274\ M^{0.29})(349\ M^{0.67}) = 9.6 \times 10^5\ M^{0.96}, \tag{9-9}$$

which scales in the same proportions as the potential lifetime basal metabolism:

$$\text{mammals:}\quad (t_{ls,max})(\dot{E}_{sm}) = (4{,}236\ M^{0.20})(282.5\ M^{0.76})$$
$$= 1.2 \times 10^6\ M^{0.96}. \tag{9-10}$$

From this it appears that about 8% of the potential lifetime maintenance is used before mating can begin. The energy cost for maintenance until mating in birds would probably scale similarly, for Western and Ssemakula (1982) calculated the time to first reproduction (t_r) to be

$$\text{birds:} \quad t_r = 850 \, M^{0.23}. \tag{9-11}$$

For further information on the allometry of the reproduction energetics of birds, see Ricklefs (1974). From the data of Bonner (1965), the length at reproduction of organisms from bacteria and protozoa to whales and sequoias is related to reproduction time in years as

$$t_r = 0.011 \, l^{0.82 \pm 0.04}. \tag{9-12}$$

If geometric similarity $(l \propto M^{1/3})$ exists across that size and phylogenetic range, this t_r would be proportional to $M^{0.27}$; if elastic similarity prevailed, it would be $M^{0.20}$, exponents within the ranges of those for other physiological times.

There is nothing in the above to support the idea that large animals devote proportionately less energy to reproduction, or to the attainment of the capacity to reproduce, than do small animals. In Chapter 11 the allometry of population dynamics will show that, in fact, the smaller mammal will not on the average produce any more offspring than will the larger mammal. Many of the allometric pieces used in these chapters are based on limited sampling. However consistent the patterns seem to be, all should be verified from larger data bases; it is my hope that this preliminary synthesis will stimulate such an effort.

Life Span of Gametes

The impression of a similarity in scaling of physiological, reproductive, and growth times is compatible with Bertalanffy's (1957) theory of "a definite and strict connection between metabolic types and growth types." It appears that with some additional data the connection could be traced further. An example is the duration in days over which spermatozoa (t_{fert}) stored in the avian oviduct remain fertile. Lake (1975) tabulated values from the literature, and these values have been combined with values (from other sources) for body mass for these species:

$$t_{fert} = 2.7 \, m^{0.18} \qquad (r^2 = 0.245; p > 0.1). \tag{9-13}$$

Lake's eight species included two Anseriformes (domestic duck and goose) that have distinctly shorter-lived spermatozoa as well as distinctly higher adult standard metabolic rates (Zar, 1968). If we rationalize from this characteristic the exclusion of these two data points, an exploratory regression for five species of Galliformes and one dove yields a significant correlation ($r^2 = 0.664$; $0.02 < p < 0.05$):

$$t_{fert} = 1.6 \ m^{0.29}. \tag{9-14}$$

Finally, if the altricial dove is excluded, we have $0.001 < p < 0.01$ for the five Galliformes:

$$t_{fert} = 1.1 \ m^{0.34} \qquad (r^2 = 0.852). \tag{9-15}$$

The small sample size and large changes in exponents that occur when one or two species are removed from the regression weakens the case for relating life span of spermatozoa either to growth rates (embryonic or postnatal) or to adult life span. However, it seems worthwhile to encourage further study. While Lake concluded that "the reasons for the interspecific differences are not clear," it is possible to be more definitive. Since body mass accounts for 66% of the variance, with an exponent close to ¼, some of the interspecific differences in spermatozoan life span may be related to the physiological time scale.

More about Bird Eggs

Rahn and co-workers (1979) described the bird egg as a "self-contained life-support system for the developing embryo," with all the nutrients, minerals, chemical energy, and water necessary for embryonic development incorporated before the shell is formed. Still to be supplied by the parent are heat and the "crucial requirement: oxygen, which drives the metabolic machinery of the embryonic cells so that they can execute the complex maneuvers of development." While bird eggs serve the purpose of reproduction, the shells have two subservient roles—as an exoskeleton and as a respiratory surface. Consideration of only the reproductive aspect would bypass one of the exciting stories of allometry. In the late 1800s and early 1900s the collecting of bird eggshells was an important hobby that doubtless had some psychological similarity to collecting postage stamps, matchbook covers, or beverage containers. In addition to yielding descriptions of the colors, sizes, and shapes of eggs of different species, oological collections have also provided a pre-DDT, pre-PCB reference collection against which effects of environmental pollution can be compared. However valuable these spin-offs have been, they apparently provided neither curiosity nor insight into the function of the avian egg in a comparative context. Our knowledge of embryology has, of course, come largely from the easily obtained eggs of *Gallus domesticus,* which from every practical consideration was the logical choice in that its full exploitation had no adverse effects on the conservation of wild birds. While one of the earliest applications of allometry to biology was Huxley's study of egg sizes (Chapter 3), until recently the method has been infrequently applied to descriptive details. The log-log relationship between incubation period and egg size was graphed by Needham (1931) and by Worth (1940), and Brody (1945) showed that egg size bore the same scaling relationship to body mass in birds as did metabolic rate.

Modern functional oology has two roots, one a study of eggshell permeability in the respiratory physiology laboratory of Hermann Rahn (Wangensteen et al., 1970), the other the field studies of Drent (1970), which form the basis and standards for the analysis of incubation. Drent showed that the rate of water loss from eggs in natural nests scaled as $M^{3/4}$. What has followed represents one of allometry's most extensive and successful stories, one that has not only reduced masses of data to neat equations and led to the conferring of the Coues Award of the American Ornithologists Union on Ar, Rahn, and Paganelli, but has yielded dimensionless design criteria for bird eggs in general using the allometric cancellation technique of Stahl (1962). For example, the product of water loss per day ($\dot{M}_{H_2O}$) and incubation time (t_{inc}),

$$\frac{(\dot{M}_{H_2O})(t_{inc})}{\text{mass of egg}}$$
$$= \frac{(0.013\ m_{egg}^{0.75})(11.48\ m_{egg}^{0.24})}{1.0\ m_{egg}^{1.0}} = 0.15\ m_{egg}^{-0.01} \tag{9-16}$$

showed that there was a size-independent fraction of 15% $\pm$ 2.5% s.d. of the egg mass at laying that is lost by evaporation before hatching (Rahn and Ar, 1974; Ar and Rahn, 1978). We have considered already the allometry of egg size, shell mass and strength, and incubation time; we shall consider growth in Chapter 10.

The recent triumphs of allometric oology fill a 358-page collection of papers (Rahn and Paganelli, 1981), so even to attempt an outline here may be unrealistic. The terminology of a "self-contained life-support system" suggests analogy to a spaceship, but the egg is the more amazing of the two. The spaceship lifts off when "all systems are go," whereas the egg is launched

9-1 The pores in an eggshell are dramatically evident in this photograph by Dennis R. Atkinson. Air was injected under pressure into the air cell at the blunt (downward) end of this egg while it was submerged in water. The bubbles are forming at the pores where the air is exiting. (Courtesy of Hermann Rahn; from Rahn and Paganelli, 1981.)

with *no* systems in visible operation. That is because our eyes cannot see the molecular schematics by which the egg will develop, and the system of pores through which the respiratory exchanges must occur is only barely discernible to the naked eye when India ink is rubbed across the surface, or when a fresh egg is lowered into a pot of water that is then rapidly heated. The gases within the egg are driven out of solution and thermally expanded, forcing them out as bubbles through the pores (Figure 9-1).

By such artificial means, or in better detail via electron microscopy, we know that the pores exist (Figure 9-2). In the natural situation, as oxygen is used in the metabolism of the developing embryo and carbon dioxide is produced, partial-pressure differences build up across the membranes and pores. These gases are driven from greater to lower pressures according to Fick's Law, as discussed previously:

$$\dot{V}_{O_2} = -DA_p \frac{\Delta p_{O_2}}{x_{sh}}.$$

(5-16)

9-2 A pore in the shell of an egg of a mute swan *(Cygnus olor)*, as seen in the radial face of a break in the shell. (Scanning electron micrograph by Cynthia Carey.)

In this application, the area A_p is the total functional pore area (number of pores times mean pore cross-sectional area) rather than the area of the eggshell, and x_{sh} is the length of the pores, which is the thickness of the shell.

For practical measurement, three terms are lumped together as the conductance (G_{O_2} for oxygen, G_{H_2O} for water vapor):

$$G_{O_2} = -D_{O_2}A_p/x_{sh}, \tag{9-17}$$

$$G_{H_2O} = -D_{H_2O}A_p/x_{sh}. \tag{9-18}$$

Substituting Eq. (9-17) into Eq. (5-16), we have:

$$\dot{V}_{O_2} = G_{O_2}\Delta p_{O_2}, \tag{9-19}$$

which is recognizably a version of Ohm's Law, that the flow is proportional to the conductance (reciprocal of resistance) times the potential difference. Similarly,

$$\dot{V}_{H_2O} = G_{H_2O}\Delta p_{H_2O}. \tag{9-20}$$

Ask a nonbiologist when an egg weighs most. Often he or she will guess that the egg is heavier just before hatching, when the embryo has grown as much as it can, than when it has been freshly laid. Consider the mass exchanges that could affect the egg mass: oxygen diffuses in while carbon dioxide and water vapor diffuse out; there are no other material arrivals or departures. The molecular weight of oxygen is 32; of carbon dioxide, 44. The main energy substrate is the lipids of the yolk. The respiratory quotient (R.Q.), or ratio of carbon dioxide molecules produced to oxygen molecules consumed, is about 0.71 for lipids. The product of R.Q. and the ratio of molecular weights is 0.98; in other words, the mass of more oxygen entering at a lower molecular weight is about the same as the mass of less carbon dioxide leaving at a greater molecular weight. Consequently, the decrease in mass during incubation is not the result of respiration but can be attributed entirely to water loss. If fresh eggs are kept at temperatures below those required for development in a desiccator, both $\dot{V}_{H_2O}$ and Δp_{H_2O} can be determined, and Eq. (9-20) is solved for G_{H_2O}. Furthermore, since the A_p and x_{sh} in Eq. (5-16) are descriptive of eggshell geometry and are indifferent to the kinds of molecules passing through, the G_{H_2O} and G_{O_2} have the same proportions as the diffusion coefficients D_{O_2} and D_{H_2O}, the actual diffusion paths for all gases—oxygen, carbon dioxide, and water vapor—being the same (Rahn and Paganelli, 1981).

The preceding relationships have opened the mysteries of eggshell function to the possibility of understanding via simple measurements: to determine the conductances of eggshells to both water vapor and oxygen, one need only record the decrease in egg mass in a desiccator at 25°C. To calculate the mean nest humidity and prehatching p_{O_2} for eggs in nature, what is needed is data on mass changes of eggs in the field and conductances calculated in the laboratory (as above) for eggs of that particular species from the local population. The effective pore area (A_p) can be calculated with Eq. (9-18), G_{H_2O}, and micrometer measurements of shell thickness, x_{sh}.

From these procedures the following general relationships have been established (Ar and Rahn, 1978, 1980):

$$G_{H_2O} = 0.38 \, m_{egg}^{0.81} \text{ (in mg/torr)}, \tag{9-21}$$

or $\quad G_{H_2O} = 0.51 \, m_{egg}^{0.81} \text{ (in mg/mbar)}. \tag{9-21a}$

The daily water loss for 93 species (in milligrams per day) is

$$\dot{M}_{H_2O} = 13.24 \, m_{egg}^{0.75 \pm 0.078}. \tag{9-22}$$

For eggs from different species with the same egg mass, there is a fair amount of variability in $\dot{M}_{H_2O}$, G_{H_2O}, and the incubation period (t_{inc}), but if t_{inc} is longer than the allometric prediction, G_{H_2O} and $\dot{M}_{H_2O}$ are lower in compensation. This results in tighter correlations with less variance for the products ($\dot{M}_{H_2O}$)(t_{inc}) and (G_{H_2O})(t_{inc}). Since $G_{O_2} = 1.07 \, G_{H_2O}$ at 37°C (Rahn and Paganelli, 1981), it follows that the maximum rate of oxygen uptake, before the chick pips, scales as do the rate of water loss and the respective conductances, in about the same way as do adult metabolic rates, $\propto M^{-3/4}$. Thus allometric analysis of eggs shows that there is a continuity in scaling from embryo to adult. In addition to the size-independent design for total water loss per gram of fresh egg during incubation, a size independence for total embryonic oxygen consumption per gram has been derived as well, amounting to 105 ml O_2/g, ± 20 s.d., equivalent to 2.09 kJ/g.

The actual rate of oxygen consumption by the developing embryo is minuscule during the first few cell divisions, increasing sigmoidally as the mass of the embryo increases (Figure 9-3). It is this maximum rate that would scale in parallel with G_{H_2O}. The extent of porosity of the shell is that necessary for such a maximum prepipping level. Another case of symmorphosis?

As noted above (Eqs. 9-17 and 9-18), the conductance, G, of the eggshell is directly proportional to the ratio of effective pore area, A_p, to shell thickness, x_{sh}. The breaking force, a measure of eggshell strength, was seen to be proportional to the square of x_{sh} (Eq. 4-25). If the thickness is scaled for adequate strength and the conductance is scaled appropriately, we can rearrange Eq. (9-18) to solve for A_p:

$$A_p \propto G_{H_2O} x_{sh} \propto (m_{egg}^{0.81})(m_{egg}^{0.448}) \propto m_{egg}^{1.25}. \tag{9-23}$$

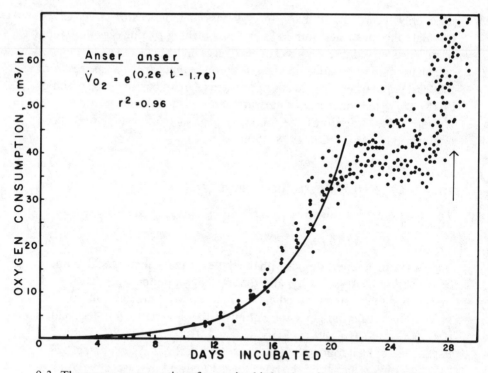

9-3 The oxygen consumption of a growing bird embryo increases sigmoidally to the maximum, a rate scaled like the conductance of the pores and membranes to oxygen and water vapor. Hatching on day 28 *(arrow)* of a domestic gosling was preceded by pipping on day 27 and accompanied by a rise in $\dot{V}_{O_2}$ from the plateau level of 40.3 cm³/hr. (From Vleck et al., 1979; with permission.)

For predicting functional pore area in square centimeters (Ar and Rahn, 1978):

$$A_p = 9.72 \times 10^{-5} \, m_{egg}^{1.25}; \qquad r^2 = 0.979. \tag{9-24}$$

In summary, four general principles have been elucidated in a decade of studies by Ar, Paganelli, Rahn, and their associates. First, gas transport via the pores of the avian eggshell is according to Fick's first law of diffusion. Second, regardless of egg size and incubation period, most eggs will lose about 15% of their initial mass by evaporation. The relative water content or proportion of water stays the same, evaporation being offset by metabolic water production and molecular recombination of the original solids. The 15% loss of total egg mass leaves an air cell at the large end of the egg. Third, the metabolic rate increases to a maximum as the embryo grows; at the maximum it is directly proportional to (and has the same scaling as) G_{H_2O}, resulting in the attainment of the same p_{O_2} and p_{CO_2} in the air cell prior to pipping, independent of egg size and incubation period. Fourth, the total metabolism, expressed as cumulative oxygen consumption or heat production per gram of initial mass, is also independent of egg size and incubation period.

10 Growth

The concept of different rates of growth of parts of the body relative to that of the body as a whole, or *allometry* as it has been called since 1936 . . . , has now been shown to have many more effects on animal (and plant) form than I mentioned in my book on relative growth.

—JULIAN HUXLEY, *PROBLEMS OF RELATIVE GROWTH* (1972)

THE USE of allometry had its origin in applications to ontogeny. But since the simple allometric equation was introduced by Julian Huxley as the heterogony formula, its primary use has been to compare adult animals of different sizes, within or between phylogenetic groups. The allometric study of mammals from mouse to elephant or of birds from hummingbird to ostrich, for example, provides insight into differential relative growth on an evolutionary scale rather than on an individual scale. Allometry, in fact, seems to have taken root more readily in the study of phylogeny and evolutionary development than it did in its starting field of ontogeny; I never heard a single word about allometry when I was an undergraduate—or even, later, as a graduate teaching assistant in embryology and developmental biology.

As Huxley, and before him D'Arcy Thompson (1917), stressed, "all organic forms, save the simplest such as the spherical or the amoeboid, are the result of differential growth." The newborn marsupial mammal is one of the most dramatic examples; it is still essentially an embryo, but the cephalic region and forelegs are much more developed than the posterior regions so that the newborn can crawl to the pouch, where it completes development

—and where the allometry of development can be studied quite profitably. The present chapter is limited to the information available from research on birds and eutherian mammals.

Our concern here will be with the total growth of an individual (in dimensions of mass and time, as analyzed allometrically within a phylogenetic group), including the costs of and provisions for this growth. Those who want a detailed look at relative ontogenetic growth should read Huxley's classic, *Problems of Relative Growth,* also Calow (1978, chap. 5) and Cock (1966). On the relationship between ontogeny and phylogeny see Gould (1971), Alberch et al. (1979), and Cheverud (1982).

The growth of an animal occurs in two major periods: embryonic and postembryonic. What are the proportions of time spent and the proportions of mass added to the developing animal in each of these periods? Is it better for the mother to drop the weight of the fetus early and have a greater postpartum rearing responsibility, or to carry the weight longer and give birth to further-developed neonates? Should a bird put more energy into a larger egg that requires longer incubation, or minimize the egg investment and spend more energy on the increased demands of a less-developed nestling? Each of these alternative reproduction strategies is actually observed, and allometry can be useful in sorting out the associated patterns.

The Scaling of Embryonic Growth

Let us first examine the sizes of birds and mammals at the dividing point between embryonic and postembryonic growth, the date of birth (see Figure 10-1). The scaling of the (eutherian) mammalian neonatal mass is slightly less than linear (Blueweiss et al., 1978, from data tabulated by Leitch et al., 1959, and Sacher and Staffeldt, 1974):

$$m_{neonate} = 0.097 \ m_{adult}^{0.92} \qquad (r^2 = 0.94). \tag{10-1}$$

From this equation we would expect the smallest mammal's neonate to weigh slightly under 10% of adult mass, and progressively larger mammals to have proportionately smaller neonates. The smallest and largest species tabulated by Sacher and Staffeldt (1974) confirm this trend, although the mass of their neonates is slightly greater than predicted. Since all neonates eventually grow to adult size ($1.0 \ m^{1.0}$), the larger animals must have slightly more (95% to 96%) of their growing to do after birth than the smaller ones (approximately 90%). Because no confidence interval for the exponent was given, it is possible that the difference between the derived $m^{0.92}$ and $m^{1.0}$ is

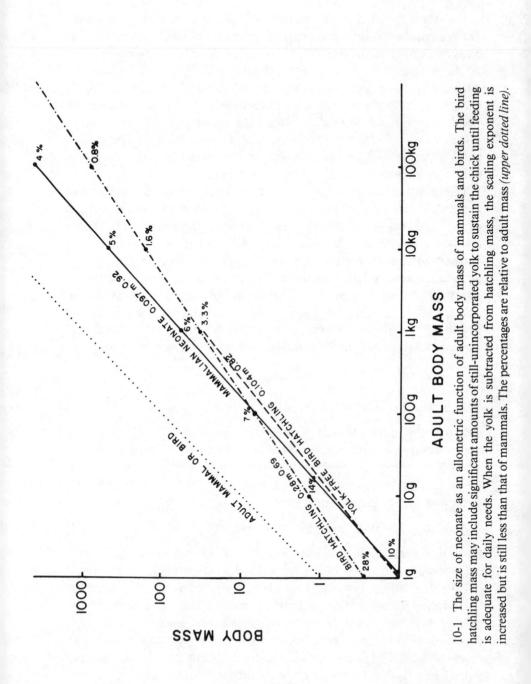

10-1 The size of neonate as an allometric function of adult body mass of mammals and birds. The bird hatchling mass may include significant amounts of still-unincorporated yolk to sustain the chick until feeding is adequate for daily needs. When the yolk is subtracted from hatchling mass, the scaling exponent is increased but is still less than that of mammals. The percentages are relative to adult mass (*upper dotted line*).

not statistically significant; in any case, the difference is small. Millar (1977) obtained a smaller exponent for the scaling of mass at birth to maternal mass:

$$m_{neo} = 0.20 \, m_{adult}^{0.71}. \qquad (10\text{-}1a)$$

The standard error of the exponent was 0.03, but the sample size was 95, compared to approximately twice as many in the Blueweiss study (Eq. 10-1).

For birds the scaling of neonate (hatchling) mass (m_{neo}) to adult mass (m_{adult}) is much different from that for mammals according to Blueweiss et al. (1978):

$$m_{neo} = 0.28 \, m_{adult}^{0.69} \qquad (r^2 = 0.86). \qquad (10\text{-}2)$$

However, this is indistinguishable from Millar's scaling equation for mammals. Still, we need to remember that hatchlings of smaller species have less growing to do after hatching than hatchlings of larger species. For example, a hatchling that will weigh 10 g as an adult has 86% of its growth ahead of it, a hatchling of a 100-g species has 93%, and a hatchling of a 100-kg species (such as the ostrich) has 99% to go before reaching adult size.

Does the difference between avian and mammalian neonate scaling reflect different constraints on oviparity and viviparity, or just a need for finer tuning of the estimating equations? First, egg size, like metabolic rate, is proportional to $M^{3/4}$; if one egg is laid each day regardless of hen size, the accomplishment is in approximately the same proportion to the other energetic demands for both large and small birds. The energy content, and therefore the cost, of producing an egg depends upon whether the egg will produce a precocial or an altricial neonate, since precocial species produce eggs with a higher percentage of energy-rich yolk. The energy content (in kilojoules) of an egg scales with adult female mass as follows (units converted from Ar and Yom-Tov, 1978):

$$\text{precocial:} \quad E = 2.26 \, m_{egg}^{0.729}; \qquad (10\text{-}3)$$

$$\text{altricial:} \quad E = 1.14 \, m_{egg}^{0.731}. \qquad (10\text{-}4)$$

Thus the energy investment is proportional to metabolic rates, a slightly higher scaling than daily energy budget; the only difference between precocial and altricial birds is that the former invests more energy in the process at the start as a trade-off against parental duties later. Furthermore, the mass of the hatchling would be expected to parallel these $M^{0.73}$ investments if the conversion efficiency is independent of size. Vleck and colleagues (1980) calculated that the cost of embryonic growth in birds is 1.23 kJ/g, and independent of egg size.

When separate allometric regressions are run, the scaling of neonate mass is different for precocial and altricial species. Not all of the yolk is used before precocial species hatch; significant amounts are retained as a reserve until the chick is capable of feeding itself. The yolk-free hatchling mass of 11 species listed by Vleck is scaled to adult mass from other sources as

$$m_{neo,yolk\text{-}free} = 0.104 \; m_{adult}^{0.82\pm0.068} \qquad (r^2 = 0.941). \qquad (10\text{-}5)$$

Yolk-free hatchling mass can also be related to fresh egg mass obtained (from other sources) for ten of these species:

$$m_{neo,yolk\text{-}free} = 0.63 \; m_{egg}^{0.96\pm0.018} \qquad (r^2 = 0.997). \qquad (10\text{-}6)$$

An exponent of less than 1.0 would be expected for yolk-free hatchling mass, because the smaller species are mostly altricial and the larger are mostly precocial, and because shell mass increases in proportion to egg mass (Anderson et al., 1979):

$$m_{sh} = 0.048 \; m_{egg}^{1.13\pm0.09} \qquad (r^2 = 0.951). \qquad (10\text{-}7)$$

Another constraint on oviparity is that the surface area for gas exchange is limited geometrically. The egg surface area (in square centimeters) is related to egg mass as follows (Paganelli et al., 1974):

$$A_{egg} = 4.835 \; m_{egg}^{0.66}. \qquad (10\text{-}8)$$

However, the proportion of that area devoted to the pores through which oxygen and carbon dioxide must pass increases out of proportion to egg size (Ar and Rahn, 1978):

$$A_{pores} = 9.72 \times 10^{-5} \; m_{egg}^{1.25}. \qquad (9\text{-}24)$$

Since, from Eq. (3-4), $m_{egg} \propto m_{adult}^{0.77}$,

$$A_{pores} \propto (m_{adult}^{0.77})^{1.25} \propto m_{adult}^{0.96}. \qquad (10\text{-}9)$$

As body and egg size increase, the eggshell must be strong enough both to contain the weight of the contents and to support the incubator. The yield-point force of an eggshell is proportional to the square of shell thickness (x_{sh}). Shell thickness scales as follows (Ar et al., 1979):

$$x_{sh} = 54.06 \; m_{egg}^{0.448}. \qquad (4\text{-}17)$$

The gas conductance (G_{O_2}) of the egg is proportional to the total area of the pores and inversely proportional to the length of the pores (diffusion distance, or shell thickness):

$$G_{O_2} \propto \frac{m_{egg}^{1.249}}{m_{egg}^{0.448}} \propto m_{egg}^{0.80}. \tag{10-10}$$

The maximum metabolic rate ($\dot{V}_{O_2}$) just before pipping is directly proportional to the G_{O_2} (Ar and Rahn, 1978). Measured directly, maximum metabolic rate (in milliliters of oxygen per day or joules per day) has a body-mass exponent somewhat smaller than 0.75, but it is still approximately proportional to adult metabolic rates (Hoyt and Rahn, 1980):

$$\dot{V}_{O_2} = 28.9 \, m^{0.714}, \tag{10-11}$$

or $\quad \dot{E}_{egg,max} = 120.9 \, m^{0.714}. \tag{10-12}$

Thus, during the embryonic growth period there appears to be a fairly good correlation among fresh egg mass ($\propto m_{adult}^{0.77}$), adult metabolic rate ($\propto m_{adult}^{0.72}$), and hatchling mass ($\propto m_{adult}^{0.69}$ to $m_{adult}^{0.82}$) and among egg energy content ($\propto m_{egg}^{0.73}$), egg maximum metabolic rate ($\propto m_{egg}^{0.71}$), and egg oxygen conductance ($\propto m_{egg}^{0.80}$). Note, however, that the scaling similarity between these two groups is only apparent, for when all are stated in terms of adult body mass, the egg metabolic rate, oxygen conductance, and egg energy content have exponents of $m_{adult}^{0.51}$, $m_{adult}^{0.58}$, and $m_{adult}^{0.53}$.

What is the energetic cost of viviparity? Brody (1945) defined the total heat increment of gestation (ΔE_{gest}, converted to kilojoules) as the difference between the resting heat production during gestation and the resting heat production when the female is not carrying young, and found this difference to be related to the mass of the neonate:

$$\Delta E_{gest} = 18,410 \, M_{neo}^{1.2}. \tag{10-13}$$

Rahn (1982) recalculated for 21 observations ($r = 0.99$; here converted to kilojoules):

$$\Delta E_{gest} = 17,129 \, M_{litter}^{1.24}. \tag{10-14}$$

Equation (10-13) can be divided by the gestation period as a function of neonatal mass (from Sacher and Staffeldt, 1974) to yield the mean daily cost increment (in kilojoules per day):

$$\frac{E_{gest}}{t_{gest}} = \frac{18,410 \, M_{neo}^{1.2}}{147 \, M_{neo}^{0.282}} = 125 \, M_{neo}^{0.92} = 0.22 \, m_{neo}^{0.92}. \tag{10-15}$$

On the other hand, Blueweiss and coworkers (1978) related m_{litter} (in grams) and t_{gest} (in days) to size as follows:

$$m_{litter} = 0.55 \, m_{adult}^{0.82}; \tag{10-16}$$

$$t_{gest} = 11 \, m_{adult}^{0.26}. \tag{10-17}$$

Therefore, the mean daily growth per litter ($\bar{m}_{litter}$, in grams per day) is

$$\bar{m}_{litter} = \frac{m_{litter}}{t_{gest}} = \frac{0.55 \, m_{adult}^{0.82}}{11 \, m_{adult}^{0.26}} \tag{10-18}$$

$$= 0.05 \, m_{adult}^{0.56} = 2.39 \, M_{adult}^{0.56}.$$

We can rearrange Eq. (10-1) to solve for adult mass, therefore:

$$m_{adult}^{0.92} = 10.31 \, m_{neo}, \qquad m_{adult} = 12.63 \, m_{neo}^{1.087}. \tag{10-19}$$

The mean daily growth per litter as a function of neonate mass is

$$0.05(12.63 \, m_{neo}^{1.087})^{0.56} = 0.05(4.138 \, m_{neo}^{0.6087})$$

$$= 0.207 \, m_{neo}^{0.609}. \tag{10-20}$$

If the derivations from Brody (Eqs. 10-13 and 10-15) and Blueweiss and colleagues (Eqs. 10-16 to 10-20) are compatible, the energy cost of embryonic growth (in kilojoules per day) must be

$$\text{energy cost} = 0.22 \, m^{0.92}/0.207 \, m^{0.609}$$

$$= 1.05 \, m_{neo}^{0.31}, \text{ or } 8.93 \, M_{neo}^{0.31}. \tag{10-21}$$

There is no obvious reason why the mean energetic cost per gram of growth should not be size independent. This scaling ($M^{0.31}$) suggests that either the heat increment of gestation must scale as a lower fractional power of neonate mass or that the allometry of mean daily growth should be recomputed. For example, Eq. (10-21) predicts that the cost per kilogram of calf embryo (26.9 kJ/g) is seven times the cost per kilogram of rat neonate. Upon combustion one gram of protein yields, on the average, 23.6 kJ of heat energy. If we assume that 60% of growth represents an increase in tissue water and 40% represents an increase in protein, the energy stored per gram of flesh is 9.46 kJ. According to Eq. (10-21), it would take 1.05 kJ to build a 1-g shrew neonate, when in fact that 1 g would represent 9.46 kJ worth of tissue, clearly impossible since no living system can violate the first law of thermodynamics and create the difference of 8.41 kJ. The 35-kg neonatal calf, on the other hand, would have a reasonable mean efficiency of 35%. We would expect efficiency to be somewhat less over longer gestations because of a prolonged fractional maintenance cost, but the discrepancies suggest that the energy costs of mammalian embryonic growth are in need of further allometric study if reproductive strategies are to be fully understood.

Brody's data (1945, table 14.6) represent different sample sizes from seven species of domestic mammals. Reducing his data to mean values in kilojoules for each species, we get:

$$E_{gest} = 20,560\ M_{neo}^{1.17} \qquad (r^2 = 0.99). \tag{10-22}$$

The gestation period for those seven species is

$$t_{gest} = 110.8\ M_{neo}^{0.29} \qquad (r^2 = 0.93). \tag{10-23}$$

Thus the average expenditure (in kilojoules per day) is

$$\dot{E}_{gest} = \frac{20,560\ M_{neo}^{1.17}}{110.8\ M_{neo}^{0.29}} = 185\ M_{neo}^{0.88} \tag{10-24}$$

and the cost (in kilojoules per day) per kilogram of neonatal mass is

$$\dot{E}_{gest} = \frac{185\ M_{neo}^{0.88}}{M^{1.0}} = 185\ M_{neo}^{-0.12}. \tag{10-25}$$

(This is equivalent to $424\ m_{neo}^{-0.12}$ in joules per gram per day.) Rahn (1982) calculated these values on the basis of litter mass rather than M_{neo} and derived the relationship (units converted to kilojoules per kilogram per day):

$$\dot{E}_{gest} = 133\ M_{litter}^{-0.08}. \tag{10-26}$$

He assumed that the exponent, -0.08, was not significantly different from zero, so that 133 kJ/(kg·day) was taken to be a size-independent constant.

How does oviparity compare in terms of total cost (exclusive of adult heat input) per kilogram of neonate produced? It is not yet clear, because of differences in exponents. Rahn (1982) found (kilocalories converted to kilojoules):

$$E_{tot} = 15,983\ M_{neo}^{1.00}. \tag{10-27}$$

This is 13% less for bird than for mammal (from Eq. 10-13) in the 1-kg neonate size, but 3.5 times as much for a 1-g neonate bird as for a 1-g neonate mammal. Rahn has identified the kinds of allometric expressions that are necessary for any understanding of vertebrate reproductive strategies, so we know what kind of information must be acquired if the comparison is to be improved.

Size and the Pace of Growth

The dynamic process by which a fertile zygote becomes an adult is complex. It simplifies matters to treat the various aspects separately: differentiation, development of individual organs and functions, growth in absolute size, and final maturation of physiological capacities (such as reproductive maturation). The ultimate sizes of the various organs in the adults are scaled differently, occupying different proportions of the body mass in small and large species (see Chapters 2 and 3, and Table 3-4). Within a species the organs do not all develop at the same rates, the cephalic region of the body developing faster than the posterior, for example. I shall leave to developmental biologists any analysis of the resulting complexities of differential growth and the allometry of individual organs. The focus here will be on total body growth, the increase in body size with time.

The fundamental dimensions of this total growth are time and mass, or time and a linear dimension. Simple as these are, when collected from species spanning a wide size range and subjected to allometric analysis, they can reveal patterns of interest to both the student of the evolution of life-history "strategies" and those who use animal models to study growth dynamics.

The idea that there is "a definite and strict connection between metabolic types and growth types" originated with Bertalanffy (1957). He separated these types according to the mass exponents for metabolic rates and the shape of the growth curve. The correlations incorporated wider phylogenetic and physiological diversity (invertebrates, poikilotherms) than I have space to deal with in this chapter. Bertalanffy did not compare body-mass exponents for growth with those for metabolism, probably because there were not enough data available at the time. However, biological data gathered and analyzed since then permit such comparison. The quantitative similarity in the scaling of physiological, reproductive, and growth times, as is evident in the similarity of body-mass exponents, is compatible with Bertalanffy's theory.

Previous studies have either perceived the allometry of physiological time as only a crude approximation, or they have overlooked it entirely. The pace of total body growth has been dealt with on a comparative basis in two series of frequently cited papers by Huggett and Laird and their associates (Huggett and Widdas, 1951; Laird, 1965; 1966a,b; Laird et al., 1968; Barton and Laird, 1969; Frazer and Huggett, 1974; Frazer, 1977. The nonlinear consequences of body or embryo size were recognized by Huggett and Widdas (1951) and by Laird (1966b). While their various "weighing factors," used to show common features of growth, appear consistent with the observed curves, they cannot suggest more than the initial assumptions by which they were applied. Empirical relationships such as the allometric summaries, on the other hand, give us something less arbitrary to attempt to explain. The quantitative deficiencies of arbitrary $M^{1/3}$, M^{-1}, and M^{-2} factors are apparent from the rough consistency of $M^{1/4}$ exponents that predominate in Table 10-1. Reanalysis of the growth studies cited above confirmed this pattern (Calder, 1982b).

The boss-to-rump length of mammalian embryos gave Huggett and Widdas (1951) a "reasonably linear" plot against time from conception. They assumed that embryo mass was proportional to volume or the cube of linear dimensions (Eqs. 4-2 to 4-4) and plotted $M^{1/3}$ as a function of time from conception (t_o):

$$M^{1/3} = u(t - t_o). \tag{10-28}$$

The growth constant u is the slope of the plot and has the dimensions of velocity:

$$u = \frac{M^{1/3}}{(t - t_o)} \propto \frac{l}{t}. \tag{10-29}$$

It was therefore called the "specific foetal growth velocity," a term the authors found useful in the study of comparative growth. Their values of t_o show a significant $M^{0.19}$ scaling (see Table 10-1), while u shows no consistent size dependence. (For a discussion of the inconsistencies that arise from assumption of $t \propto M^{1/3}$ see Calder, 1982b.)

Laird (1965) stated that "the allometric equation does not include a time parameter." Of course, it can: the allometric equation is merely an empirical correlation and can have whatever dimensions we wish to correlate, including the variable of time.

In Chapter 6 the principle of "synchrony of times," Stahl's (1962) extension of Hill's "physiological time," was applied to the functioning of adult animals. The allometric examination of growth, both embryonic and postnatal, shows great similarity to the $M^{1/4}$ scaling of physiological time in adults. In fact, Laird's (1966b) values for the "standard doubling time" during the approximately linear rapid-growth stage, expressed as a function of the asymptotic body mass at cessation of growth, provide yet another allometric regression with the $M^{1/4}$ characteristic of many physiological time scales (Taylor, 1965, 1968; Lindstedt and Calder, 1981; Calder, 1982b; see Table 10-1).

Growth is, after all, a physiological process. Embryonic and postnatal growth, maturation, maintenance, and aging are becoming more generally regarded as parts of a continuum (Timiras, 1978; Lindstedt and Calder, 1981; Rahn, 1982; Calder, 1982b; Western and Ssemakula, 1982; Western, 1983). If this is indeed the case, it should be confirmed in the scaling of time.

Table 10-1 Growth times and size, expressed as mass of egg (M_{egg}), of neonate (M_{neo}), of litter or clutch (M_{litt}), of adult (M_{adult}), or of asymptotic growth mass (M_{asymp}). (Adapted from Calder, 1982, with permission of the Zoological Society of London.)

Time (days)	Mass (kg)	Mass (g)	r^2	n	95% C.I. for scaling exponent b	Sources of data or equations
Mammals (Eutheria)						
Embryonic						
t_o = conception to onset of rapid growth[a]	$26.9\,M_{neo}^{0.19}$	$7.25\,m_{neo}^{0.19}$	0.98	7	—	Huggett and Widdas, 1951; Calder, 1982b
Gestation	$147\,M_{neo}^{0.28}$	$21.33\,m_{neo}^{0.28}$	—	91	—	Sacher and Staffeldt, 1979
Gestation	$130\,M_{litt}^{0.30}$	$16.37\,m_{litt}^{0.30}$	0.69	95	—	Rahn, 1982
Gestation	$66.2\,M_{adult}^{0.26}$	$10.99\,m_{adult}^{0.26}$	—	91	—	Sacher and Staffeldt, 1979
Postembryonic						
Incremental growth[b]	$177.8\,M_{adult}^{0.28}$	$25.7\,m_{adult}^{0.28}$	0.92	167	—	Case, 1978
Standard doubling[c]	$21.0\,M_{asymp}^{0.25}$	$3.73\,m_{asymp}^{0.25}$	0.77	7	—	Laird, 1966b; Calder, 1982b
50% growth	$130.5\,M_{adult}^{0.25}$	$23.2\,m_{adult}^{0.25}$	—	—	—	Stahl, 1962
98% growth	$447\,M_{adult}^{0.26}$	$74.3\,m_{adult}^{0.26}$	—	—	—	Stahl, 1962

Reproductive maturity	$274\,M_{adult}^{0.29}$	$37.2\,m_{adult}^{0.29}$	0.57	56	±0.069	Hafez et al., 1972; Economos, 1981b; Eisenberg, 1980; Lindstedt and Calder, 1981
Maximum longevity, captive	$4{,}234\,M_{adult}^{0.20}$	$1{,}064\,m_{adult}^{0.20}$	0.59	63	±0.042	Sacher, 1959
Mammals (Marsupialia)						
Eyes open	$86.1\,M^{0.16}$	$28.5\,m^{0.16}$	0.66	44	0.036	Russell, 1982
Pouch life—all	$108.5\,M^{0.27}$	$16.8\,m^{0.27}$	0.78	59	0.039	Russell, 1982
Pouch life—Dasyuridae	$72.7\,M^{0.13}$	$29.6\,m^{0.13}$	0.67	16	0.051	Russell, 1982
Pouch life—Macropodidae	$116.4\,M^{0.29}$	$15.7\,m^{0.29}$	0.73	22	0.083	Russell, 1982
Weaning—all	$163.2\,M^{0.22}$	$35.7\,m^{0.22}$	0.67	56	0.042	Russell, 1982
Weaning—Dasyuridae	$136.3\,M^{0.12}$	$59.5\,m^{0.12}$	0.73	17	0.043	Russell, 1982
Weaning—Macropodidae	$143.0\,M^{0.37}$	$11.1\,m^{0.37}$	0.82	18	0.091	Russell, 1982
Birds						
Embryonic						
Incubation	$53.2\,M_{egg}^{0.22}$	$11.64\,m_{egg}^{0.22}$	0.74	90	—	Ar and Rahn, 1978
Incubation	$59.1\,M_{neo}^{0.22}$	$12.93\,m_{neo}^{0.22}$	0.74	444	—	Rahn, 1982
Incubation	$28.9\,M_{adult}^{0.17}$	$9.11\,m_{adult}^{0.17}$	—	—	—	Rahn et al., 1975

Table 10-1 (continued)

Time (days)	Mass (kg)	Mass (g)	r^2	n	95% C.I. for scaling exponent b	Sources of data or equations
Postembryonic						
Standard doubling[c]	$14.7\,M_{adult}^{0.18}$	$4.24\,m_{adult}^{0.18}$	0.34	5	—	Laird, 1966b; Calder, 1982b
Standard doubling[d]	$13.2\,M_{adult}^{0.34}$	$1.26\,m_{adult}^{0.34}$	0.58	148	±0.052	Calculated from Ricklefs, 1979
Hatch—fledge, altricial	$69.7\,M_{egg}^{0.23}$	$14.2\,m_{egg}^{0.23}$	—	27	—	Calculated from Ar and Yom-Tov, 1978; differences between altricial and precocial not significant
Hatch—fledge, precocial	$112.4\,M_{egg}^{0.26}$	$18.7\,m_{egg}^{0.26}$	—	27	—	
Hatch—fledge, altricial	$38.1\,M_{adult}^{0.18}$	$11.0\,m_{adult}^{0.18}$	—	27	—	
Hatch—fledge, precocial	$44.2\,M_{adult}^{0.20}$	$11.1\,m_{adult}^{0.20}$	—	27	—	
Age at first breeding	$1{,}048\,M_{adult}^{0.23}$	$214\,m_{adult}^{0.23}$	0.40	40	±0.102	Western and Ssemakula, 1982
Maximum longevity, wild	$6{,}424\,M_{adult}^{0.20}$	$1{,}614\,m_{adult}^{0.20}$	0.61	152	±0.020	Lindstedt and Calder, 1976
Maximum longevity, captive	$10{,}330\,M_{adult}^{0.14}$	$2{,}780\,m_{adult}^{0.14}$	0.49	58	±0.04	Lindstedt and Calder, 1976

a. Extrapolated from essentially linear slope for rapid growth phase by Huggett and Widdas, 1951.
b. Total growth divided by rate during linear phase from 5% to "about 30 to 50% of adult mass" (Case, 1978).
c. Minimum time for doubling of body mass during rapid growth phase in Gompertz plot. Mammals ($p < 0.01$) and birds ($p < 0.02$) from Laird, 1966b.
d. Minimum time for doubling of body mass during rapid growth phase in a logistic plot.

The length of each of these periods is proportional to body size in approximately the same way (mean exponent for M_{adult}: 0.24 ± 0.04 for mammals, 0.23 ± 0.02 for wild birds). This means that the fractions of a lifetime required for the stages of growth are size independent, even though the absolute times vary widely as a function of size. Furthermore, if the total life span and the fraction of life devoted to growth scale in similar fashion, the remaining plateau of the prime of life—and the aging process that terminates the prime—must also scale in parallel; if the pace of growth speeds up in the small animal, the pace of aging does too (as will become evident in the next chapter).

Allometric Comparisons of Growth

The condition or state of development of the neonate may be either altricial (naked, blind, and helpless) or precocial (furred or feathered, eyes open, and capable of some coordinated movement on its own behalf), or somewhere between, in both birds and mammals. The length of the incubation period (t_{inc}) and the length of the fledging period (from hatching to fledging) are positively correlated, while the posthatching growth rate is inversely related to incubation period (Drent, 1975). Ar and Yom-Tov (1978) separated precocial and altricial birds to derive:

$$\text{precocial:} \quad t_{fledging} = 1.064 \, t_{inc}^{1.167}; \quad\quad\quad (10\text{-}30)$$

$$\text{altricial:} \quad t_{fledging} = 1.053 \, t_{inc}^{1.061}. \quad\quad\quad (10\text{-}31)$$

These relationships do not differ significantly, but will be used later in other speculative derivations.

To limit the variability due to family differences in degree of precociality or altriciality, Eisenberg (1981) ran separate regressions for log (gestation + eye-opening time) for different families of mammals. Excluding four of his relationships with r^2 less than 0.5, the body-mass exponents had a mean of 0.16 ± 0.09 s.d., not distinctly different from the exponent for Eutheria data combined.

Comparison of specific fetal growth velocities (u in Eq. 10-29) has given equivocal results. For birds, u ranges from 0.06 to 0.21; for mammals from 0.063 to 0.20. This overlap makes no distinction between the two major reproductive strategies of viviparity and oviparity. Stahl's allometric cancellation technique, however, facilitates such a comparison and in addition may yield some of the general design criteria of reproduction and growth.

We can begin by examining fetal growth rate as the mean increase in embryonic mass per day of gestation or incubation; all equations will be stated in terms of adult body mass in grams. The body mass of hatchling birds contains a variable amount of unincorporated yolk, especially in precocial species. The data on yolk-free hatchling mass are fewer, but remove this source of variability. Vleck and coworkers (1980) have tabulated values with incubation times, for 11 species with an adult size range from 11 g to 3,425 g. The average growth rate ($\overline{\dot{m}}_{neo}$, in days) is

$$\overline{\dot{m}}_{neo} = 0.010 \ m_{adult}^{0.70} \qquad (r^2 = 0.946; \ p < 0.001). \qquad (10\text{-}32)$$

When the total growth represented by neonatal mass (Eq. 10-1) is divided by gestation time (Table 10-1, line 4), the average embryonic growth rate of mammals is

$$0.097 \ m_{adult}^{0.92} / 10.99 \ m_{adult}^{0.26} = 0.0088 \ m_{adult}^{0.66}. \qquad (10\text{-}33)$$

Equations (10-32) and (10-33), with close similarity in exponents, may now be combined to compare growth rates of the two classes:

$$\text{birds/mammals} = 0.010 \ m_{adult}^{0.70} / 0.0088 \ m_{adult}^{0.66}$$
$$= 1.13 \ m_{adult}^{0.04}. \qquad (10\text{-}34)$$

Rahn (1982) has also compared growth rates in terms of hatchling mass (uncorrected for unincorporated yolk) of birds and litter mass of mammals. Converting his allometric equations to units of gram mass, we obtain a ratio equation similar to (10-34):

$$\text{birds/mammals} = 0.078 \ m^{0.78} / 0.061 \ m^{0.70} = 1.27 \ m^{0.08}. \qquad (10\text{-}35)$$

When comparison is made at the 1-kg intercept rather than at 1 g, Rahn's (1982) larger residual mass exponent of 0.08 yields a ratio of 2.2 for avian vs. mammalian growth rates, compared to a ratio of 1.49 if Eq. (10-34) is solved for a 1-kg bird. In any case, avian oviparous growth has the advantage of being more rapid than mammalian viviparity, but the quantitative difference needs to be resolved with greater precision and additional data. We could generalize by comparing at some representative midpoint in the distribution of body or neonatal masses of the two classes. From Rahn's graphs it appears that the median hatchling mass is about 25 g. If we take the predictions from Eqs. (10-34) and (10-35), the growth rate of this median bird embryo is 1.3 to 1.6 times that of a 25-g mammalian embryo.

Let us now compare the postnatal growth of birds and mammals. Postnatal growth has been plotted either according to the Gompertz equation

(Laird, 1966b),

$$m_t = m_o e^{(\Lambda_o/\alpha)(1-e^{-\alpha t})}, \tag{10-36}$$

or the logistic equation (Ricklefs, 1979),

$$m_t = m_{asymp}/[1 + e^{-K(t-t_i)}]. \tag{10-37}$$

In both, m_t = mass at time t; m_o = computed initial mass; m_{asymp} = asymptotic mass at completion of growth; Λ_o = initial specific growth rate; α = rate of decay of Λ_o that has the effect of progressively reducing the slope of the curve until a horizontal asymptote (m_{asymp}) is reached; K = the logistic rate constant; and t_i = the time of maximum growth rate (point of inflection in curve). These two equations have roughly the same effect, that of generating a sigmoid curve which approximates the observed growth curve of an animal.

Using these equations, the original measurements, and/or allometric summaries such as those of Ricklefs for 148 species of birds:

$$K = 1.10 \; m^{-0.34} = 0.105 \; M^{-0.34} \tag{10-38}$$

($r^2 = 0.58$; $p < 0.001$; s.e. of exponent $= -0.34 \pm 0.026$). One can then calculate the allometry of the standard doubling time (t_{2x}), the period required for a young animal to double its mass, from ⅓ to ⅔ of the asymptotic or fledging mass, centered around the point of inflection and thereby encompassing the time of most rapid growth. The derived equations appear in Table 10-1. From these doubling times it appears that birds require much less time for postnatal as well as prenatal growth — one-third as much time at the 1-g intercept, or at the 1-kg size:

$$\text{bird/mammal } t_{2x} = 13.2 \; M^{0.34}/21.0 \; M^{0.25} = 0.63 \; M^{-0.09}. \tag{10-39}$$

The interclass difference is even greater if times to reach adult size are compared, for the young birds of most species attain adult mass (or may even exceed it with stored fat) before fledging. The time to fledging, t_{fl}, can be restated from Table 10-1 in terms of adult body mass:

$$t_{fl, precocial \; birds} = 11.1 \; m_{adult}^{0.20}, \tag{10-40}$$

$$t_{fl, altricial \; birds} = 11.0 \; m_{adult}^{0.18}. \tag{10-41}$$

One might expect t_{fl} to be shorter for precocial birds that have already developed somewhat by the time they hatch, but the equations are not significantly different.

For birds, as for mammals, the body-size-dependent times for embryonic and postnatal development and growth are related in a way that is exponentially similar to their physiological time scales. Because of this attainment of approximately adult mass by fledging time, that time should be comparable to the time, in days, for growth of mammals to 98% of adult mass:

$$t_{98\%} = 74.3 \ m^{0.26}. \tag{10-42}$$

The postnatal growth of mammals is much slower than that of birds, taking 6.8 times as long at the hypothetical 1-g size and relatively longer at larger sizes if the differences in exponents reflect a significant difference in the influence of body-size increase. This faster growth of birds would enable them to fend for themselves sooner; however, because sexual maturation precedes attainment of 98% of adult mass in mammals, they are undergoing perhaps more qualitative change while the birds are emphasizing quantitative change for the early youth stage.

The ratios of postnatal to prenatal development times are similar for precocial and altricial birds, but once again we see quantitatively the appreciably slower postnatal development of the eutherian mammals:

$$\text{precocial birds} = t_{fl}/t_{inc} \quad = 1.22 \ m^{0.03}, \tag{10-43}$$

$$\text{altricial birds} = t_{fl}/t_{inc} \quad = 1.21 \ m^{0.01}, \tag{10-44}$$

$$\text{eutherian mammals} = t_{98\%}/t_{gest} = 6.76 \ m^{0.00}. \tag{10-45}$$

The small residual-mass exponents suggest that postnatal growth is on an allometrical continuum with prenatal growth, and that the quantitative differences between oviparity and viviparity continue beyond birth. Generalizations about oviparity and viviparity should therefore include both embryonic and postembryonic growth, just as generalizations about either or both phases of growth should include the physiological time scale.

The postnatal growth rates during the linear phase of growth midway through development toward adult body mass in several vertebrate taxa have been analyzed allometrically (Case, 1978; Drent and Daan, 1980). The results are summarized in Figure 10-2. Clearly, the growth rates of poikilotherms are an order of magnitude smaller than those attained by the Mammalia, analyzed collectively, and are of the order of eutherian mammals analyzed separately, a parallel to the differences in metabolic rates of these groups. This led Case to conclude that "the evolution of endothermy was a key factor in lifting physiological constraints upon growth rates," a correlation noted first by Bertalanffy (1957).

Furthermore, just as the metabolic rates of the marsupials are interme-

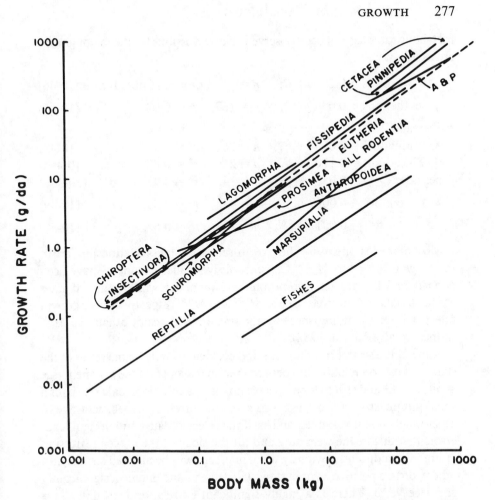

10-2 Postnatal growth rates of vertebrates during the linear phase of rapid growth, from the data of Case (1978). The relative positions of the eutherian *(dashed line)*, marsupialian, reptilian, and fish regressions are in the same qualitative order as their metabolic regressions. The respective ratios of growth to adult metabolism (g/da per W) are 1.95 $m^{-0.04}$, 0.22 $m^{0.08}$, 2.94 $m^{-0.10}$, and 1.09 $m^{-0.09}$. A & P = Artiodactyla and Perissodactyla.

diate between reptiles and eutherians but closer to those of the latter (Figure 10-2), the allometry of growth rates of marsupials is also intermediate (0.015 $m^{0.82}$ vs. 0.039 $m^{0.72}$ for Eutheria). Birds of the orders Pelecaniformes (pelicans, gannets, boobies, cormorants), Charadriiformes (gulls, terns, puffins), and Falconiformes (vultures, hawks) develop about five times as rapidly during this linear growth phase, although a parallel increase in metabolic rates has not been observed. The regressions for the major groups are as follows:

$$\text{birds} = 0.203 \, m^{0.75}, \quad n = 81 \, (3 \text{ orders}; \, r^2 \text{ not given}); \quad (10\text{-}46)$$

$$\text{eutherians} = 0.039 \, m^{0.72}, \quad n = 162, \quad r^2 = 0.92; \quad (10\text{-}47)$$

marsupials:

$$\text{per individual} = 0.015 \, m^{0.82}, \quad n = 26, \quad r^2 = 0.97 \quad (10\text{-}48)$$

$$\text{per litter (all)} = 0.170 \, m^{0.57}, \quad n = 26, \quad r^2 = 0.87 \quad (10\text{-}48a)$$

$$\text{per litter} \, (>1) = 0.088 \, m^{0.75}, \quad n = 14, \quad r^2 = 0.92 \quad (10\text{-}48b)$$

$$\text{reptiles} = 0.0047 \, m^{0.67}, \quad n = 43, \quad r^2 = 0.86; \quad (10\text{-}49)$$

$$\text{fish} = 0.0012 \, m^{0.61}, \quad n = 10, \quad r^2 = 0.62. \quad (10\text{-}50)$$

Most of the exponents for the growth-rate/body-mass relationships are in the range of $M^{\sim 2/3}$ to $M^{\sim 3/4}$. Conspicuously smaller exponents have been derived for fish, primates, lagomorphs, and ungulates, while the growth rates of marsupials and squirrels scale as $M^{0.82}$ and $M^{0.87}$ respectively. The growth rate per litter excluding species with single births scales exactly as $M^{3/4}$ in marsupials (Russell, 1982).

Some of the scatter in exponents and coefficients may be the result of the statistical inverse relationship between the coefficient and the exponent (see White and Gould, 1965, for a review and discussion). If the data set spans a small sample size or size range, or has a low correlation coefficient due to other independent variables, and/or if the selected data contain significant errors that distort the weighting and tilt the slope of the log-log regression around its midpoint, obviously the y-intercept will be affected strongly. In other words, a positive error in slope will decrease the intercept significantly and vice versa. There is a highly significant correlation ($p < 0.001$, $r = 0.797$) between the intercept (log of coefficient a) and the slope (exponent b) of 16 equations relating growth rate to size in groups of eutherian mammals listed by Case (1978, appendix 3). Again, correlation does not establish causation. Nevertheless, one must be aware of the possibility that because the analysis is limited to narrower taxonomic ranges, the body size range is usually reduced, and in the process the precision in determination of the exponents and coefficients is also reduced. There is therefore a problem in deciding what is artifact and what is growth strategy.

For biological or statistical reasons, the primates and ungulates showed a conspicuously different allometry of growth rates than the rest of the eutherians. If we arbitrarily exclude these groups, the mean of exponents for five orders is 0.69, close to the all-eutherian 0.72. The exploratory manipulations that follow will be based on the "all-eutherian" growth allometry.

The average embryonic growth rates of eutherian mammals are summarized in Eq. (10-33), which can be compared to maximum postembryonic growth as follows:

$$\text{postembryonic/embryonic} = 0.0389\ m_{adult}^{0.72}/0.0088\ m_{adult}^{0.66} = 4.42\ m_{adult}^{0.06}. \tag{10-51}$$

I have attempted to relate growth time scaling to the scaling of physiological time in general. For further comparisons it is necessary to convert growth rates (in grams per day) to the reciprocally related growth times, as the minimum times required for an increment of total growth:

$$\text{total growth/growth rate} = \text{minimum growth time,} \\ m^{1.0}/0.0389\ m^{0.72} = 2.57\ m^{0.28}. \tag{10-52}$$

This represents the relatively linear phase of growth from 5% to "30 to 50 percent of adult body weight" in Case's (1978) study, which is in good agreement with Stahl's equation (Table 10-1, line 7) for time to attainment of 50% of adult size.

Stahl also gave an equation for 98% growth time (Table 10-1, line 8), the reciprocal of which must approximate the average postembryonic growth rate ($\bar{m}$):

$$\bar{m} = 0.0135\ m^{0.74}. \tag{10-53}$$

This can now be compared to the average embryonic growth rate:

$$\text{postembryonic/embryonic} = 0.0135\ m^{0.74}/0.0088\ m^{0.66} = 1.53\ m^{0.08}. \tag{10-54}$$

Taken together, these equations suggest that the average growth rate increases postpartum by a minimum of 53% (if we ignore diverging exponents as probably insignificant). This would seem to point toward selection for an early, altricial birth, other things (like security for the neonate) being equal.

Provisions for Growth

There are several aspects of embryonic growth that have what are probably size-related needs. The developing embryo must obtain both oxygen and nutrients from the placental circulation, another step in the respiratory cascade. It is well known that fetal hemoglobin has a higher affinity for oxygen than maternal hemoglobin, which makes it possible for fetal blood to attain a greater fraction of saturation at the lower oxygen partial pressure that exists at the bottom end of the maternal respiratory cascade. In adults the oxygen affinity is inversely related to size (Schmidt-Nielsen and Larimer, 1958):

$$p_{50} = 34.7 \; M^{-0.054}, \tag{10-55}$$

where p_{50} is the partial pressure of oxygen (torr) necessary to attain half-saturation of hemoglobin with oxygen. This is explained as providing easier off-loading of oxygen for the more intense specific metabolic rates of the smaller mammals. Is there a parallel allometry of fetal p_{50}? To my knowledge, this question has not been answered.

In the other direction, mammals living at high elevations tend to have blood with stronger oxygen affinity to aid in uptake of oxygen from the alveolar air to the pulmonary circulation. Since the animal cannot "have it both ways," how is the oxygen affinity of the hemoglobin adapted to the needs of a pregnant masked shrew (Sorex cinereus) and her embryos, living at elevations of over 3,800 m in the Rocky Mountains (and perhaps in even more hypoxic situations than those of my casual observations)?

As we have seen, growth times are roughly proportional to the sort of mass dependency exhibited by physiological times in general, $M^{-1/4}$. Sivertsen (cited in Bourlière, 1964) showed an inverse correlation between the number of days needed for doubling of neonate mass and the protein content of the milk. However, that correlation was based on only 9 species. On a log-log regression for 21 species of terrestrial eutherians, the correlation disappeared (data from table 5 of Bourlière, 1964, excluding Pinnipedia and Cetacea, which have milk concentrations that are characteristically higher than those of other mammals). Case (1978) also found no significance in the correlation between growth rates and milk protein levels.

In view of the parallel scaling between growth times and physiological time scales, one might expect that the milk intake of the sucklings would scale in similar fashion to the food intake of the adults. Food intake of the adults should be scaled, like metabolic rates, to $M^{-3/4}$ (see Table 5-9). Milk intake of the young must be proportional to milk yield of the mother, which is, in kilograms per day:

$$\dot{M}_{milk} = 0.0835 \; M_{adult}^{0.765 \pm 0.022}. \tag{10-56}$$

This regression is based on data from 15 species of eutherian mammals not artificially selected for milk yield (22 data points, mouse to camel; Hanwell and Peaker, 1977). The scaling of the equivalent milk energy output per day does not appear to be significantly different (95% confidence limits of exponents overlap). In kilojoules per day this is

$$\dot{E}_{milk} = 532 \ M_{adult}^{0.694\pm0.041}. \tag{10-57}$$

To put this into perspective, the energy density of milk of naturally selected lactators is 2.24 times that of the artificially selected domestic cow's milk. Furthermore, if Eq. (10-57) is compared with Eq. (3-2), (3-13), or (3-13a) converted to common units, it can be seen that the energy output in milk from a nursing mother averages 1.8 times her basal metabolic requirement. Although this increment represents energy transported from the rest of her body, the final secretion is accomplished by a fantastic set of mammary organs, which amount to only 1% to 8% of total body mass from elephant to mouse size:

$$M_{mammary} = 0.045 \ M^{0.819\pm0.026}. \tag{10-58}$$

For the sample size of 14, the 95% confidence limits overlap the exponent for milk yield. However, a slightly higher scaling for organ mass would not be surprising, as it would reflect a need for proportionately more nonsecretory supporting tissue with greater organ mass.

Conclusions

Some past models of growth have incorporated weighting factors that were arbitrary, or at best approximate, to describe the scaling of growth rates and times to animal size. A more accurate description seems to result when Huxley's empirical heterogony or allometry is applied to a phylogenetic (as contrasted to an ontogenetic) size sequence. Of greater importance is our realization that just as growth is a physiological process, the pace of growth is an aspect of the same physiological time scale that marks the rest of the life history.

However, a state of delusion can arise from the amassing of numbers and formulas. Can they be applied to real life? What are their limitations? Do they have value? If so, the merits will lie in our ability to generalize about the patterns and to predict baseline quantities. It is clear, for example, that among the homeotherms oviparity is faster than viviparity, incubation taking 44 ($M^{-0.09}$)% as long as gestation in Eutheria. Postnatal growth rates are also faster, but reproductive maturation and aging are slower in birds after the initial rapid growth. Whether the embryonic growth is oviparous or viviparous, it accounts for about the same proportion of total growth from

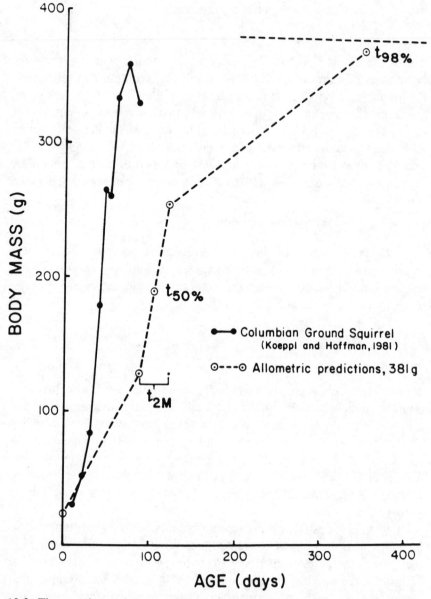

10-3 The actual growth curve of the Columbian ground squirrel compared with allometric predictions for its 381-g body mass.

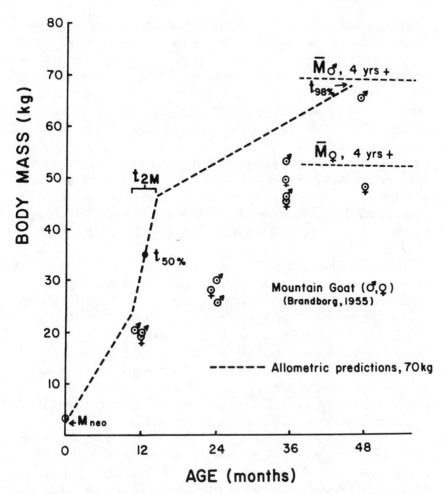

10-4 The actual growth data of the mountain goat compared with allomet-ric predictions for its 70-kg body mass.

zygote to adult, at least at the small end of the size range; the exponents diverge as species size increases.

What sort of predictive performance does the growth allometry offer? We can consider as examples two mammals whose geographic ranges overlap (see Figures 10-3 and 10-4). The actual growth curve of the Columbian ground squirrel *(Spermophilus columbianus)* rises considerably faster than a curve sketched from allometric predictions for neonatal mass, 50% and 98% growth times, and standard doubling time, taking only 40% as much time as the prediction for attainment of 50% of adult mass. Growth of the mountain goat *(Oreamnos americanus)* is, in contrast, considerably slower than predicted, taking over twice as long as predicted to reach 50% of adult mass. Perhaps the rapid mass gain of the ground squirrel is more fat for hibernation than additional flesh and bone. To speculate further, sparser food supply available for the mountain goat could be the cause of slower growth rates. A still larger Jersey cow takes only half as long as the prediction for attainment of 50% of adult body mass, because it is no doubt fed optimally for agricultural purposes (Crampton and Harris, 1969). The standard doubling time for postnatal growth was derived from data on only five domesticated species (Table 10-1) and should be reexamined using the abundant literature on mammalian growth. Recall also that allometric equations represent only the qualitative trends over several orders of magnitude in body mass, so a prediction that is off by a factor of two is not unlikely (see Figure 3-2).

We have examined only total growth, not the differential growth rates of components or the changes in form associated with internal differences of component growth. It would be interesting to see how body composition scales ontogenetically, as compared with the phylogenetic size progression described in Chapter 2. Although ontogenetic scaling is beyond the scope of this book, one example may provide a useful perspective. In most mammals brain mass increases in linear or slightly greater proportion to fetal mass, but postnatal increase in brain mass is proportional to a small fractional exponent of body mass, about $M^{0.1}$. Prenatal growth of the brain in humans and macaques *(Macaca mulatta)* is the same. After birth, whereas the macaque moves to a typical mammalian hypoallometric brain growth, the human brain continues to grow at the prenatal scaling until the infant is about two years old (Holt et al., 1975; Gould, 1977). Thus differences in eventual brain size and intellectual capacity, as well as the degree of child care and the type of parental behavior, are reflected in the ontogenetic allometry.

When I began to review the relationship between size and growth for this chapter, I had only a vague notion that there must be some sort of correlation. As with the next chapter, I found much more than I had expected. I came away convinced that a better understanding of the processes of reproduction and growth will emerge from wider application of allometry.

11 Life History and Body Size

It seems likely that an organism is an integrated system by virtue of the fact that none of its properties is entirely uncorrelated, but that most are demonstrably interlinked.

—E. F. ADOLPH (1949)

Because size scales the main life history parameters of mammals it should also be a central theme in ecology from the individual to the community level of organization.

—DAVID WESTERN (1979)

THE DEVELOPMENT of an allometric perspective on ecology has begun only recently. Before extrapolating into ecological allometry, we might find it useful to consider just what it is that the ecologist attempts to measure and explain: numbers of organisms, the space they occupy, the dynamics of their pace of living, and the energy for which they compete in order to support life and its replacement. The complex interactions which originate in these dimensional factors have stimulated a considerable body of theory. There are many abstract concepts such as niche widths, fitness, competitive exclusion, patchiness, and hyperspace that are not measured with calibrated devices in the M.K.S. or S.I. systems. Eventually, when the difficult tasks of accurately describing ecology have been mastered, we will be talking quantitatively about how numbers of animals of different kinds subdivide the available resources of space, time, nutrients, and energy. Qualitative matters such as aspects of behavior at the species level, strategic "decisions" on which food plant or habitat to select, and coevolution may not be within allometric reach for the time being.

However, there are many aspects of ecology that *are* readily amenable to allometric treatment, an opportunity that has been long overlooked: "Few branches of biology have attracted more analytical mathematical treatment than has the study of populations" (Cole, 1954). Despite three additional decades of intense interest, the simplicity of allometry and the predominance of size have not been brought to bear on this vital topic until recently.

The heart of population biology is the logistic equation

$$\frac{dN}{dt} = \frac{rN(K-N)}{K} = r_{max}N - r_{max}N\left(\frac{N}{K}\right). \tag{11-1}$$

Here N is the number of individuals in the population, r_{max} is the intrinsic rate of increase (or maximum rate in absence of constraints), and K is the carrying capacity. The concept of r- and K-selection represents a relative evolutionary "choice" along a continuum of life-history strategies. At one end is rapid population growth at rates approaching the maximum or intrinsic r; this is an opportunistic strategy for populations in variable or unpredictable, but often unsaturated, environments. At the other end of the continuum is a stable, slowly reproducing pattern for equilibrium populations (at carrying capacity K), where competitive ability is necessary to hold one's place. A number of characteristics have been correlated with these alternative strategies (Murray, 1979; Pianka, 1983). Among them are small body size in r-selection and larger body size in K-selection. Of the remaining 14 correlates listed by Murray, 6 can be put on allometric scales: rate of development, time to first reproduction, r_{max}, survivorship, longevity, and extent of sociality. Most of the remaining characteristics (except the climates favoring r- or K-selection!) can be reasoned from these allometric scalings (see Chapter 12). Whether the available niche calls forth the size with the appropriate life history, or the available size determines the niche, body size is the single most important characteristic of an organism.

Life in nature is far more complex than the life of a sample held captive in a laboratory. The ecologist cannot control the variables as a physiologist is able to do. It is to be expected that the patterns of the estimating equations obtained as least-squares fits from data gathered in the wild will not be so clear as the simple "Y vs. X" in the lab. However diffuse or however obscured by the uncontrolled variables that make nature esthetically appealing, the patterns of size dependence are there; size still says more about what the animal needs than any other feature. Thus the allometry of space, time, and energy is as basic to the currently popular study of "life-history strategies" as it has been to the study of animal design in physiology and functional morphology.

As we make this progression from organismic to environmental allometry, we are using the same sort of attack that foresters use on large forest fires. They start building the fire line at a relatively quiet base of the fire, where control can be imposed most readily — not at the leading edge where the fire is raging out of control. We have managed to secure an allometric line around anatomy, physiology, and reproduction; we may now proceed in an attempt to bring ecology under control as well. While this analogy may exaggerate a bit, we are starting with a framework gained expediently in the controlled situation of the laboratory and seeing if it can be extended to the real world.

In some cases size-dependent patterns are apparent without $Y = aM^b$. A beautiful example is the relation of body size to life history dramatized in the biology of African antelopes and buffalo. Africa has more than 70 species of these animals, amounting to 64% of the living species of Bovidae in the world. This amazing diversity is not manifested in a randomness of life histories. Jarman (1974) has distinguished five classes of social organization, which are related to five classes of feeding styles:

Class A. Animals are found singly or in pairs on one small home range throughout the year, mostly browsers that are quite "particular" about the parts of plants they will eat, although utilizing a diversity of species.

Class B. Animals live in groups of 1 to 12, most commonly 3 to 6, on one home range throughout the year, entirely grazing or entirely browsing, "very selective" for plant parts of high nutritive value.

Class C. Animals occur in groups of 6 to 200, most commonly 6 to 60, feeding "rather selectively" on a range of grasses and browsing a range of vegetation types within a "fairly large home area."

Class D. Animal groups of 6 to a few hundred occupy a "poorly defined home area" and migrate seasonally. Their food is largely grasses, unselective for species, but more selective for parts or growth stages; food is of low nutritive value.

Class E. Permanent herds numbering many hundreds to 2,000 move seasonally within a large home area, feeding unselectively on forage of low nutritive value.

The pattern in this diversity is one of body-size dependence (see Figure 11-1). The correlation with body size of both size of social group and type of feeding is obvious. Generalized into a continuum, the following tendencies go with increased body size: (1) larger social groups, (2) larger home range, (3) more likely migration, (4) less selective feeding, (5) lower nutritive value in food.

There are a number of interesting questions here. Can these semiquantitative generalizations be quantified allometrically? If group size, home range, and body size all increase together, what is the combined effect on

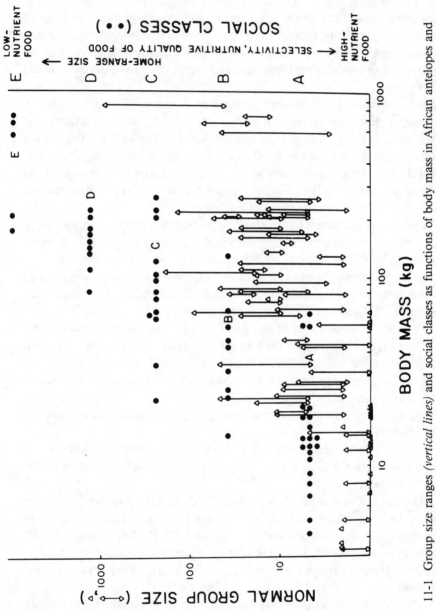

11-1 Group size ranges (*vertical lines*) and social classes as functions of body mass in African antelopes and buffalo. Large body size is correlated with large group size, large home range, and less selectivity in the diet. (Modified from Jarman, 1974, *Behaviour* 48:226, 240; with permission.)

biomass of large herds of large animals? Was the evolution of feeding behavior and body size not a highly interdependent process? Is there an ecological parallel to the physiological relationship of size, rate, and time scale?

A hectare of habitat can produce food at a rate dependent upon climate, soil, and kinds of plants. What will be the composition of the animal community that depends, in turn, upon the food resources? The plants may produce a variety of foods: seeds, fruits, roots, buds, leaves, bark, stems, sap, and nectar. Some of these may be "offered" in order for the plants to utilize the mobility of the animals as pollinators or disseminators of seeds, as is the case with nectar, pollen, nuts, and other fruits. More often there is no benefit, and there may even be harm to the plant, from the grazing, browsing, or uprooting of plant parts by the animals.

In any case, there is a finite limit to how much the plants can produce, in terms of microliters of nectar per hour or vegetative growth per season. Overgrazing would not only limit the foraging efficiency of an animal, but it could damage the plants and reduce their productivity. Within the plant the energy density and nutritive value of the various parts and growth stages varies considerably. Per gram of food, seeds contain considerably more energy than leaves or stalks. As a consequence (perhaps a coevolved consequence involving both plants and animals), there has been an adaptive radiation of body size. Thus we see small nectar feeders and seed harvesters and large grazers and browsers, as well as a diversity of sizes of their predators. The characteristics of the community of consumers depend on both external and internal relationships—not only type of food, seasonal availability, and rate of production by the primary producers, but size of animal, metabolic rate, and life history of consumer. All of these characteristics are interrelated.

In evaluating the functional relationships of animal and environment, we can classify the organismic properties as those of structure, rate, and time, all of which have been correlated with body mass in preceding chapters. The structural characteristics include linear dimensions proportional to $M^{1/4}$ to $M^{1/3}$ and organ masses and capacities from $M^{2/3}$ to $M^{\sim 1.1}$. Rates are generally functions of $M^{3/4}$, while times are related to $M^{1/4}$ (and frequencies of timed or cyclic events tend to be scaled to $M^{-1/4}$).

Given that life is scaled in these patterns at the organ and organismic levels, it is reasonable to expect such scaling to be evident in the life history of the animal in the community, and for the community structure to reflect these characteristics. In this chapter we consider animal ecology as it relates to animal size and internal function. Size and time are seen to be the principal characteristics of life history and ecology.

While ecological allometry is in its initial stages of development, the dimensions must be used clearly and precisely. Too often in the ecological literature the entities of interest and their dimensions are not clear, resulting in meaningless equations which may incorporate factors that cannot be added legitimately because they do not have the same dimensions. For example, a trap into which more than one ecologist has fallen is energy + energy/time = ? This can be avoided by rigorous dimensional analysis. The dimensions on both sides of an equality as well as all additive and subtractive terms must be consistent. This is facilitated by use of standard symbols for all terms having the same dimensions, for instance, $M =$ mass in kilograms, or $m =$ mass in grams, $V =$ volume, $A =$ area, $s =$ speed or $u =$ velocity, $a =$ acceleration, $F =$ force, $p =$ pressure, $E =$ energy, $\dot{E} = dE/dt =$ power, with subscripts to distinguish different terms with the same dimension. Then it will be clear immediately that dimensional consistency prevails (Calder, 1982a).

Since allometric relationships are merely empirical correlations that do not establish cause and effect (although they may be instructive in pursuit of that goal), dimensional consistency will not be apparent on both sides of the allometric equation. The coefficient a (in $Y = aM^b$) may be thought of as containing or summarizing "hidden dimensions" (Riggs, 1963). Recall that when the mass term is governed by an exponent other than the isometric 1.00, conversion of an expression for mass M in kilograms to m in grams does not simply change the coefficient a by a factor of 10^{-3}, but by $(10^{-3})^b$.

Home on the Range

Kleiber, and a host of physiologists who followed him, have confirmed that metabolic rates (at rest and in activity, homeothermic and poikilothermic) are proportional to $M^{3/4}$ (Kleiber, 1961; Bartholomew, 1982). Logically, the internal functions that support the metabolism, such as ventilation and circulatory flow rates for delivery of oxygen and elimination of wastes, also scale approximately as $M^{3/4}$. Since the internal metabolic support systems are matched to metabolic needs, ecological variables relating to food supply should, in turn, also be functions of $M^{3/4}$. Thus we are biased toward finding $M^{3/4}$ patterns and quite legitimately look for them. It is exciting to find another $M^{3/4}$, and tempting to imagine one.

The scaling of home ranges of mammals was one of the first ecological

Table 11-1 Home ranges and territories of terrestrial vertebrates.

Group	Area (ha)		r	n	p	References
Mammals						
Primary con-sumers	Home range =	$4.7\,M^{1.02}$	0.934	32	0.001	Calder, 1974[a]
	=	$2.3\,M^{1.02}$	0.866	28	0.004	Harestad and Bunnell, 1979
Secondary consumers	=	$66.8\,M^{1.22}$	0.956	8	0.001	Calder, 1974[a]
Omnivores	=	$34.0\,M^{0.92}$	0.949	7	0.001	Harestad and Bunnell, 1979
Carnivores	=	$13.2\,M^{1.36}$	0.900	20	0.001	Harestad and Bunnell, 1979
Birds						
Passerines and nonpasser-ines	Home range ∝	$M^{1.16}$	—	75	0.001	Schoener, 1968
	Territory ∝	$M^{1.09}$	—	61	0.001	Schoener, 1968
	Both A_{hr} and territory	$98.6\,M^{1.15}$	0.773	77	0.001	Schoener, 1968
Herbivores	Territory ∝	$M^{0.70}$	—	3	n.s.	Schoener, 1968
Omnivores	Home range ∝	$M^{0.51}$	—	26	n.s.	Schoener, 1968
	Territory ∝	$M^{0.35}$	—	19	n.s.	Schoener, 1968
Carnivores	Home range ∝	$M^{1.39}$	—	46	0.001	Schoener, 1968
	Territory ∝	$M^{1.31}$	—	39	n.s.	Schoener, 1968
Lizards	Home range	$4.8\,M^{0.95}$	—	—	—	Turner et al., 1969

a. Citing Buskirk, unpublished.

variables thought to fit this pattern. In separate regressions for "croppers" (≈herbivores) and "hunters" (≈carnivores), home ranges appeared to scale as $M^{0.63}$. Statistically this exponent did not differ significantly from ¾, and the scaling of home range to metabolic needs seemed logical. This was accepted with perhaps too-hasty enthusiasm. Later, however, a more extensive data base showed that mammalian home range actually scales linearly or hyperallometrically ($M^{b \geq 1.0}$; see Table 11-1; Buskirk, cited in Calder, 1974; Harestad and Bunnell, 1979).

Home ranges of birds also seem more or less to parallel the mammalian pattern described above. A preliminary analysis of bird data yielded mass exponents equal to or exceeding 1.0 for territories and home ranges

(Schoener, 1968). A regression line based on the data in Schoener's table 1 for all birds predicts avian home ranges greater than those for herbivorous mammals of equal body weight and home ranges greater than those for omnivorous mammals, when body weight exceeds 10 g. These predictions are consistent, at least qualitatively, with the fact that in flight, as opposed to while running, higher speeds are attainable and energy costs are lower (the ratio of metabolic costs of running and flying is 2.16 $M^{0.05}$). Home-range size approaches a linear scaling for lizards as well ($\propto M^{0.95}$; s.e. and c.i. not given; Turner et al., 1969).

The advantages cited allow birds to patrol and forage over larger home ranges. However, when Schoener separated birds according to diet, herbivorous and omnivorous birds were found to have mass exponents less than one ($M^{0.35}$ to $M^{0.70}$). These correlations were not statistically significant, the omnivorous birds being of limited size range (12.2 g to 622 g) and the herbivorous birds encompassing not only a small size range (48 g to 1,050 g) but also a small sample size ($n = 3$). The foraging area is of a more pronounced hyperallometry ($M^{1.22}$, $M^{1.36}$, $M^{1.31}$, $M^{1.39}$) for carnivorous birds and mammals than for avian and mammalian vegetarians and omnivores.

Home-range area (A_{hr}) for 20 species of primates, from 1.1 kg to 160 kg body mass, is proportional to (average daily metabolic needs, $\dot{E}_{tot}$)$^{2.28}$, but simple models relating the two variables did not predict the observed relationships (Harvey and Clutton-Brock, 1981). The primates can be compared with other eutherian mammals by combining Eq. (8-40) or (8-40a) and the first equation in Table 11-1:

$$\dot{E}_{tot} \propto M^{0.67}; \qquad M \propto \dot{E}_{tot}^{1/0.67=1.5}; \tag{11-2}$$

$$A_{hr} \propto M^{1.02} \propto (\dot{E}_{tot}^{1.5})^{1.02} \propto \dot{E}_{tot}^{1.53}. \tag{11-3}$$

While this exponent for $\dot{E}_{tot}$ is within the wide 95% confidence interval (1.27 to 6.15) of the $\dot{E}^{2.28}$ for primates, this correlation (at least as a first attempt) does not yield any new insight. However, the approach is certainly valid and should be useful in the future. The expectation for A_{hr} to scale with metabolic needs has not been supported by empirical evidence (Harestad and Bunnell, 1979; Mace and Harvey, 1983), an enigma that should be one of the major challenges in ecological energetics. Why do larger mammals hold more land than their energy requirements suggest would be necessary? One idea is that as area increases, it encompasses proportionally more marginal habitat, which could be uneconomical to harvest. What has been overlooked is the fact that larger animals on larger areas tend to live in larger social units, so that more animals are sharing the same home range.

Home Range and Population Density

What were the prerequisites or predisposing factors involved in the evolution of sociality? Sociality has been defined as the "state of group formation when members of a population of differing sex and age structures have the same space, *i.e.*, have markedly overlapping home ranges" (Armitage, 1981). A principal-component analysis of life-history traits in ground squirrels did not yield a correlation between a "sociality index" and body size. However, when a much larger body-size range is examined allometrically, the results for this intriguing biological problem are more promising, placing Armitage's definition of sociality within the "central theme" urged by Western, in the quotation at the beginning of this chapter.

The present analysis is limited to herbivorous eutherian mammals, since data acquisition and allometric analysis have proceeded farthest for this group. Population density (ρ_n) is inversely dependent upon body size in herbivorous mammals (Damuth, 1981a):

$$\log \rho_n = -0.75 \,(\log m) + 4.23, \tag{11-4}$$

where density is number of animals per square kilometer, $r = -0.86$, $S_b = 0.026$. Converted for comparative purposes, this equation reads:

$$\rho_n = 95.5 \, M^{-0.75}. \tag{11-5}$$

The carnivores show a parallel relationship with a lower absolute population size for any given body mass but the same proportionate decrease of home-range area with increase in body size (Mohr, 1940).

On the basis of reciprocal dimensions we might expect population density $(\rho_n$, animals per square kilometer) and home-range size (hectares per animal) to have the same numerical exponents but of opposite sign. The inverse of population density has the following dimensions of area per animal (in km^2 and ha respectively):

$$\rho_n^{-1} = 0.0105 \, M^{0.75}, \tag{11-6}$$

or $\qquad \rho_n^{-1} = 1.05 \, M^{0.75}. \tag{11-7}$

Can we reconcile values for area per animal based purely on population density (Eq. 11-4) with values for home range (Table 11-1) such as the following?

$$A_{hr} = 2.3 \, M^{1.02}. \tag{11-8}$$

If both Eqs. (11-7) and (11-8) approximate the truth, herbivorous mammals that are larger than 55 g body mass must actually range over an area greater than their per capita share of the local habitat, and their home ranges must overlap in some fashion (Damuth, 1981b; see Table 11-2). The degree of overlap can be expressed in terms of the ratio of $A_{hr}/\rho_n^{-1}(=A_{hr}\rho_n)$:

$$A_{hr}\rho_n = 2.3\ M^{1.02}/1.05\ M^{0.75} = 2.19\ M^{0.27}. \tag{11-9}$$

This means that a body-size increase by a factor of 10^4 should result in a home-range overlap about 10 times as great. Might this equation predict approximate herd or harem size for an animal of a given body size?

Depending on which expression for home range is used, the allometry of population density and home-range size indicates that for animals up to the size of a hare (2 kg), only one to five home ranges must fit in the same per capita share of available space calculated from population density. These smaller mammals would therefore be expected to live as solitary individuals, in pairs, or in small family groups. (Did you ever run into a herd of mice?) From the size of a marmot (*Marmota flaviventris,* 3.6 kg) to the size of the wapiti (*Cervus canadensis,* 300 kg), even greater home-range overlap means that larger numbers of individuals are expected to occupy the same area (resulting in groups of 3 to 6 individuals for marmots and 10 to 21 individuals for wapiti; Figure 11-2). Table 11-2 compares social group sizes actually observed in species of different body sizes with group sizes predicted from the allometry of home-range and population density. Another size-dependent factor that facilitates a social group is the longevity of experienced animals from which the young can learn (the "old-timer index" in Table 11-2 and the "shared patrol" in Table 11-3 later).

The validity of the relationship appearing in Eq. 11-9 was confirmed in Damuth's tabulation of data on body mass, home-range area, and population density. He derived an empirical relationship directly for 18 species of herbivorous eutherian mammals (*M* range: 0.030 kg to 380.2 kg):

$$A_{hr}\rho_n = 0.44\ m^{0.34} = 4.6\ M^{0.34}. \tag{11-10}$$

The correlation coefficient ($r = 0.84$) is highly significant ($p < 0.0001$) and the 95% confidence interval of the exponent 0.34 ± 0.11 included the exponent 0.27 of Eq. (11-9).

The greater the product $A_{hr}\rho_n$, the greater the intensity of competition and the greater the desirability of social systems that contribute to mutual toleration within these groups. Instead of wasting energy on confrontation and physical conflict, members of social systems receive common benefits such as predator alert systems and "baby-sitting" services. (Our own species could profit from this lesson!)

Table 11-2 Allometric predictions for herbivorous mammals, showing the advantages of body size for sociality.

Animal size equivalent	Body mass (kg)	Home range (ha)		Per capita space (ha)	Overlap (individuals/home range)			Old-timer index[f]	
		a	b	c	a ÷ c	b ÷ c	Observed	$100(t_{1/2} \div t_{max})$	$100(t_{sc} \div t_{max})$
Mouse	0.02	0.043	0.088	0.056	0.77	1.57	Solitary, pairs, families?	24	6
Vole	0.05	0.108	0.221	0.111	0.97	1.99		28	7
Bunny	1.0	2.3	4.7	1.05	2.19	4.48		44	9
Hare	2.0	4.7	9.5	1.77	2.64	5.38		48	10
Marmot	3.6	8.5	17.4	2.74	3.10	6.35	6–8[d]	53	10
Pronghorn	50	124.4	254.1	19.74	6.3	12.9	4–9[e]	79	13
Wapiti	300	773	1,580	75.7	10.2	20.9	13–61[e]	104	15

a. $A_{hr} = 2.3\ M^{1.02}$ (Harestad and Bunnell, 1979).

b. $A_{hr} = 4.7\ M^{1.02}$ (Buskirk, in Calder, 1974).

c. $p/n = 1.05\ M^{0.75}$ (calculated from Damuth, 1981a).

d. *Marmota flaviventris* (Armitage, personal communication).

e. (Cahalane, 1961; Burt and Grossenheider, 1964; Bourlière, 1964).

f. See Eqs. (11-38) and (11-59).

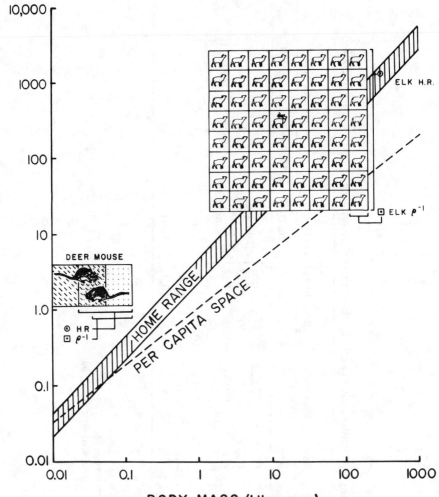

11-2 Home-range areas of herbivorous mammals are linearly proportional to body mass, exceeding the scaling of per capita space calculated from population density and leading to progressively greater home-range overlap for larger animals. The upper limit of the A_{hr} plot is from the Buskirk regression, the lower from Harestad and Bunnell (see Table 11-1).

One could argue that since additional advantages would accrue to each member of a large social group, and since the cost of transport is relatively lower for a large animal, larger animals could expand their home ranges beyond the area required to provide sufficient food in order that their home ranges would overlap with the home ranges of more individuals. However, the concept of relatively cheaper transport for larger animals is an artifact of transportation costs' being customarily expressed *per unit body mass*. Taylor and colleagues (1982), for example, expressed the incremental cost of running ($\$_{run}$) as

$$\$_{run} = 10.7 \ M^{-0.316}. \tag{7-36}$$

Certainly the negative exponent indicates an inverse relationship between the cost of running per unit mass and body mass. Meanwhile, the *total mass* that the animal must transport is, of course, $M^{1.0}$. Therefore, if $\$_{run}$ is expressed in joules per meter (instead of in joules per kilogram per meter, as in Eq. 7-36):

$$\$_{run} = (10.7 \ M^{-0.316})(M^{1.0}) = 10.7 \ M^{0.68}. \tag{11-11}$$

The total cost of crossing the home range is the product of cost of running and distance. The distance (l_{hr}) or perimeter of the area is proportional to the square root of home-range area (Table 11-1); if the A_{hr} according to Harestad and Bunnell were a square,

$$l_{hr} \propto (23,000 \ M^{1.02})^{1/2} \propto 152 \ M^{0.51}. \tag{11-12}$$

The total cost of crossing would then be

$$\$_{run} = (\text{J m}^{-1})(\text{distance}) = (10.7 \ M^{0.68})(152 \ M^{0.51}) = 1,622 \ M^{1.19}. \tag{11-13}$$

This equation clearly shows that the cost of crossing home ranges in the observed size range does *not* decrease as body size increases; in fact it increases faster than body size. However, if the necessary patrolling were shared by several members of a social unit, a decrease per individual could be accomplished.

Size, Competition, and Energy Conversion

There are several interesting implications to the fact that population density is proportional to the inverse of $M^{3/4}$ (Damuth, 1981a,b). First, the basal rate of energy consumption of a whole population of density ρ_n is independent of body size:

$$(\rho_n)(\dot{E}_m) \propto (M^{-0.75})(M^{0.75}) = M^{0.0}. \tag{11-14}$$

This means that no herbivorous species of mammal can outstrip another species, energetically, merely by virtue of its size. Second, since the efficiency of conversion of plant material to primary consumer biomass seems to be size independent (Kleiber, 1961; Humphreys, 1981; Schroeder, 1981), the energy available to the secondary consumer or predator is likely to be independent of prey size. Third, the standing crop biomass (M_{sc}) of a herbivorous species is positively correlated with size, being equal to the population density times the average mass of individuals of that species:

$$M_{sc} = (\rho_n)(M) \propto (M^{-0.75})(M) = M^{0.25}. \tag{11-15}$$

From the size independence of both population metabolic rate and conversion efficiency it can be concluded: "Species of small mammals are able to produce, on average, the same amount of biomass over time as do species of larger mammals, whereas at a given moment their standing-crop biomass is considerably less, because the population turnover rates and individual growth rates per unit weight of small species are much greater" (Damuth, 1981a). Damuth's analysis confirms the observation that "with larger body size there are fewer species and fewer species per genus, but not necessarily less participation in the energy flow" (Van Valen, 1973). The smaller species have a smaller standing crop but are living at a faster pace, the two factors essentially canceling out. In fact, the biomass turnover per square kilometer per day of mouse to wapiti-sized herbivore decreases by only 30% as body size increases by 1,500,000%. This biomass turnover rate is calculated by allometric cancellation of biomass (Eq. 11-15) and turnover time (to be discussed later in this chapter).

Overlooked in this analysis is the fact that there tend to be more species of smaller-sized mammals in a local fauna. For example, the vicinity of the Rocky Mountain Biological Laboratory in Gothic, Colorado, has (or has had) seven species of mouse-sized rodents (10^1 g to 10^2 g body mass), six species of rodents and lagomorphs (10^2 g to 10^3 g), four herbivore species (10^3 g to 10^4 g), three species of 10^4 g to 10^5 g (beaver, deer, bighorn sheep), and only one over 10^5 g (American wapiti). This can be expressed allometrically as:

number of species per order of magnitude $= 4.36 \ M^{-0.20}$ (11-16)

($r^2 = 0.88$). The allometric species diversity of local faunas should be examined elsewhere as well.

While the rate of energy consumption *by a species* is independent of its size, the rate for all the herbivorous mammals in the community may be higher at the small end of the scale, and the biomass may scale as something less than the $M^{1/4}$ indicated in Eq. (11-15). However, the corrections would not scale as steeply as suggested by Eq. (11-16), because of differences in relative abundance and lack of complete overlap in the "fine grain" of the environment. Obviously there is opportunity, and need, for collaboration between the field biologist and the allometrician. Another qualification regarding Eq. (11-14) is that the daily metabolic rates of wild animals do not necessarily scale as $M^{3/4}$, but appear more likely to scale as $M^{2/3}$ (Table 8-3).

Crossing and Patrolling the Home Range

Matters are simplified if we treat the home range as a square or circle. The linear distance across width or diameter is proportional to the square root of that area. Travel speeds of mammals appear to scale as $M^{-1/4}$ (see Eqs. 7-3 and 7-12). Another expression of travel speed, in effect, is the distance mammals move in a day (Eq. 7-14). Dividing this into distances traveled per hour through a 12-hr active day length, the average speed (meters per hour) from Eq. (7-14) would be

$$\dot{L}_{dm} = 86.5 \, M^{0.25}. \tag{11-17}$$

Excluding the Carnivora, which range farther than the herbivores, the equation for the other mammals, mostly omnivores or herbivores (Garland, 1983b), is

$$\dot{L}_{dm} = 72.9 \, M^{0.22}. \tag{11-17a}$$

The scaling exponent for travel speed is essentially independent of gait or urgency.

The time required for a herbivorous mammal to cross its home range would be

$$\text{time} = \text{distance/speed} \propto (A_{hr})^{1/2}/u. \tag{11-18}$$

Substituting (11-8) and (11-17a) into (11-18) and converting to units of meters and hours, an average crossing time ($\bar{t}$) might be

$$(\bar{t} = (2.3 \times 10^4 \, M^{1.02})^{1/2}/72.9 \, M^{0.22} = 2.08 \, M^{0.29}. \tag{11-19}$$

Traveling at the greater speed of the transition from a trot to barely galloping, the crossing time could be predicted by substituting (11-8) and (7-3) into (11-18):

$$t_{t-g} = (2.3 \times 10^4 \, M^{1.02})^{1/2}/5,500 \, M^{0.24} = 0.028 \, M^{0.27}. \tag{11-19a}$$

The similarity of the scaling factors for this travel and other physiological times suggests that the size of the home range is related to the animal's size and physiological schedule in a fashion such that crossing one's home range is a "similar movement" performed in a similar physiological time.

Proceeding from known values as "calibration," Pennycuick (1979) used theoretical arguments to derive the allometry of walking speed u (meters per second) and foraging radius r (meters) of ungulate mammals of the Serengeti. These differ in exponents from those of Eqs. (11-12) and (11-17) above.

$$u = 0.50 \ M^{0.13} \tag{11-20}$$

$$r = 2,000 \ M^{0.40}. \tag{11-21}$$

The time for crossing that would be predicted from these relationships would be twice the radius divided by the speed (converted to hours and kilometers):

$$t = 2(2.0 \ M^{0.40})/1.8 \ M^{0.13} = 1.11 \ M^{0.27}. \tag{11-22}$$

Essentially the same scaling factor is obtained in either case, but the home range (πr^2 or $4r^2$ if circular or square, respectively) would not scale as $M^{1.02}$, determined for North American mammals by Harestad and Bunnell (1979), but rather as $M^{0.80}$. Is this a geographic or phylogenetic difference, or does it merely reflect the crudity of our preliminary analysis? Here is another opportunity for further study.

How much distance must a mammal travel daily to secure its territory or home range? Garland's empirical descriptions of L_{dm} (Eqs. 11-17 and 11-17a) include feeding as well as any turf disputes and patrolling, but presumably the visual presence of the feeder and its awareness of encroachers helps to advertise or exercise the claims. Hence, predictions from Eq. (11-17a) should provide a reasonable approximation for our present consideration, if we exclude the more wide-ranging carnivores (see Table 11-1). The larger species can be assumed to be overlapping in a social unit of $2.19 \ M^{0.27}$ animals (Eq. 11-9). The aggregate travels of all would be the product of Eqs. (11-9) and (11-17a):

$$L_{dm,total} = (2.19 \ M^{0.27})(72.9 \ M^{0.22}) = 160 \ M^{0.49} \tag{11-23}$$

in meters per hour, or in a 12-hr day, 1.92 km. For much of this travel the animals would be bunched together, but the interesting curiosity is that if there were any tendency to split up or spread out, the aggregate L_{dm} for the group would resemble, in exponent, the scaling of a linear dimension of the home range, $A_{hr}^{1/2}, \propto M^{1/2}$. Or view it the other way around: if each animal were responsible for a characteristic linear distance scaled as $M^{1/2}$, why are the L_{dm} scalings derived by Garland all close to $M^{1/4}$?

Finally, what if home ranges were scaled at a metabolically sufficient $M^{3/4}$, without social overlapping and mutual defense of larger $M^{1.0}$ home ranges? Characteristic linear dimensions (diameter or perimeter) being proportional to the square root of the area, the patrol distance might then scale as $(M^{3/4})^{1/2}$, or $M^{3/8}$. This exponent is greater than that for the observed daily movement distances that are adequate for social animals, so as animal size and home-range size scale up, proportionately more patrolling would be necessary for the solitary owner of the hypothetical $M^{3/4}$ home range.

Territory Size in Birds

Territory size for 61 species of birds (not separated into feeding types) is proportional to $M^{1.09\pm0.11}$ (Schoener, 1968). As in the preceding treatment of home ranges of herbivorous mammals, the territory is assumed to have the shape of a square or a circle. The distance across the territory (width or diameter), and therefore the patrol distance around the perimeter, is proportional to the square root of the area (Table 11-1).

How long does it take for a flying bird to cross its territory? The approximate speed of flying birds scales as $M^{0.17}$ (Eqs. 7-15 and 7-16). The time required to cross the territory should then scale as follows:

$$t_{crossing} = \frac{\text{distance}}{\text{speed}} \propto \frac{(M^{1.09})^{1/2}}{M^{0.17}} \propto M^{0.38}. \tag{11-24}$$

When Schoener's 61 species of birds were categorized by diet, however, body-mass exponents for territorial area (A_t) for the three feeding types varied widely (Table 11-1): $M^{0.70}$, $M^{0.35}$, and $M^{1.31}$. Using these mass exponents, but continuing to assume that the speed for all three groups scales to $M^{0.17}$, we find that according to Eq. (11-24), the times for crossing the territory would scale as $M^{0.18}$, $M^{0.01}$, and $M^{0.49}$. Only the last is based on a significant scaling exponent for A_t; this exercise is merely a suggested approach for future studies of territory scaling.

More than mammals, which need to patrol the boundaries of their home ranges in order to maintain their claim, songbirds can rely in part on the option of substituting a broadcast song for physical presence at the territory boundary. They are frequently able to save themselves the added energetic expense of flying across the territory at a rate five or more times the metabolic rate while perched. Let us assume that the power output of the song is directly proportional to the respiratory flow rate (milliliters of air per minute), which scales as $M^{0.74}$ (Brackenbury, 1977, 1980; see also Chapter 5). The volume of the song attenuates with the square of the distance from the source of the song. (Attenuation characteristics of different types of vegetation in the path of the sound waves also differ, but for simplicity's sake this additional, complex variable is assumed to be size independent.)

Heuwinkel (1978) found that the sound pressure level of the reed warbler (*Acrocephalus scirpaceus*) was adequate to mark territories acoustically. On the basis of the allometry for power output and assuming size-independent auditory sensitivity, a song will be effective at the other side of a bird's territory if the diameter of the territory is proportional to $M^{0.37}$, which would seem to mean that the area of the territory should be proportional to $M^{0.74}$. This is not the case for carnivorous or omnivorous birds, and the scaling of territory size for the herbivorous birds in Schoener's study ($n = 3$) is not statistically significant although it corresponds to the prediction.

Power output is not the only size-dependent factor related to bird song. High frequencies attenuate more rapidly than low frequencies. Since the resonant frequency of a vibrating membrane (such as the syrinx and the tympanic membrane of the middle ear) is inversely proportional to membrane length, song frequency should bear an inverse relationship to body size. The correlation is not quite that simple; still, there is some evidence for such a relationship in the warblers of the genus *Sylvia* (Bergmann, 1976). Thus it is possible that a larger songbird's vocal advertisement travels farther not only because of his larger respiratory capacity, but also because of lower frequencies in the song. Actually, bird songs contain a mixture of high and low frequencies, and differences in their attenuation can serve as a measurement of distance from the source (Richards, 1981). The speculative nature of presumed relationships among body mass, territory size, and bird song points out the need for additional study in this field.

Time Scales in Ecology

An animal's life, in the simplest ecological terms, is composed of genetic inheritance, an allotted span of time, the space controlled and the energy acquired from the use of genes and time, and the reconversion of energy surplus to maintenance requirements to n multiples of the original genetic information (where n = number of offspring). Of these, the terms that are size dependent or presently suitable for allometric analysis are time, space, energy, and number of offspring. In this section we consider the dimension of time.

Only a static view of an ecological unit or system can come from counting or weighing the population or biomass. The understanding of ecological *processes* must be dynamic. Static and dynamic views differ by the dimension of time. It is not enough to consider the amount of food collected, without regard to the time in which it is harvested; in the long run it is the *rate* of energy input that must keep up with *rate* of metabolism. The lifetime basal energy requirement per kilogram of animal is essentially constant, independent of size (Stahl, 1962). The mouse and the moose differ primarily in the pace of their lives, the physiological time scale which is approximately proportional to $M^{1/4}$. This means that the turnover times for nutrient and energy flow through the animal are size dependent. It would seem to follow that physiological time would have great significance too in the ecology of animals.

Several ecological time scales can be found (or derived from the literature) that have similar body-mass dependencies in mammals: (a) crossing or circling the home range, $t \propto M^{0.26}$, (b) foraging times, (c) population doubling at r_{max}, $t \propto M^{0.26}$ to $M^{0.36}$, (d) growth and maturation times, $t \propto M^{0.18}$ to $M^{0.29}$, (e) population cycling time, (f) maximum life span $t \propto M^{0.20}$, (g) turnover of the standing crop biomass, $t \propto M^{0.29}$ or $M^{0.33}$.

These exponents are similar to those for standing crop biomass ($M_{sc} \propto M^{0.25}$) and degree of overlap in home range ($A/\rho^{-1} \propto M^{0.27}$), for which there is a plausible explanation (see below). Thus, the absolute or environmental times are progressively longer for the larger animals, which also have greater standing crop biomasses and greater overlap in use of space. These are calculated in Table 11-3 for a series of hypothetical herbivorous mammals ranging in size from mouse to wapiti.

FORAGING TIME AND FREQUENCY

The survival of any individual — and, in turn, the species — depends upon the acquisition of enough food to fuel the metabolism. It is not surprising, then, that a wide array of life-history characteristics such as home range, territory, distribution, life span, and population dynamics are directly or indirectly related to the rate of procurement of adequate amounts of food. Furthermore, as is the case at the physiological level, size and other physical dimensions must be considered if comparative accounts or theoretical models are to be meaningful.

The size dependencies of mammalian foraging variables such as the frequency and amount of travel necessary have been derived by Garland

Table 11-3 Allometric predictions from ecological time scaling for herbivorous mammals.

| | | | Speed | | Ecological times | | | | |
| | | | | | Crossing home range | | | | |
Animal size equivalent	Body mass (kg)	Home-range width (m)[a]	$\bar{u}$ (m/hr)	u_{l-g} (m/min)	Gallop (min)	Average (hr)	Shared patrol (hr)[b]	Population doubling, minimum (da)	Standing crop turnover (da)
Mouse	0.02	23	31	36	0.64	0.75	0.97	79	83–119
Vole	0.05	37	38	45	0.83	0.98	1.01	101	113–155
Bunny	1.0	171	73	92	1.87	2.35	1.07	219	303–371
Hare	2.0	244	85	108	2.25	2.87	1.09	263	381–453
Marmot	3.6	329	96	125	2.64	3.40	1.10	306	462–538
Pronghorn	50	1,258	172	234	5.37	7.30	1.16	607	1,102–1,153
Wapiti	300	3,137	255	360	8.70	12.27	1.20	967	1,938–1,990
Scaling:		$M^{0.51}$	$M^{0.22}$	$M^{0.24}$	$M^{0.26}$	$M^{0.29}$	$M^{0.02}$	$M^{0.26}$	$M^{0.29,0.33}$
Text equation:		11-13	11-17	7-3	11-18b	11-18a	11-9, 11-18a	11-36a	11-34, 11-35

a. For distances or times around the perimeter of a circular home range, multiply by $\pi = 3.1416$. Distances and times are based on Harestad-Bunnell A_{hr}; if Buskirk's A_{hr} (see Table 11-1) is used, they are increased about 50%.

b. Shared patrol time calculated as time for one individual to travel at average daily movement speed (previous column) divided by ("Overlap $a \div c$" from Table 11-2 = number of individuals).

(1983b) on the following basis:

foraging bout distance (L_{for}) = radius of home range (A_{hr});

$$\text{frequency of foraging } (f_{for}) = \frac{\text{daily consumption } (M_{food})}{\text{stomach capacity}}$$

(the stomach is assumed to be filled to capacity in each foraging bout);

daily foraging distance $(\dot{L}_{for}) = (L_{for})(f_{for})$

From published and newly derived allometric relationships (without a distinction between herbivores and carnivores) the equations for the above are, in units of kilograms, kilometers, and days,

$$L_{for} = 0.436 \, M^{0.54}, \tag{11-25}$$

$$f_{for} = \frac{0.152 \, M^{0.74}}{0.050 \, M^{1.0}} = 3.04 \, M^{-0.26}, \tag{11-26}$$

$$\dot{L}_{for} = (0.436 \, M^{0.54})(3.04 \, M^{-0.26}) = 1.32 \, M^{0.28}. \tag{11-27}$$

This derived expression of distance moved per day for foraging is remarkably similar to one that Garland (1983b) derived from published estimates of daily movement distances of 76 species of mammals (half from primates):

$$\dot{L}_{dm} = 1.038 \, M^{0.25}. \tag{7-14}$$

When Garland separated the wide-ranging Carnivora from the other mammals, $\dot{L}_{dm}$ for both groups scaled as $M^{0.22}$, the carnivores traveling 4.4 times as far as other mammals of the same size. Thus, if Garland had derived $\dot{L}_{for}$ separately according to food habits, the coefficients would have been changed more than the exponents.

Like heartbeat, breathing, and frequency of gut contraction, foraging frequency is also proportional to $M^{-1/4}$. Since frequency is the reciprocal of time, there is in foraging an evident continuity or consistency with the scaling of physiological time.

TURNOVER TIME OF THE STANDING CROP

The living state is part of an open system dependent upon energy intake and replacement of worn parts and expired members. The stable individual must achieve energy, fluid, and mineral balance. The stable population must, over the long run, attain a birth rate equal to the mortality rate. In an animal population there should therefore be a direct relationship between productivity $(\dot{E}_p)$ and respiration or metabolic utilization $(\dot{E}_m)$. For both poikilotherms and homeotherms the productivity is essentially linearly proportional to respiration (McNeill and Lawton, 1970; Humphreys, 1979; May, 1979; Schroeder, 1981).

For example, for homeotherms (mammals from mice to the 60- to 100-kg Uganda kob, plus a 20-g bird),

$$\dot{E}_p = 0.0167 \, \dot{E}_m^{1.014}. \tag{11-28}$$

Since basal metabolic rates are approximately proportional to $M^{3/4}$ and conversion efficiency is size independent, it follows that productivity per animal should also be scaled as $M^{3/4}$. The rate of productivity per unit of habitat area is the size-independent product of population density, individual metabolic rate, and production/respiration efficiency (Eq. 11-14; see Damuth, 1981a). This pattern is also revealed by multiplying the productivity/biomass ratios of herbivorous mammals from Beland and Russell (1980) or Banse and Mosher (1980) by the standing crop biomass from Damuth (1981b), thus solving for productivity per area:

$$\dot{E}_p/A \propto (M^{-0.29})(M^{0.25}) \propto M^{-0.04}, \tag{11-29}$$

$$\dot{E}_p/A \propto (M^{-0.33})(M^{0.25}) \propto M^{-0.08}. \tag{11-30}$$

Considering that ecological measurements are fraught with many more uncontrollable variables and possibilities of error than are physiological measurements in the laboratory, Eqs. (11-29) and (11-30) appear to be reasonable approximations in an ecological context to the "dimensionless design criteria" of Stahl (1962). However, the scaling product for population metabolic rate according to Damuth (1981a) does not take into consideration the nonbasal scalings of activity and thermoregulation, which have higher and lower exponents (see Table 8-4). Consequently, we must anticipate some necessary refinements.

Kleiber (1975) contributed the concept of the turnover rate for chemical energy in the animal body. He used its reciprocal as the metabolic turnover time (t_{mt}) for the energy (E) contained in body fat and protein:

$$t_{mt} = (E_{fat} + E_{protein})/\dot{E}_m. \tag{11-31}$$

Just as there is a turnover of the materials within an organism, there is a turnover of organisms within a population. The survival of the species depends upon reproductive replacement for the losses, just as the survival of the individual depends upon energy balance.

The productivity/biomass ratio is the quotient of production rate divided by standing crop or mean population biomass. As used by ecologists, it has the dimensions $\dfrac{kg/(m^2 \cdot yr)}{kg/m^2}$, or $\dfrac{kcal/(m^2 \cdot yr)}{kcal/m^2}$. Either of these reduces to

yr^{-1} by canceling the mass or energy units. For herbivorous mammals (Beland and Russell, 1980) we get:

$$\frac{\text{annual productivity}}{\text{standing crop biomass}} = 7.28 \; m^{-0.29} = 0.98 \; M^{-0.29}. \tag{11-32}$$

A similar analysis avoided a heavier weighting of commonly measured species by using means of reported values (Banse and Mosher, 1980). The productivity/biomass ratio (units converted) for herbivorous mammals derived by these authors was

$$\frac{\text{annual productivity}}{\text{mean biomass}} = 1.20 \; M^{-0.33}. \tag{11-33}$$

The confidence interval for this exponent, $M^{-0.26}$ to $M^{-0.39}$, includes the exponent in Eq. (11-32).

The reciprocal of these relationships is biomass or standing crop divided by productivity rate, which reduces to turnover time for standing crop (t_{sc}, in years). From Eq. (11-32) this would be

$$t_{sc} = 0.14 \; m^{0.29} = 1.02 \; M^{0.29}. \tag{11-34}$$

This is similar in exponent, but about three-fifths of the time predicted for generation length (average age of females when giving birth to offspring, $1.74 \; M^{0.27}$; Millar and Zammuto, 1983). From Eq. (11-33) we would have

$$t_{sc} = 0.83 \; M^{0.33}. \tag{11-35}$$

The scaling exponents in either case are within the range of those for physiological, reproductive, and growth time scales. It is interesting to note that the scalings of t_{sc} are also similar to the reciprocals of annual birth rate scalings (Table 9-1). The product of birth rate and turnover time, births per standing crop, suggests size independence. The product of the coefficients would ideally reach the replacement value of 2, assuming a one-to-one sex ratio and steady-state population size. My derivations from Eisenberg's (1981) tables yield values that are high (2.9 to 4.0), while Western's allometry of birth rates falls short of replacement (1.0 to 1.3). One can only hope that the potential for matching allometries will stimulate progress in this area.

MINIMUM TIME FOR POPULATION DOUBLING

The rate of population growth was described in the logistic equation (11-1). Its maximum under ideal conditions is r_{max} (days^{-1}). The carrying capacity, K, represents the maximum population that the environment can support.

For five species of mammals spanning four orders of magnitude, this intrinsic rate of increase scales as

$$r_{max} = 6.3 \times 10^{-3} \, M^{-0.26} \qquad (r^2 = 0.96). \tag{11-36}$$

From values of r_{max} one can calculate a theoretical minimum time required for a population to increase in number by integrating Eq. (11-1) and solving for time (Lindstedt and Calder, 1981):

$$t = \frac{1}{r_{max}} \ln \frac{N_t(K - N_o)}{N_o(K - N_t)} = 159 \, M^{0.26} \ln \frac{N_t(K - N_o)}{N_o(K - N_t)}. \tag{11-36a}$$

The time required for a population to increase in number from N_o/K to N_t/K is thus roughly proportional to the fourth root of body mass. It would actually take longer for a population to double in number than calculated by Eq. (11-36a), since r_{max} is a theoretical number attainable only under optimal conditions seldom found in nature. While the values presented are low, they should be proportional to the actual slower population growth rates, not skewed toward either large or small body size.

Equation (11-36a) can be solved for a population doubling ($N_t = 2N_o$) that is centered around the point of inflection ($\frac{1}{2}K$) in a logistic (sigmoid) curve from zero to the carrying capacity $K(N_o = \frac{1}{3}K; N_t = 2N_o = \frac{2}{3}K)$:

$$t_{2N_o} = 159 \, M^{0.26} \ln \frac{2N_o(3N_o - N_o)}{N_o(3N_o - 2N_o)} = 159 \, M^{0.26} \ln(4)$$

$$= 220 \, M^{0.26} \text{ days} = 0.602 \, M^{0.26} \text{ yr.} \tag{11-36b}$$

This t_{2N_o}, as an index of reproductive capacity, can be combined with the t_{sc} (Eq. 11-34 or 11-35) to yield a dimensionless design criterion of mammalian ecology:

$$t_{sc}/t_{2N_o} = 1.02 \, M^{0.29}/0.602 \, M^{0.26} = 1.7 \, M^{0.03}, \tag{11-37}$$

or $\qquad t_{sc}/t_{2N_o} = 0.83 \, M^{0.33}/0.602 \, M^{0.26} = 1.4 \, M^{0.07}. \tag{11-37a}$

If this relationship is confirmed when t_{sc} and t_{2N_o} have been derived from larger samples, we can say, as a size-independent rule of thumb, that standing crops of mammals are replaced by a reproduction rate that is 1/1.7 to 1/1.4 or 59% to 73% of the maximum or intrinsic rate. (The slighter of the two residual exponents, $M^{0.03}$, results in a range of ratios of 1.5 at 20-g body mass to 1.7 at 1 kg, and 2.0 at 300-kg mass.)

Recent estimates of r_{max} from life-history data in nature confirm the $M^{-0.26}$ scaling (Eq. 11-36) from a much larger sampling of 44 species of mammals (Hennemann, 1983). However, the coefficient was only 35% of that above, yielding a minimum t_{2N_o} longer than either expression of t_{sc} from field data in Eqs. (11-34) and (11-35).

Critically reviewing the published information, Caughley and Krebs (1983) found a higher scaling for r_{max} (in yr^{-1}; $r^2 = 0.92$, $n = 8$ from 0.03 to 3,700 kg):

$$r_{max} = 1.5 \ M^{-0.36}, \qquad\qquad\qquad (11\text{-}38)$$

from which can be derived (as in Eq. 11-36a):

$$t_{2N_o} = 0.924 \ M^{0.36}. \qquad\qquad\qquad (11\text{-}38\text{a})$$

This would give t_{sc}/t_{2N_o} ratios of 1.1 $M^{-0.07}$ to 0.90 $M^{-0.03}$, a proximity to unity that suggests that the difference between population doubling and population replacement lies in extrinsic factors, with the same reproductive pace and effort in either case.

The Trade-off between Time and Space

We have noted a similarity between the exponents for physiological and ecological time, on the one hand, and standing crop biomass and the degree of overlap in home range, on the other. What are the reasons for the similarities in scaling exponents of these two ecological variables to body size? As a general statement, the smaller the herbivorous mammal, the more rapidly it reproduces, grows, lives, and dies: its role in the population average or ecological turnover time is shorter. The larger the animal, the greater the standing crop biomass, for the size per individual increases with a higher body-mass exponent than that with which the population density decreases ($M^{1.0}$ vs. $M^{-0.75}$). This greater standing crop biomass results in a higher degree of overlap between individual home ranges and is often manifested in social groupings such as leks or herds. It is as if there were a trade-off between time and space in the life-history "strategy," as the following exercise demonstrates.

Equation (11-14) indicated that the rate of foraging per unit of habitat area should be essentially size independent (Damuth, 1981a), while the degree of overlap by different individuals in the same space was shown in Eq. (11-9). The body mass for which home-range overlap results in hypothetical groups of four individuals occupying the same home range can be obtained by solving Eq. (11-9) in reverse:

$$2.19 \ M^{0.27} = 4; \qquad 0.27 \log M = \log (4/2.19) = 0.2615; \qquad (11\text{-}39)$$

$$\log M = 0.2615/0.27 = 0.9686; \qquad M = 9.30 \text{ kg}. \qquad (11\text{-}40)$$

The turnover time for a standing crop of four herbivores of this size could be predicted with Eq. (11-34):

$$t_{sc} = 1.02(9.30)^{0.29} = 1.95 \text{ yr.} \qquad (11\text{-}41)$$

Compare this turnover time with that of an individual whose home range does not overlap any other home range belonging to members of his own species. Equation (11-9) is then solved for one:

$$2.19 \, M^{0.27} = 1; \qquad M = 0.0548 \text{ kg.} \qquad (11\text{-}42)$$

The turnover time predicted for this animal would be

$$t_{sc} = 1.02(0.0548^{0.29}) = 0.44 \text{ yr.} \qquad (11\text{-}43)$$

The turnover time for this solitary individual is approximately one-fourth (22.5%) that for the four herbivores with overlapping home ranges. That is, in 1.95 years a home range shared by four individuals of a hypothetical 9.3-kg species would turn over once; in the same 1.95 years the home range of a single individual of the hypothetical 54.8-g species would turn over about four times. Thus, while the four individuals of the larger species occupy the same home range simultaneously, the four individuals of the smaller species occupy the same home range consecutively. In both instances the same number of individuals would occupy the home range over the course of two years. In effect, the life history of a small herbivore is one of more self-reliance and the possibility of a more rapid response to environmental changes, while the larger herbivore has the benefits of a social system. If the second derivation of t_{sc} from Eq. (11-35) had been used, the solitary animals could have turn over five times as fast as the four larger animals grouped on a mutual home range.

Population Dynamics and Size

The potential physiological life span, or maximum recorded longevity, is clearly size dependent, as is the average life span implied from the standing crop turnover time (see Table 6-1). Since larger animals live longer, the annual rates of aging and dying must be less than those of smaller animals and these differences should be reflected in the survivorship curves. Surprisingly, the scaling of population characteristics such as survivorship and mortality has been largely neglected. We need to determine whether survivorship, mean longevity, life expectancy, and other intermediate portions of the potential life span exhibit the size dependence that the life-span allometry leads us to expect. If so, allometry can be a tool for the study of population biology, as it is for other aspects of life history.

Let us first examine a problem in the survivorship of birds. When the avian literature becomes inadequate, we shall turn to mammalian information, which permits a quantitatively satisfying allometric reconcilation "from the cradle to the grave."

Hummingbirds are the smallest birds, a fact that makes them interesting in many respects — heat, water, and salt exchanges, for example. In the process of handling birds for such studies, several colleagues and I collaborated in a bird-banding project, gaining data on the histories of individual birds; survivorship curves for the small end of the size range became possible (Calder et al., 1983). How does the survivorship of hummingbirds compare with that of other kinds of birds? Allometry is the inevitable next step.

It had been concluded that mortality rates of adult birds stay essentially the same throughout life, with about the same fraction of the survivors in each age class disappearing each year (Deevey, 1947; Farner, 1955; Ricklefs, 1973). The hazards of migration, predation, and disease appeared to take their toll before the vulnerability and decline of aging could occur.

In a given age-class, the number (n) of individuals that survive from the previous year should decay exponentially if the mortality is age independent:

$$n = n_o e^{-qt}, \tag{11-44}$$

where n_o is the original number at the start (first-year adults in this case, or 100 if n is expressed as a percentage), e is the base of natural logarithms, q is the annual disappearance or mortality rate, and t is time in years. From this exponential decay equation one can calculate the time t_{n/n_o} that elapses until the survivors constitute a fraction n/n_o of the original first-year adults:

$$t_{n/n_o} = \frac{\ln (n/n_o)}{-q}. \tag{11-45}$$

If the mortality of birds is age independent, t_{n/n_o} should scale in the same fashion as the maximum life span, $t_{ls,max}$.

The question of age dependence or independence in avian mortality was reexamined by Botkin and Miller (1974), who tabulated published mortality rates and $t_{ls,max}$ of 15 of the best-studied species of birds ranging in size from 11 g to 8,500 g. Potential natural longevities for survival of 10^{-3}, 10^{-4}, and 10^{-5} of the original cohorts were calculated on the assumption of age-independent mortality. These authors indicated that "there is one chance in 1000 that a royal albatross . . . breeding in New Zealand today was 25

Table 11-4 Adult disappearance, longevity, and life expectancy of selected birds. Data mostly from table 1 of Botkin and Miller (1974) with body-mass data from other sources; hummingbird data from Calder et al. (1983); albatross longevity from New Zealand Wildlife Service.

Species	Body mass (g)	Average adult mortality (yr^{-1})	Maximum life span (yr)	Life expectancy (yr)	Potential natural longevity $t_{0.368}$ (yr)	Potential natural longevity $t_{0.001}$ (yr)
Broad-tailed hummingbird, *Selasphorus platycercus*	3.5	0.43	8	1.3	1.43	8.2
Blue tit, *Parus caeruleus*	11.4	0.72	9	0.9	0.79	5.4
Robin, *Erithacus rubecula*	16.3	0.62	12	1.1	1.03	7.1
Redstart, *Phoenicurus phoenicurus*	14.5	0.56	—	1.3	1.22	8.4
Common swift, *Apus apus*	40.2	0.18	21	5.1	5.04	34.8
Starling, *Sturnus vulgaris*	76.3	{ 0.52 / 0.63	20	1.4	{ 1.37 / 1.01	{ 9.4 / —
Blackbird, *Turdus merula*	90.2	0.42	7	1.9	1.84	12.7
Alpine swift, *Apus melba*	101.7	0.18	16	5.1	5.04	34.8
Lapwing, *Vanellus vanellus*	220	0.34	16	2.4	2.41	16.6
Fulmar, *Fulmaris glacialis*	734	0.06	10	16.2	16.16	111.6
Sooty shearwater, *Puffinus griseus*	787	0.07	27	13.8	13.78	95.2
Herring gull, *Larus argentatus*	946	{ 0.09 / 0.04	36	24.5	{ 10.60 / 24.50	{ — / 169.2
Gray heron, *Ardea cinerea*	1,508	0.31	24	2.7	2.70	18.6
Gannet, *Sula bassana*	2,895	0.06	17	16.2	16.16	111.6
Royal albatross, *Diomedea epomophora*	8,500	{ 0.09 / 0.03	49	32.8	{ 10.60 / 32.85	{ — / 226.8
Allometric summary:		$1.38\,m^{-0.36}$	$6.60\,m^{0.18}$	$0.41\,m^{0.46}$	$0.43\,m^{0.43}$	$2.53\,m^{0.47}$
r:		−0.825	0.698	0.836	0.826	0.713
S_b:		0.071	0.052	0.083	0.083	0.083

years old when Captain Cook made his first visit to the island in 1769!" They had to conclude that mortality does increase with age, although the rate of the age-dependent increase is comparatively small and therefore difficult to observe.

To go into this allometrically, from other sources we can add to the data of Botkin and Miller's table 1 the body masses of the species (except the penguin, deleted for lack of a mass value), data on the broad-tailed hummingbird, and a longer record for longevity of the royal albatross (Table 11-4). The resulting allometric relationship is

$$\text{maximum life span} = 6.60 \; m^{0.18} = 22.9 \; M^{0.18}, \tag{11-46}$$

which can be compared with $17.6 \; M^{0.20}$ from Lindstedt and Calder (1976) for 152 species. Equation (11-46) was derived from data on the most extensive studies to date, those most likely to yield extreme longevities for wild birds — hence the 30% greater coefficient. The important point is that the scaling is essentially the same. If mortality is indeed age independent, then the survivorship curve on a semilogarithmic plot should be a straight line, and we would expect the scaling for survival times of any fraction of the original cohort, such as e^{-1} (36.8%) or 50%, to be parallel to that for physiological maximum life span (see Figure 11-3). To the contrary, for expressions of life span shorter than the records for physiological maximum, the scaling exponents are greater:

$$\text{life expectancy} = 0.410 \; m^{0.46} = 9.84 \; M^{0.46}; \tag{11-47}$$

$$\text{survival of } e^{-1} = 0.432 \; m^{0.46} = 10.36 \; M^{0.46}. \tag{11-48}$$

Both of these are highly significant correlations ($r^2 = 0.699$ and 0.682, respectively; $p < 0.001$). If Eqs. (11-47) and (11-48) are both true, and also Eq. (11-46), the difference in exponents can only mean that a greater proportion of the population survives to the maximum potential life span, that is, that there are more senior citizens in populations of larger species of birds. In fact, the ratio of life expectancy to maximum life span for these species increases dramatically with size:

$$t_{exp}/t_{ls,max} = 0.410 \; m^{0.46}/6.60 \; m^{0.18} = 0.062 \; m^{0.28} = 0.429 \; M^{0.28}. \tag{11-49}$$

For the 11.4-g blue tit, the observed life expectancy is 10% of the potential maximum; for the albatross, it is 67%. (The allometric summary, Eq. 11-49, predicts 23% and 78%, respectively.) Thus from either the allometric generalization or the specific values tabulated by Botkin and Miller, the former assumption of age-independent mortality in birds is clearly wrong.

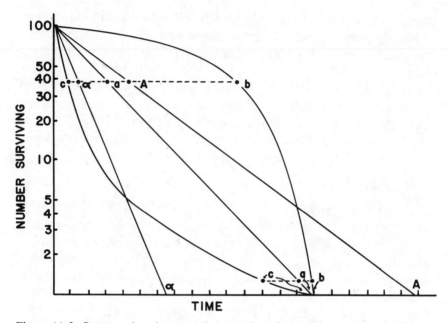

Figure 11-3 Suppose that there are three species of hypothetical birds of different sizes, α, a, and A, with maximum life spans appropriate to their respective sizes, as plotted along the *x*-axis. If mortality is age independent, survival should follow a straight line for simple exponential disappearance on this semilogarithmic plot, resulting in the same scaling (seen as similar triangles) for survival of intermediate fractions of the starting numbers. If survival is age dependent, the curves would curve, as for species *b* and *c*. As maximum longevity is approached, the curves converge, blurring this distinction. (From Calder, 1983d; with permission. Copyright Academic Press.)

We turn now to the mammalian data, which permit analysis of age-dependent mortality.

Survivorship with Age-Dependent Mortality

In view of the foregoing considerations our assumption of age-independent mortality of birds is in question. The exponent q, or disappearance coefficient, as used in the previous section, is not only size dependent, it is time dependent. Such a time or age dependence is incorporated in the Gompertz equation (as in Eq. 10-36):

$$n = n_o e^{(q_o/\alpha)(1-e^{\alpha t})} \tag{11-50}$$

in which α is the Gompertz "constant" for the actuarial aging rate, or the annual rate of increase in the mortality or disappearance of individuals from a cohort, and q_o is the initial annual mortality rate, or initial vulnerability (Sacher, 1978; Rosen et al., 1981). Since larger animals live longer, we might expect to find a size dependence for q_o and α in Eq. (11-50). The information is available for eutherian mammals, so let us examine the allometry:

$$q_o = 0.0124\ M^{-0.56} = 0.594\ m^{-0.56}; \qquad r^2 = 0.710; \qquad (11\text{-}51)$$

$$\alpha = 0.709\ M^{-0.27} = 4.45\ m^{-0.27}; \qquad r^2 = 0.622. \qquad (11\text{-}52)$$

(See Table 11-5 and Calder, 1982d.) Units of q_o and α are yr^{-1}.

We can obtain something a little more familiar and comparable with expressions in the dimension of time, which underlies much of the biology emphasized in this book, by taking a reciprocal function of Eq. (11-52), the mortality doubling time (t_{2q_o}):

$$\begin{aligned} t_{2q_o} &= 0.693\ \alpha^{-1} = 0.693\ (0.709\ M^{-0.27})^{-1} \\ &= 0.98\ M^{0.27} = 0.156\ m^{0.27}. \end{aligned} \qquad (11\text{-}53)$$

Thus we see that the body-mass exponent for mortality doubling time is indistinguishable from the $M^{1/4}$ scaling common for physiological and ecological time scales (Lindstedt and Calder, 1981). In recent years the process of aging has come to be regarded as a fundamental part of the continuum on a physiological time scale that extends from embryonic and postnatal growth, to normal adult physiological time, and on to maximum physiological life span. The consistency of body-mass exponents is further suggestion that this continuity exists.

Does the age dependence incorporated in the Gompertz equation resolve the sort of discrepancy noted between the scalings of life expectancy and maximum longevity (Eqs. 11-46 through 11-49)? To answer this and pursue the application of physiological time further, we solve Eq. (11-50) for time:

$$t = \frac{\ln\left[1 - \dfrac{\alpha}{q_o}\left(\ln \dfrac{n}{n_o}\right)\right]}{\alpha}. \qquad (11\text{-}54)$$

Now, if we want to predict life expectancy, or the time for survival of one in 1,000 $(t_{0.001})$ we can substitute body-mass predictions for q_o and α, from Eqs. (11-51) and (11-52), into (11-54):

$$\begin{aligned} t_{0.001} &= \frac{\ln\left[1 - \dfrac{(0.709\ M^{-0.27})}{(0.0124\ M^{-0.56})}(\ln 0.001)\right]}{0.709\ M^{-0.27}} \\ &= \frac{\ln\left[1 - (57.2\ M^{0.29})(-6.908)\right]}{0.709\ M^{-0.27}}. \end{aligned} \qquad (11\text{-}55)$$

Table 11-5 The allometry of population dynamics for eutherian mammals.

Size of mammal (kg)	n		$t_{ls,max}$ (yr)	$t_{1/2}$ (yr)	q_0 (yr)	α (yr)$^{-1}$	t_d (yr)	r^2 values	References
	Data	Species							
0.008–2,500	63	—	$11.6\,M^{0.20}$	—	—	—	—	0.59	Sacher, 1959
0.023– 453	18	11	$6.9\,M^{0.28}$	—	—	$1.11\,M^{-0.36}$	$0.62\,M^{0.36}$	0.77, 0.63[a]	Rosen et al., 1981
0.021– 484	9	—	—	$4.44\,M^{0.32}$	$0.0124\,M^{-0.56}$	$0.52\,M^{-0.25}$	$1.33\,M^{0.25}$	0.73, 0.71[a], 0.67	Sacher, 1978
0.021– 484	27	17	—	$5.08\,M^{0.35}$	$0.0124\,M^{-0.56}$	$0.71\,M^{-0.27}$	$0.98\,M^{0.27}$	0.67, 0.71, 0.62[a]	Calder, 1982d, 1983a

Definitions: α = Gompertz constant for rate of increase in death rate with age in $n = n_0 e^{(q_0/\alpha)(1 - e^{\alpha t})}$.
q_0 = initial mortality rate at time $t = 0$.
t_d = time for doubling of mortality rate = 0.693 α^{-1}.
$t_{ls,max}$ = maximum recorded longevity in captivity, assumed to approximate physiological maximum.
$t_{1/2}$ = mean or expected life span.

a. Calculated from $t_d = 0.693\,\alpha^{-1}$.

This equation has become a bit unwieldy, so I resort to the expediency of an allometric approximation. I substitute a series of body-mass values, 0.01, 0.1, 1, 10, and 100 kg, into Eq. (11-55), and then run an allometric regression on the values obtained. This yields:

$$t_{0.001} = 8.3 \ M^{0.32}. \tag{11-56}$$

Similarly, one can calculate an average life expectancy as the survival time for 50% of the cohort $(t_{1/2})$:

$$t_{1/2} = 5.08 \ M^{0.35}. \tag{11-57}$$

This is comparable to the equation obtained from regression of life expectancy values directly from Sacher's (1978) table 1:

$$t_{1/2} = 5.44 \ M^{0.32}. \tag{11-58}$$

The ratio of life expectancy to maximum observed longevity (from Table 6-1) is

$$t_{1/2}/t_{ls,max} = 5.08 \ M^{0.35}/11.6 \ M^{0.20} = 0.44 \ M^{0.15} = 0.16 \ m^{0.15}. \tag{11-59}$$

From this we see that, even if we take the age dependence of mortality into consideration, the average 1-kg mammal's life expectancy is a larger fraction of the physiological maximum than it would be for a smaller mammal. This being the case, the effects on the ecology of mammals are profound. There are at least three interesting consequences of the relation of life expectancy to body mass that can be inferred from Eq. (11-59).

First of all, the population–age distribution should reflect the higher proportional life expectancy. The physiological maximum is an extreme value, whereas life expectancy is actuarial, a statistical probability or average. The greater this average relative to the maximum life span, the better the chances for individuals to attain senior-citizen status, and therefore the greater proportion of these "golden-agers" in the total population. However, a caveat is in order for Eqs. (11-56), (11-57), and (11-59). Although the q_o in Eq. (11-51) is based on a wide range from shrew to horse, increasing the sample size from 9 would probably improve the equation. Equation (11-52) for the Gompertz α is based on data from 15 species. Values calculated for q_o and α in Eq. (11-56) can give a general appreciation of the effect of size (Figure 11-4). However, like any allometric relationship, it may do a poor job of predicting the pattern for outlying taxa, such as *Peromyscus,* as can be anticipated from differences between observed and predicted values (Table 12-3 later).

The shape of the population–age distribution histograph supposedly tells

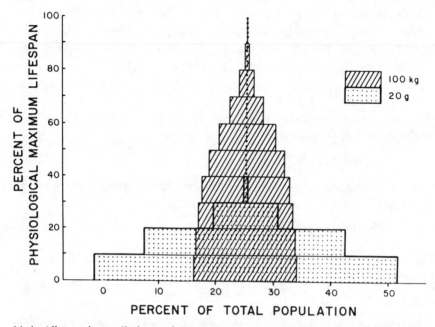

11-4 Allometric predictions of population–age distribution for a 20-g (mouse-sized) mammal and a 100-kg (deer-sized) mammal. Note the similarity of the small mammal's age distribution, wide at the bottom and sparse after attaining 30% of maximum recorded life span, to that of a human population in a developing country *(dotted area)*. The age distribution of the large mammal, more evenly distributed *(slashed area)*, is like that of a developed, stable population with zero growth.

whether a population is growing, stable, or declining, depending on the relative widths of the younger and older age groups. In Figure 11-4 the allometrically generalized prediction of population–age distribution of the 100-kg mammal suggests an almost stable population size, whereas the distribution for a 20-g mammal looks like something from Third World human demography. The differences in this case actually are due to body size: the moral is that body size needs to be considered in demographic studies of animals.

A second implication bears on the allometric description of sociality. As was noted previously, the greater scaling factor for home range than for per capita space based on population density results in, or stems from, the positive scaling of group size ($\propto M^{0.27}$ to $M^{0.34}$, Eqs. 11-9 and 11-10) for herbivorous mammals. Sociality would be furthered by historical continuity and vice versa. The population–age distribution of larger mammals includes a higher proportion of senior citizens in the herd, from whom the young could learn. This should result in a positive allometry for cultural diffusion and the benefits of experience. Alternatively, there might have been natural selection for these benefits that would have favored evolution of the larger mammals. It will be interesting to see how A_{hr} and p, survivorship and social structure, fit together when the missing allometric descriptions can be made!

The third consequence to consider is the allometric relationship between reproduction and survivorship. My impression is that despite the importance to evolutionary ecology of the concept of reproductive fitness, it has been expressed only in relative terms, thereby limiting consideration to a species-by-species patchwork and retarding progress toward general principles. A quantitative baseline should facilitate the development of a broader comparative ecology. The body-mass exponents for life expectancy and mean annual birth rates suggest the possibility of deriving ecological design constants or similarity criteria comparable to what Stahl (1962) proposed for physiology.

As a first approximation in Table 9-1, the allometry of maximum life span was combined with Western's (1979) expression of mean annual birth rate (births per year, in percentage of whole population) as an average for reproductive value ("age-specific expectation of future offspring"; Pianka, 1983). However, most animals fail to attain the maximum longevity recorded for their species, so a more realistic relationship might be the product of animal birth rate and the life expectancy derived in Eq. (11-57):

$$(1.26\ M^{-0.33})(5.08\ M^{0.35}) = 6.4\ M^{0.02}. \tag{11-60}$$

The scaling exponents just about offset one another to produce a size-independent average of 6.4 offspring per member of the population (12.8 per monogamous pair; see Figure 11-5). The reciprocal of 6.4 is a reproductive-success or replacement efficiency of 16%, which persists to maturity.

Theoretically, the reproductive success should be twice as good, equal to e^{-1} or 37%; limited data (30 species of birds, 5 of mammals) back this with no size-dependent trend (Alerstam and Högstedt, 1983). Expressed as in Eq. (11-60), this would be 2.7 M^0 young per lifetime. With some refinement the two approaches seem capable of agreement. The birth rate component was derived from 21 species of mammals, mostly artiodactyls and no rodents, while the q_o used to derive life expectancy was based on only 9 species, including rodents and our own species—which conforms to few typical mammalian patterns.

This need for refinement in the allometric description of lifetime repro-

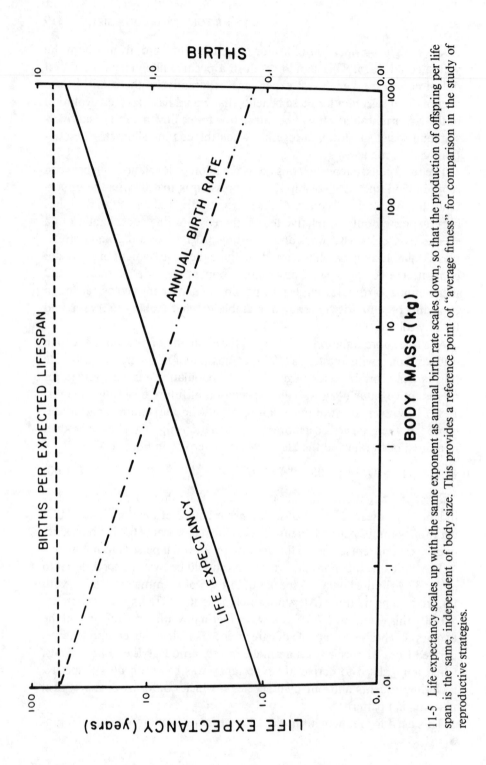

11-5 Life expectancy scales up with the same exponent as annual birth rate scales down, so that the production of offspring per life span is the same, independent of body size. This provides a reference point of "average fitness" for comparison in the study of reproductive strategies.

duction will be seen again in Figure 14-2, which exaggerates survivorship to times in excess of predicted maximum life span. No doubt the discrepancy is due in part to the fact that the q_o used to derive Eq. (11-57) was from captive animals not exposed to predation and other hazards of nature.

The analysis of life tables of wild mammals has produced additional scaling relationships for t_{exp} at birth. For 35 species (including one marsupial) the relationship is indistinguishable from Eq. (11-34) for t_{sc} when units are converted from Damuth (1982) for comparison ($r^2 = 0.81$):

$$t_{exp} = 1.02 \, M^{0.30 \pm 0.026}. \tag{11-61}$$

This suggests that t_{exp} at birth is cut 80% (compared to captivity) by the hazards of life in the wild. Further, this would produce, in the manner expressed in Eq. (11-60), a life total of only 1.27 $M^{-0.03}$ offspring per individual (2.6 per monogamous pair and corresponding to 78% reproductive success, twice the theoretical value). A similar analysis of 29 species (38% of which overlap with those used in Eq. 11-61) yielded $M^{0.24}$ scaling ($r^2 = 0.76$; Millar and Zammuto, 1983). Combination of the 52 eutherian species used by these two independent analyses and resolution of disagreement in M and t_{exp} values (cited differently from the same primary references) would fine-tune the relationships. Something ecologically useful to avian population dynamics should come from this approach as well.

Population Cycles and Irruptions

One learns in introductory physics that the period of a pendulum's swing is proportional to the square root of its length. Suspecting that either the laws of physics or my memory might have changed in the past 30 years, I suspended weights on three strings 50, 100, and 200 cm long, displaced them each 30 degrees from the vertical, and released them simultaneously. Their frequencies were 43, 30, and 21 swings per minute, and the respective periods were 1.4, 2.0, and 2.9 seconds per swing. Someone who had never taken physics, entering the lab 15 seconds after I had released the pendulums, might have wondered why each was swinging at a different rate. Although they were not used in this experiment, there were three hammers on the lab bench; a small one used by a physician to test the patellar reflex, a carpenter's hammer, and a rubber mallet used for beating dents out of a fender. The observer might have hypothesized that different swinging frequencies were imparted by blows delivered by hammers with different characteristics. He or she might have decided to concentrate on just one type of cycle, say the one with the 2.0-second swing, and tried to figure out which hammer made it swing that fast. In fact, analysis of the behavior of all three pendulums might have expedited an explanation. This, I feel, is analogous to what has occurred in the study of population cycles.

One learns in ecology about the phenomena of regular cycles of boom and crash in populations of several species of mammals and birds in the northern tundra and taiga, and about the spectacular population explosions followed by mass die-offs following introduction of large ungulates to new locations, or removal of natural population control through elimination of predators or disease. Voles and lemmings (subfamily Microtinae) cycle at fairly regular intervals with peaks occurring every three or four years, while the snowshoe or varying hare *(Lepus americanus)* and grouse (family Tetranoidae) cycle with periods of eight to ten years (Figure 11-6).

Those who have studied these population cycles have tended to consider the longer cycles in isolation from the shorter ones, thereby overlooking the possibility that the two could be related. For example, Keith (1963) specialized in the study of eight- to ten-year cycles. He reviewed the evidence and hypotheses that might explain such a periodicity: sunspot cycles, cycles in ozone accumulation and ultraviolet penetration, availability of key nutrients or the general food supply (perhaps secondary effects of meteorologic cycles), and synergistic effects of oscillations in more than one environmental factor. The three- to four-year cycles were specifically excluded from this search for explanations; yet if a sunspot cycle, for example, had such a profound effect on the population of varying hares, would it not get the voles in step with the same cadence as well?

In the other camp, Krebs and Myers (1974) reviewed the topic of the approximately four-year cycles of the small mammals, addressing four specific questions to guide the search:

A. What prevents unlimited increase in the population?

B. What causes the cyclic periodicity of three to four years?

C. What produces synchrony of populations over large areas?

D. What determines the amplitude of the function?

Of these, A and D both pertain to amplitude, an unlimited increase going to an unlimited amplitude, so they can be grouped as one question. Furthermore, it seems necessary to know the cause before attempting to limit the effects, so the combined question should be secondary to B. Question B, however, may fail to make a necessary distinction between *cause of cycle*

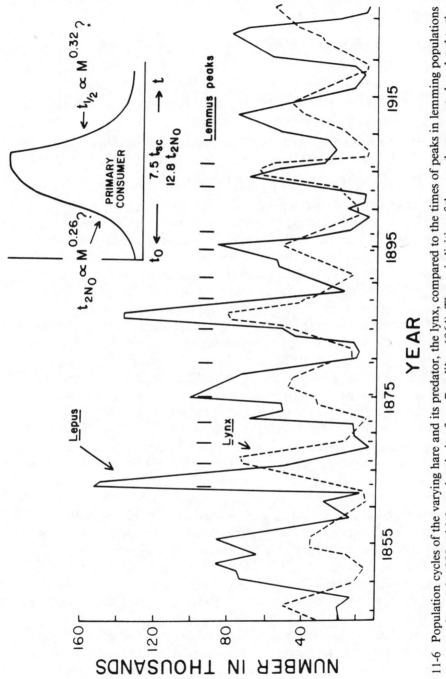

11-6 Population cycles of the varying hare and its predator, the lynx, compared to the times of peaks in lemming populations (redrawn from Itô, 1980, with lemming timing from Bourlière, 1964). The periodicities of these cycles appear to be related to the scaling of reproduction, aging, and turnover times of primary consumers (see inset, upper right, symbols as in text).

and *cause of periodicity,* so it should be split into two. Our agenda of questions might then read:

A. What causes a population to cycle from boom to bust?

B. What imparts the characteristic three- to four-year or eight- to ten-year periodicities to such oscillations?

C. What prevents unlimited increase in the population, thereby determining the amplitude of the function?

D. What produces synchrony of populations of microtine rodents in the same body-size range over large areas without synchronizing the cycles of varying hares to the same forces?

We turn now to allometry to see if it is possible that the previous studies of oscillation made the distinction in the wrong place, that is, between small mammals with shorter cycles and medium-sized mammals with longer cycles, rather than between environmental causes and perhaps endogenous periodicities such as are observed for the pendulums.

Food supply, radiation, climate and other atmospheric conditions, predation, and disease certainly have profound influences upon reproduction and population size, either directly or indirectly. All must be considered as potential stimulators of population oscillations and controllers of the amplitude of changes, whether or not they affect the periodicity. The absolute time scale of the environment, by which we measure the population changes, is something to which the animal must be entrained on daily and annual bases, in order that feeding and reproduction coincide with the best possible conditions of light, climate, and food supply. However, as allometry has shown repeatedly, the significance of an hour, day, or year in the life of an animal depends upon its physiological time scale, which is size dependent and endogenous. In the absence of immigration and emigration, a population can grow only when births exceed deaths, and decline only when mortality exceeds natality. The minimum time for population doubling, life expectancy, and standing crop turnover time are size dependent, and scale in a manner approximately similar, quantitatively, to physiological and growth times — that is, to $M^{1/4}$. The period of population irruption would seem particularly closely related to the rate at which a species can reproduce and double its population. Do population cycle times, dependent on reproduction and mortality, scale in a parallel manner? If there is a significant endogenous component to the periodicity of population oscillations, the temporal correlation between a population cycle and an environmental factor may be only a coincidence that does not show us the true cause of the oscillation.

Cycles in the abundance of primary consumers seem to lead to cycles of similar periods in their predators. On the tundra, where lemming cycles peak every 3.5 to 4 years, a similar cycle length is exhibited by red and arctic foxes *(Vulpes fulva* and *Alopex lagopus).* However, in the taiga, the red fox and the lynx *(Lynx canadensis)* prey principally on the varying hare and have approximately 10-year cycles like their prey, as seen in Figure 11-6 (Keith, 1963; Itô, 1980). Since the basic periodicities originate at the primary consumer level, it is appropriate to consider the herbivores exclusively in this preliminary allometric exploration.

The most complete records for microtine cycles of those tabulated by Bourlière (1964) are for the lemming *(Lemmus lemmus)* in southern Norway in the last century. The peaks occurred 3.5 ± 0.81 (s.d.) years apart. The lemming has a body mass of about 50 g. The varying hare varies its population in cycles of 8.4 years (Keith, 1963; Itô, 1980) and has a body mass of about 1.36 kg. If log cycle period is plotted against log body mass for these two species, the allometric line connecting the two data points is

$$t_{\sim} = 7.7 \ M^{0.27}, \tag{11-62}$$

where $t_{\sim}$ is cycle time in years. Significant or coincidental, this scaling is essentially the same as that for standing crop turnover time $(\propto M^{0.29})$ and minimum population doubling time $(\propto M^{0.26})$, although the coefficient 7.7 is considerably larger than for either t_{sc} or t_{2N_o}, reflecting the fact that the population cycle must involve more than one generation.

At the very least, these scaling parallels encourage further consideration of animal size as a quantitative factor in population cycling. We anticipated a connection between the scaling of reproduction and mortality rates and other variables in population dynamics on the one hand, and oscillations generated in populations (by whatever means) on the other, and the correlations seem to support this.

If the period of oscillation in populations is a consequence of size-dependent reproduction and population growth rates, perhaps the pace of boom and crash in populations of large herbivores might fall on the same scale whether or not these are sustained oscillations. Once the population of large ungulates exceeds the carrying capacity, starvation and sudden population decreases are inevitable. In many cases the data are not reliable. For example, the oft-cited case of the Kaibab deer population was reanalyzed by Caughley (1970), who concluded: "Data on the Kaibab deer herd in the period 1906–1939 are unreliable and inconsistent . . . the study is unlikely to teach us much about eruption of ungulate populations." Precise census data are also lacking for the irruptive fluctuations in populations of introduced red deer *(Cervus elaphus)* in New Zealand.

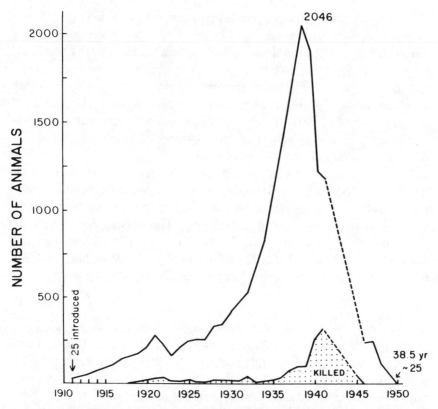

11-7 The reindeer population on St. Paul Island subsequent to its introduction in 1911. The dashed line interpolates for the period in World War II when counts were not taken. If the period of nongrowth from 1921 through 1927 is omitted from the 38.5-yr rise and fall, the "cycle" is very close to that predicted by extrapolation from lemming and hare cycle periodicity. (Data from Scheffer, 1951.)

However, the boom and crash of the reindeer *(Rangifer tarandus)* following their introduction to St. Paul Island in the Pribilofs provides a useful case for further examination of the allometric basis of population irruptions (Figure 11-7). Twenty-five reindeer were introduced to the island in 1911. Except for a two-year dip (1921–1923) and a brief leveling (1925–1927), the herd size increased exponentially to 2,046 animals in 1938, three times the estimated carrying capacity of the environment. At this point a precipitous decline began, until only 8 deer remained in 1950 (Scheffer, 1951). Although the reindeer were hunted, the kill was only 28% of the 1,176 animals in 1941, and significantly less in most years; hunting, therefore, could not have been influential. The time from introduction of 25 reindeer to reduction to an equal number (determined by interpolating between the 1949 and 1950 censuses) was about 38.5 years. A full-grown reindeer may weigh 318 kg (Walker, 1975), for which Eq. (11-62) predicts a population cycle of 36.4 years. Using 318 kg for mass, the relationship from data on small mammals extrapolates well, but 318 kg may exceed the size of animals in the Pribilofs. The mean body mass of adults over three years of age during the rapid growth phase on another Bering Sea island was 147.7 kg.

Fewer details are available subsequent to the liberation of 29 reindeer on St. Matthew Island in 1949. In 1958, there were 1,350 reindeer; in 1963, an estimated 6,000, followed by a crash to 42 the following year (Klein, 1968). This period of about 20 years is much shorter than the 38.5 years for the herd on St. Paul Island, but differences by a factor of two are relatively inconsequential to a scaling that spans several orders of magnitude in body size. One could subtract the six-year setback (1921 to 1927) and then return to exponential growth on St. Paul Island and call it a 32.5-year cycle, but I have not done so. Lemming, hare, and the two reindeer points give us:

$$t_{\sim} = 7.4 \, M^{0.24}; \qquad r^2 = 0.96. \tag{11-63}$$

Finally, if we include nine species of mammals from 18.5 g to 390 kg (Table 11-6), we have:

$$t_{\sim} = 6.53 \, M^{0.25}; \qquad r^2 = 0.891; \qquad S_{yx} = 0.0988; \qquad S_b = 0.0326. \tag{11-64}$$

The exponent does not seem to want to change appreciably; it is essentially the same as for the scaling of physiological, reproductive, and growth times.

The inclusion with the cycles of lemmings and voles of a 390-kg moose may seem a bit far-fetched. However, right after taking such allometric liberties, I was interested to read in *Audubon* magazine: "Wolf-moose relationships on Isle Royale National Park may be about to repeat a 25- to 30- year cycle . . . researchers believe they are witnessing 'a predator-prey cycle with an extended period of oscillation'" (Sayre, 1982). This was confirmed by a phone conversation with Rolf Peterson, head of the study team there. This relationship has been extended to include 39 species of mammals and birds, without changing the exponent ($r^2 = 0.83$; Peterson et al., MS):

$$t_{\sim} = 7.93 \, M^{0.25}. \tag{11-64a}$$

Another perspective is derived by expressing cycle time as a multiple of the minimum time for population doubling:

$$t_{\sim}/t_{2N_o} = 7.93 \, M^{0.25}/0.602 \, M^{0.26} = 13.2 \, M^{-0.01}, \tag{11-65}$$

Table 11-6   Cycle times (peak to peak, trough to trough, or average of both) of herbivorous mammalian population oscillations. (From Calder, 1983, with permission, from *J. Theor. Biol.* 100:278. Copyright Academic Press, Inc., London.)

Species	Body mass (g)[a]	Cycle time (yr)	Source[a]
Apodemus sylvaticus	18.5	2.17	Southern, 1970
Clethrionomys glareolus	22.0	1.33	Southern, 1970
Microtus agrestis	33.8	4	Krebs and Myers, 1974
Clethrionomys rufocanus	40.5	4	Krebs and Myers, 1974
Lemmus lemmus	50.0	3.5	Bourlière, 1964
Lemmus trimucronatus	78.3	3.2	Krebs and Myers, 1974
Lepus americanus	1,360	8.4	Itô, 1980
Rangifer tarandus	147,700	20 ⎱ mean:	Klein, 1968
Rangifer tarandus	147,700	38.5 ⎰ 29.3	Scheffer, 1951
Alces alces	390,000	20[b]	Mech, 1966

a. Cycle times are estimated from data in the references listed; body masses are from various fauna manuals and papers. Because the latter were not necessarily from the same population at location, errors may be incorporated. Still, even a 100% error will have negligible effect in a regression spanning four orders of magnitude.

b. The period for the population cycle of moose on Isle Royale now appears to be 25 to 30 yr (Peterson et al., MS).

and as a multiple of turnover time for standing crop biomass:

$$t_{\sim}/t_{sc} = 7.93\, M^{0.25}/1.02\, M^{0.29} = 7.8\, M^{-0.04}. \tag{11-66}$$

In the form of essentially size-independent design constants (Stahl, 1962), we seem to have a good case for an endogenous basis of periodicity in population cycles. This is still in a relatively crude preliminary form, with limited sample size, so the coefficients of the ratios, 13.2 and 7.8, should perhaps not be interpreted except as a semiquantitative indication that the cycle of boom and bust occurs over several generations or standing crops, and that the rate of increase is only a fraction of the intrinsic or maximum population growth rate.

Let us now return to our questions about population cycles.

   A.   The allometry says nothing about what causes some populations to oscillate rather than to damp out to some relatively constant equilibrium level. However, we do have good reason to doubt the effectiveness of trying to assign causation to an environmental factor that happens to oscillate with the same period.

B. The periodicity appears to be endogenous rather than exogenous.

C. We have not explored the determination of cycle amplitudes. That would, no doubt, require considerable measurement of food supplies and considerable knowledge of the biology of each species.

D. The synchrony of several populations in different areas is also beyond allometric insight. There appears no reason to ascribe it to endogenous factors, so it must be environmental in nature. The coefficient of variation for 22 *Lemmus* peaks from southern Norway (1862–1941) was 23% of the mean cycle duration of 3.5 years. The cycles in different Scandinavian locations and species were in synchrony with southern Norway *Lemmus* cycles 82% of the time, suggesting that an environmental factor, perhaps that stimulating or destabilizing the populations in the first place, was responsible for the synchronization. Therefore, causality and periodicity must be separated in seeking the ultimate explanation. By exposing the size dependency of periodicity, we may be on the way to discovery of the common features of population irruptions and to recognition of possible spurious correlations.

Conclusion

The variety and complexity of living organisms have necessitated specialization among biologists. Except in the most obvious areas of temperature and osmotic responses in stressful environments, energetics, and sensory perception, the fields of ecology and physiology have been pursued quite separately. Upon reflection, however, it seems that there must be considerably more coupling between the animal's internal functions and its environment than has developed between the disciplines that study these topics. Our interpretations of laboratory metabolic data on the one hand, and ecological observations on the other, have suffered from academic isolation. There are too few biologists to describe all the details of morphology, physiology, reproduction, growth, and ecology, let alone to understand how these details are functionally assimilated into a life history. Analytical ecology seems handicapped by a situation analogous to having 5% of the pieces of each of 100 jigsaw puzzles. For species A, we have measurements of variables Y_1, Y_3, and Y_6; for species B, we know Y_1, Y_2, Y_4, Y_5, and Y_7; for C, we have Y_2, Y_5, Y_6, and Y_7, and so on. How can we find the patterns that connect these fragments? If size does have an effect predominating enough that a quantitative statement can made, it is not necessary for us to have the missing data for Y_2 or Y_4 in species A, or Y_3 or Y_6 data in species B, as long as the allometric equations are representative and span a large range of body sizes.

Allometry has outgrown its restricted use in comparative physiology and morphology and can be expected to play an important role in analytical ecology. As it was stated so well by Western and Ssemakula (1982): "If the allometry is due to physiological and anatomical scaling principles, then we can expect that theories of life history strategies will be greatly simplified; if not, then we are faced with the challenge of explaining why individually adaptive traits aggregate into consistent patterns."

Internal function is linked to the ecological niche through the characteristics of body size and size-dependent physiological time scales and energy requirements. Since the animal is not a closed system, quantitative details of internal function must extend beyond the skin, to a realm where they have ecological consequences. Because of this the role of physiological ecology as a bridge for our understanding of population and community ecology may prove to be much greater than previously expected. Students of modern ecology have tended to shy away from courses in physiology and several other basic biological disciplines (see the excellent essay on this problem by King, 1980). This has been done in order to expedite concentration on ecological problems, but the apparent expediency actually may have retarded ecological comprehension.

The mark of scientific progress seems to be the ability to state a few principles that have wide general application. As a prerequisite, we must find the patterns that accurately describe what is to be explained. Stahl's size-independent "design constants" have clarified several physiological patterns. While the data base is small and the relationships yielded need refinement, it seems that there are design constants in ecology:

(population density)(individual metabolic rate) $\propto M^{-0.75+0.75} \propto M^{0.00}$;

productivity = (density)(mean body mass)(productivity/biomass)
$$\propto M^{-0.75+1.0-0.29} \propto M^{-0.04};$$

(density)(home range)(productivity/biomass) $\propto M^{-0.75+1.02-0.29} \propto M^{-0.02}$;

lifetime reproduction = (annual birth rate)(life expectancy)
$$\propto M^{-0.33+0.35} \propto M^{0.02};$$

lifetime metabolism = (life expectancy)(average daily metabolism)
$$\propto M^{0.35+0.66} \propto M^{1.01};$$

lifetime metabolism per unit mass $\propto M^{1.01}/M^{1.0} \propto M^{0.01}$;

population cycle time/standing crop turnover time $\propto M^{0.25-0.29} \propto M^{-0.04}$.

These first approximations to ecological design constants are encouraging. It appears that ecological scaling analysis can contribute to our view of evolutionary ecology and sociobiology. Of four population parameters suggested by Wilson (1975) to connect the theory of evolutionary ecology to the theory of sociobiology, two are clearly size dependent as documented in this chapter: (a) individual birth and death schedules and (b) equilibrial population densities. Furthermore, of five listed manifestations of sociobiology, two treated herein (at least for herbivorous mammals) show predictable size dependencies: (a) group size and (b) age composition. The second is true for birds as well. The use of allometry in ecology is argued further in the next chapter, with regard to the horizontal and vertical allometry of r- and K-selection.

12 Adaptive Strategies

A size-based analysis will distinguish strategies which are largely an outcome of size, the "physiological mechanism" or first order strategies, from those which vary amongst populations, and according to ecological conditions, the second order strategies.

—DAVID WESTERN (1979)

A PRESENTATION based on allometry once inspired the comment that "allometry sweeps all the interesting variability under the rug." Certainly, if we were to leave the impression that size is the only characteristic that matters, we would miss the adaptive modifications that are departures from the basic patterns — modifications that improve species fitness. Nonetheless, unless there is an allometric baseline, no objective comparison of those interesting variations is really possible.

Variations of this sort are seen as vertical displacements from an allometric regression line, traits in which an animal has a conspicuously higher or lower value than the general trend predicts. With a graphic portrayal in mind, I shall refer to the variation within a body size as "vertical allometry." This is equivalent to Western's (1979) "second-order strategies" or Smith's (1980) "narrow allometry," as contrasted with the trends across a broad size range (Western's "first-order strategies" or Smith's "broad allometry," which I prefer to call a "horizontal allometry"). Vertical allometry compares animals of the same approximate body size with the allometric predictions for that size and class of animals. As examples we can consider variations in gill area of 0.5-kg fish, in thermoregulatory response to heat stress in 40-g birds, in eggs of chicken-sized birds, in longevity of mice, and in the respiratory consequences of having a long neck. Finally, I will consider the continuum of r- and K-selection, which until Western's (1979) paper was a confusion of vertical and horizontal allometry. Before we commence such a survey, it might be worthwhile to review allometric baselines, so that the variability can be put into perspective.

After examining morphological allocation of body mass, physical support, physiology, reproduction, growth, and ecology, one cannot fail to be impressed by the repetition of patterns in scaling: $M^{1/4}$, $M^{1/3}$, $M^{1/2}$, $M^{2/3}$, $M^{3/4}$, M^{1}, M^{0}, $M^{-1/4}$, $M^{-1/3}$. True, we have had to accept, on an interim basis, data sets that are crude and assumptions and extrapolations that must be regarded with some skepticism. The exponents do not correspond perfectly to the above fractions; "$M^{-1/4}$" includes $M^{0.2}$ to $M^{0.3}$, so that r.m.e.'s of combined expressions, though small, are perhaps significant. Even if the patterns were perfect, a perfect explanation might still evade us.

The fact remains that the patterns are widespread. Scaling has not been random. For whatever reasons, natural selection has imposed constraints; otherwise the correlations with body mass would not be so strong. This implies that conformity provides some advantages, or that nonconformity carries a price. Consider again this statement: "It seems likely that an organism is an integrated system by virtue of the fact that none of its properties is entirely uncorrelated." Adolph (1949) concluded this about physiological characteristics, but we have seen through allometry that the integration does not cease at the skin, but is coupled to the environment.

Natural selection thus had to proceed with a conflict between the advantages of integration by conforming to the basic scaling rules and the competitive advantage of being able to reach a little higher, to take in a little more energy, or to get by on a little less. However, if the animal gets off the allometric line with regard to one feature, the interlinkage or integration is disturbed and other changes must follow to match or counterbalance the consequences of the initial adaptive nonconformity. The changes may be morphological, physiological, and/or acceptance of a more limited life-style.

Fish Gills

The fine divisions of fish gills, the lamellae, provide a tremendous proliferation of surface area for the diffusion of oxygen from the environmental water to the blood of a fish. The gill area may be as much as 48.5 times body-surface area in the false albacore *(Gymmosarda alleterata)* and 8 times body surface in the mackerel *(Scomber scombrus)*. The larger the fish, the more oxygen it requires, and the larger the surface area needed. We would, therefore, expect an allometric increase in respiratory-surface area for larger fish, and within a given body size an increase to meet demands of higher activity.

In a classic study of marine teleost fish, Gray (1954) compared the gill-surface area of four species having the same body size, 548 ± 12 g, but different activity patterns (see Table 12-1). Allometric equations grouping Gray's data for 31 species of marine teleosts, or for 33 species of marine and freshwater species combined with the data of Hughes (1966) and Hughes and Morgan (1973) would predict a respiratory area of 1,232 cm^2 to 2,621 cm^2 for a 548-g fish. The vertical allometry of fishes within this size range seems to support the principle of symmorphosis, that "no more structure is formed and maintained than is required to satisfy functional needs" (Taylor and Weibel, 1981). Fish gills would be logical organs for adherence to symmorphosis because the gill epithelium is the site of osmotic water loss from marine fishes or because of the swamping effect of osmotic uptake as well as salt losses by freshwater fishes. In either environment no more gill surface should be exposed than is needed for oxygen uptake. A sedentary fish like the bottom-dwelling toadfish can swim in fairly rapid bursts; but this can be done anaerobically, so the gill area can be well below the average (allometrically predicted) amount. The moderately active fish are between the two predictions, and the highly active menhaden vastly

Table 12-1 Respiratory surface area of the gills of marine fishes of the same body mass. (Data from Gray, 1954.)

Fish	Body mass (g)	Total gill area (cm^2)	Behavior
Menhaden	540	8,211	Very active
Sheepshead	544	2,542	Moderately active
Tautog	547	1,982	Moderately active
Toadfish	560	729	Sedentary
Allometric prediction	548	2,621[a]	—
		1,232[b]	

a. mm^2 = 792 $m^{0.92}$ (r^2 = 0.751; n = 31, data from table 1, Gray, 1954; p < 0.001; standard error of exponent = 0.098).

b. mm^2 = 579 $m^{0.85}$ (r^2 = 0.697; n = 33, data from Hughes, 1966, and Hughes and Morgan, 1973, excluding immatures, Antarctic fish, air breathers, and Chondrichthyes; p < 0.001; standard error of exponent = 0.101).

exceeds the average amount of gill area. The vertical allometry of gill area shows a similar correlation in decapod crustaceans (Hughes, 1983).

Thus this adaptive radiation in gill area appears to have been secondary to the adaptation to a sedentary or very active life-style. We could speculate that the menhaden might have to work harder at osmotic regulation than would the toadfish, and that if rate of water loss or salt excretion were plotted allometrically, parallel differences would be observed.

Evaporative Cooling in 40-g Birds

When the environmental temperature rises above the regulated body temperature of an animal, heat flows inward. The only way to get rid of that heat, and the heat produced metabolically, is to evaporate water. Birds do not have sweat glands, so depend upon the respiratory system and buccal surfaces for augmentation of this evaporation. In different birds this may involve panting (shallow respiratory pumping) and/or gular flutter (vertical oscillation) of the floor of the mouth and throat, which mixes local air without respiratory involvement. There are also phylogenetic differences in levels of metabolic heat production, the Caprimulgiformes having, for example, a relatively low basal metabolism. With these differences in mind, Lasiewski and Seymour (1972) selected a 40-g vertical allometry of a passerine weaverbird, quail, dove, and caprimulgiform poorwill. The data showed that it is not the rate of evaporation per se, but the balance between metabolic rate and evaporation rate that determines how effective the thermoregulation will be. There was no clear superiority for panting over gular flutter or vice versa, which helps to explain why both techniques have persisted.

The Eggs of Chicken-Sized Birds

This book has already surveyed the general allometric patterns of egg structure and function. There are, however, many departures from these patterns which are instructive as apparent examples of adaptation to particular niches, and which show that a single adaptive departure cannot be made without accompanying adjustments that move the bearer from conformity to other allometric plots (Rahn and Paganelli, 1981).

The kiwis (*Apteryx* species) of New Zealand lay what appear to be ridiculously large eggs (Figure 12-1). The chicken-sized brown kiwi (*A.*

12-1   For a chicken-sized bird, kiwis lay unusually large eggs, which are thought to
be a legacy from larger moa ancestors. The absence of risk to predators in New
Zealand served to relax the natural selection that would otherwise have restricted egg
sizes and incubation periods to the confidence limits of the normal allometry for
these life-history variables. This x-ray photograph of a female brown kiwi was taken
15 hours before the egg was laid. (Copyright 1978 by the Otorohanga Zoological
Society; used with permission.)

australis, 2.2 kg) produces eggs of 350 g to 400 g or more, overdoing it by a
factor of five or so if its egg is compared to even a grade AA jumbo egg from a
chicken-sized chicken *(Gallus domesticus).*

When Sir Julian Huxley made the first plot of egg size versus body size on
log-log coordinates, he excluded *Apteryx* from the regression line. In fact,
the kiwi egg is so far from where it should be that it provides an excellent
example of an extreme from which patterns may become apparent. If kiwis
were but dimly aware of their evolutionary history, they might wonder
about the inconsistency of natural selection: should they be bigger or
smaller? They make a poor case for orthogenesis if the widely held interpretation is correct, that their ancestors in Gondwanaland obeyed Cope's Law
of body-size increase until they were permanently grounded as overweight.
When New Zealand and Antarctica separated, the only nonflying homeo-

therms on New Zealand were of the ancestral moa-kiwi stock, from which the kiwis apparently scaled down to a mammal-like niche.

That is, they scaled down in most respects, but not with respect to the proper allometry for bird eggs (Tables 3-1 and 6-1, Eqs. 9-21, 9-21a, and 9-22). The derivations from allometric norms are naturally interdependent, but I cannot say what nature selected first and what had to follow in order for the egg to remain functional — a dilemma that is more complicated than the simple chicken-and-egg problem. Thus the chronology of reduction that follows may not be in proper order (Calder, 1978a,b, 1979):

(1) Selection of smaller body size, without predators to exert selective pressure for an "appropriately" smaller egg. (Let us temporarily bypass the vertical allometry of reproduction energetics.)

(2) Reduction in shell thickness, possible in the protected environment of the burrow, an apparent compromise (27.5 g) between the 36.6 g of shell appropriate for a 403-g egg and the 5.6-g to 9.2-g shell normally expected from a 2.2-kg hen's shell gland.

(3) Prolongation of incubation (by the male) to 74 days, 1.67 times the normal for a 403-g egg, perhaps *due to* the lower incubation temperature (35.4° vs. 37.7° C "normal") or *in order to* produce a more highly developed chick (that receives no parental guidance or feeding), permitted by the absence of predators. (The incubation period for a 2.2-kg species of bird would be 33 days, for a 403-g egg 44 days; the gestation period of a 2.2-kg mammal is about 80 days!).

(4) Reduced porosity of eggshell (G_{H_2O} = 60% of the prediction from Eq. 9-21) to reduce rate of evaporation over the 1.67-fold long incubation ($0.60 \times 1.67 = 1.002$, back to normal). The total loss over incubation is about 17% of fresh egg mass, within the range of so-called normal eggs. This calculated porosity reduction has been confirmed by SEM measurements (Silyn-Roberts, 1983).

(5) Equivalent reduction in peak oxygen consumption to 0.63 of normal, as a consequence of reduced conductance.

(6) Juvenile and adult standard metabolic rates are 65% and 61% of allometric predictions also, so that the same proportional step from embryo to bird occurs in kiwis as in normal birds.

The example of the kiwi illustrates some principles, but the allometry makes a general case for these principles as trade-off or compensation, established convincingly by Ar and Rahn (1978). The total water loss during incubation is, from Eqs. (9-16) and (9-20), the product

$$(G_{H_2O})(p_{H_2O})(t_{inc}) \propto (M_{egg}^{0.81})(M_{egg}^{0})(M_{egg}^{0.22}) \propto M_{egg}^{1.03}, \qquad (12\text{-}1)$$

in linear proportion to fresh egg mass, or

$$(G_{H_2O})(t_{inc}) = \text{a constant, } c. \tag{12-2}$$

The constant c, obtained by allometric cancellation in Eq. (12-2), is 5.94 mg/(g · mbar). Ar and Rahn assembled the data for G, t_{inc}, and M_{egg} for 90 species of birds, eggs ranging from 0.92 g to 1,500 g and found a constant, in units of mg/(g · mbar):

$$\frac{(G_{H_2O})(t_{inc})}{M_{egg}} = 6.84 \pm 1.15 \text{ s.d.} \tag{12-3}$$

Then what about all the interesting adaptive variability in the respiration of birds' eggs? It can still be enclosed by size-independent dimensional design criteria in the manner of Stahl (1962, 1963, 1967), as demonstrated by Eq. (12-3).

Reproduction Energetics

Not only is this 12.5-cm-long egg of the kiwi huge, it also has proportionately one of the largest yolks — 62% to 66% of the volume of the fresh egg. Yolk contains about 6 times as much energy as albumen per gram, amounting to 92% of the total energy in this egg. Thus the total energy content of the kiwi egg is over 11 times that which would be adequate for successful reproduction by a domestic hen of the same body size (Reid, 1971a,b, 1977; Calder et al., 1978; Calder, 1979).

In this example of vertical allometry, egg size spans the range from the proportionately small eggs of the Galliformes to a size which natural selection elsewhere would have ruled absurd. Perhaps only in the context of an isolated island continent lacking predators and offering a vacant niche could the flightless ratite stock lead to a scale-down of adult size from moa to kiwi without a scale-down of egg size (Cracraft, 1974). Such extremes raise questions about reproductive strategies: How should the available energy be allocated between self-maintenance and reproduction? How should the reproductive allocation be subdivided among offspring? How valid is "the assumption that the evolutionary resolution of problems involving mutually incompatible stresses or limited resources tends towards optimality" (King, 1980; see also Stearns, 1976)?

When we go beyond mere comparison of egg and yolk sizes to consider clutch sizes and incubation periods, a more complete picture emerges. The Galliformes lay larger clutches (chickens lay six or more eggs per clutch, the total mass of which is similar to the one or two eggs in a kiwi's clutch). Apparently, maintenance of a stable population in the face of losses to disease and predation necessitates *not* "putting all your eggs in one egg"; to have this insurance, the chickens spread their "premiums" over a series of smaller "deposits." For the kiwis, on the other hand, such insurance is not necessary, because predation is not a factor and a "lump sum" is a good investment.

Long Necks

Consider the flamingo *(Phoenicopterus ruber)*, whose tracheal volume (V_{trds}) is twice that predicted for a typical 2.2-kg bird (Table 12-2, col. 7). At the end of expiration the trachea thus has twice as much "used air" (reduced oxygen, increased carbon dioxide) in it, which will return to the lung as the first air in the next inspiration (Chapter 5). To ensure that an adequate amount of oxygen is inspired, the tidal volume (col. 3) is also more than doubled, while the respiratory frequency is reduced (Bech et al., 1979; Table 12-2, col. 2). Recall the reciprocal relationship between natural frequency and compliance (Eq. 5-23) and assume that the density of the system is the same as in other birds. These considerations make it appear that when the "proto-flamingo" stuck its neck out, there was a necessary increase in the compliance of the respiratory system, perhaps by increasing the proportional allocation to air sacs. This would lead to a lower natural frequency; fewer but deeper breaths would tend to restore the ratio of tracheal dead space to tidal volume. The trachea of the mute swan is also very long and contains twice as much air as it supposedly should. The product of lower breathing rate and deeper tidal volume results in a "standard" minute ventilation for the bird's size.

The mammal with the longest neck, the giraffe *(Giraffa camelopardalus)*, has a trachea that is 2.5 times as long as would be typical for its 400-kg body size. Unlike birds in general, which have compensated for the effect on resistance by increasing diameter to lower resistance, the giraffe actually has a tracheal diameter 10% less than predicted from the allometric equation of Tenney and Bartlett (1967), which partially counteracts the effect of the extra length, but the dead-space volume is still twice the typical expectation. The resting respiratory rate is 20% below the prediction, suggesting that tidal volume has been increased (assuming normal minute volume and oxygen extraction). Tidal volume measurements via a pneumotachographic flow device in one nostril do not support this, however, though there is a

Table 12-2 A demonstration that those who stick their necks out should breathe slower and deeper. Here f_r = respiratory frequency; V_t = tidal volume; $\dot{V_i}$ = minute volume; $\dot{V}_{O_2}$ = oxygen consumption; V_{trds} = volume of tracheal dead space; obs = observed; and pred = predicted from the allometric equations given below.

	(1) Body mass (kg)	(2) f_r (min^{-1})		(3) V_t (ml)		(4) $\dot{V_i}$ (ml/min)	
Species		Obs	Obs/pred	Obs	Obs/pred	Obs	Obs/pred
Flamingo[b]	2.21	9.6	0.71	89.3	2.29	846	1.62
Mute swan[c]	9.77	3.1	0.37	537	2.90	1,591	1.06
Giraffe[f]	400	9.0	0.80	3,300	0.84	30,000	0.66
Allometry							
Birds		17.2 $M^{-0.31g}$		16.9 $M^{1.05h}$		291 $M^{0.74i}$	
Mammals		53.5 $M^{-0.26j}$		7.7 $M^{1.04j}$		379 $M^{0.80j}$	

a. $\dot{V}_{O_2}$ predictions for basal metabolic rate.
b. *Phoenicopterus ruber;* from Bech et al., 1979.
c. *Cygnus olor;* from Bech and Johansen, 1980.
d. To 9.77-kg body mass from data for 8.57-kg mute swan.
e. Hinds and Calder, 1971.
f. *Giraffa camelopardalus;* from Hugh-Jones et al., 1978.
g. Lasiewski and Calder, 1971.
h. Bech et al., 1979.
i. Product of cols. (2) and (3).
j. Stahl, 1967.

possibility that greater airflow in the other nostril (without interposition of the resistance of the pneumotachographic sensor) may have caused an actual increase in tidal volume to be overlooked (Hugh-Jones et al., 1978; Calder, 1981).

Longevity of Mice

In general, life expectancy and maximum life span in captivity are positively correlated with body size. We would therefore expect that two species of mice of approximately the same size would have similar life spans. The Nearctic white-footed mouse (*Peromyscus leucopus,* 22 g to 30 g), however, lives considerably longer than the Palearctic house mouse (*Mus musculus,* 18 g to 21 g), making them an ideal pair for the studies of aging

Table 12-2 *(continued)*

(5) $\dot{V}_{O_2}$ (ml/min)		(6) $\dot{V}_{O_2}/\dot{V}_i$		(7) V_{trds} (ml)		(8) V_{trds}/V_t	
Obs	Obs/pred[a]	Obs	Obs/pred	Obs	Obs/pred	Obs	Obs/pred
38	1.91	0.04	1.0	17.8	2.01	0.20	0.92
110	1.89	0.07	1.8	88.4[d,e]	2.00	0.16	0.68
—	—	—	—	1,930	2.00	0.58	1.67
11.3 $M^{0.72}$[a]		0.05 $M^{-0.05}$[g]		3.70 $M^{1.09}$[e]		0.22 $M^{0.04}$	
11.6 $M^{0.76}$[a]		0.03 $M^{-0.04}$[g]		0.82 $M^{1.18}$[e]		0.11 $M^{0.14}$	

conducted by the late George Sacher and coworkers (Sacher and Hart, 1978; Hart et al., 1979).

The actual mice used in these studies differed somewhat in size, the *Peromyscus* averaging 31% greater body mass, for which the first equation in Table 6-1 would predict a 6% difference in maximum life span. However, the *Peromyscus* outlived the prediction for its size by 50%, while the *Mus* fell short of the prediction by 34%. The maximum life span of *Peromyscus* was consequently 2.4 times that of *Mus*. The superior performance of *Peromyscus* was amazingly similar, quantitatively, in life expectancy, time for mortality rates to double (at the actuarial aging rate α), lifetime metabolism (caged daily metabolic rate times life span), and rate of DNA synthesis in response to the damage of ultraviolet radiation.

The similarities are of great interest because of suggested relationships between aging and such factors as metabolic life span, physiological time, and ability to repair somatic DNA damage. Note that in its longer life, *Peromyscus* metabolizes more energy per gram of tissue. While *Peromyscus* has a larger brain, the percentage increase is only slightly greater than the difference in body mass (brain 2.6% of body mass vs. 2.5% in *Mus*). The differences in life span are greater than the differences in gestation period, which could be taken as another index of the physiological pace of life. This leaves the capacity for repair, manifested in unscheduled DNA synthesis, as the best correlate of longevity among the variables listed (Table 12-3).

Table 12-3 A comparison of the longevity and aging characteristics of two species of mice. (From data of Sacher and Hart, 1978, and Hart et al., 1979.)

Characteristic	House mouse (*Mus musculus*)	White-footed mouse (*Peromyscus leucopus*)	Ratio *Peromyscus/Mus*
Body mass	18.1 g	23.8 g	1.31
Calculated maximum life span[a]	1,899 da	2,006 da	1.06
Observed maximum life span	1,250 da	3,000 da	2.40
Maximum, observed/calculated	0.66	1.50	2.27
Life expectancy	605 da	1,420 da	2.35
Initial death rate (q_o)	1.87×10^{-4}/da	1.16×10^{-4}/da	0.62
Actuarial aging rate (α)	3.85×10^{-3}/da	1.65×10^{-3}/da	0.43
Doubling time of mortality $(0.693\,\alpha^{-1})$	182 da	422 da	2.32
Lifetime metabolism	776 kJ/g	1,560 kJ/g	2.01
Mean unscheduled DNA synthesis after ultraviolet irradiation[b]	11 u	26.5 u	2.41
Gestation period[c]	19 da	23 da	1.21
Brain mass[c]	0.45 g	0.63 g	1.40

a. From Sacher, 1959.

b. Measured 22 hr after 10 J/m² of 254 nm irradiation; units are average grains per cell nucleus.

c. From Sacher and Staffeldt, 1974.

Growth Rates and Life Spans

Why do some species have life spans that are characteristically longer than would be predicted from size alone? An allometric plot with a reasonably tight correlation coefficient such as that for maximum life span suggests the existence of physical constraints. How do some species become exempt from the implied limitations? Would the animal that lives longer not leave more offspring, leading to natural selection against the animal that "obeyed the allometric rules"? The persistence of the short-lived as well as the long-lived suggests that there must be a trade-off in the "decision" favoring one life-style over another. Perhaps the life-style is best characterized by the physiological pace or the time scale of the life history. We anticipate living longer than would be predicted from our body size. This extreme longevity is preceded by slower-than-predicted growth and maturation. For eutherian mammals in general, life span and age at puberty (sexual maturation time, Table 10-1) correlate well (Cutler, 1978). Thus, we have general trends and correlations of the exceptions, both of which can give insight into life-history "strategies" or "pendulums" (Chapter 14).

One approach is to collect information on several variables and see if the products or ratios of two or more of these variables are more predictable from body size than when each is taken separately. Economos (1981b) extracted from the literature figures for 22 species, from shrew to elephant, species for which all of the following were available: gestation time (t_{gest},yr), age at puberty (t_{pub},yr), record life span (t_{ls},yr), and neonatal (M_{neo},kg) and adult (M_{adult},kg) body masses. These were combined into

$$\text{"rate of becoming"} = \frac{(t_{gest})(t_{pub})}{(t_{ls})(M_{neo})} = 0.298 \, M_{adult}^{-0.53}, \tag{12-4}$$

for which the correlation coefficient r was -0.992. If data for each of the factors are analyzed separately, these are the results:

$$t_{gest} = 0.19 \, M_{adult}^{0.23}, \quad r = 0.896; \tag{12-5}$$

$$t_{pub} = 0.92 \, M_{adult}^{0.30}, \quad r = 0.851; \tag{12-6}$$

$$t_{ls} = 10.5 \, M_{adult}^{0.23}, \quad r = 0.842; \tag{12-7}$$

$$M_{neo} = 0.05 \, M_{adult}^{0.83}, \quad r = 0.979. \tag{12-8}$$

The average correlation coefficient is 0.892, body size thereby appearing to control about 80% of the variability ($r^2 = 0.71$ to 0.96), with "strategic trade-offs" perhaps claiming the rest. If instead of using individual "rate of becoming" ratios, we run an allometric cancellation of Eqs. (12-5) through (12-8), we obtain the same exponent and a slightly higher coefficient:

$$\text{rate of becoming} = 0.334 \, M_{adult}^{-0.53}. \tag{12-9}$$

These calculations show that the variability manifested as departures from the allometric prediction for a single variable is only part of a trade-off; when the rest of the trade is taken into account, there is less overall variability (as expressed in r values) for Eq. (12-4) than for Eqs. (12-5) through (12-8). This check on the validity of allometric cancellation has also proved to be reassuring.

Whether or not the rate of becoming is the best expression biologically is another matter. It is not a rate, but the reciprocal of a rate, having dimensions of $t^2/(t)(m)$ which, reduced, has as units the years per kilogram of neonatal mass. Its reciprocal is the mean rate of embryonic growth (M_{neo}/t_{gest}) as set forth in Eq. (10-33), except that a correction factor (the ratio of time at puberty to maximum time until death) has been applied. Perhaps more useful would be the development of size-independent dimensional and nondimensional growth constants as proposed by Stahl (1962), rather than the $M^{-0.53}$ reciprocal of the rate of becoming, which cannot be associated directly with other allometric scalings.

r-Selection and K-Selection

The earliest description of the life cycles of r-selected animals, those selected for environments that are variably inhospitable or unpredictable, appears in the Old Testament:

> That thou givest them they gather: thou openest thine hand, they are filled with good. Thou hidest thy face, they are troubled: thou takest away their breath, they die, and return to their dust. Thou sendest forth thy spirit, they are created: and thou renewest the face of the earth. (Ps. 104:28 – 30)

These are conditions that lead to variable and nonequilibrium populations. At the other extreme are the inhabitants of relatively stable environments, wherein populations have attained equilibrium; in between is a continuum of intermediate accommodations (Pianka, 1983). A refinement of major significance to the understanding of this continuum was made by Western (1979), who distinguished between a size-dependent first-order strategy and a second-order population strategy that fine-tunes the animal to ecological conditions.

In Western's view, selection for body size is first-order strategy and can account for most of the variability in life-history characteristics used to identify r-selection or K-selection, a view reinforced by Stearns (1983a,b). I have attempted to follow up this suggestion with the allometry of mammals, as developed in the preceding chapters; the result is Table 12-4. If these scalings are adequate representations, they can then serve as baselines for evaluating second-order strategies or adaptations to specific ecological situations as well as phylogenetic legacies. For example, Western depicts reproductive rates as a vertical allometry between population strategies for r-selection, in which the reproductive rate exceeds that predicted from body

Table 12-4 The r-to-K selection continuum: an allometric interpretation of the correlates for mammals. (Adapted from Murray, 1979, and Pianka, 1983. See text for sources of scaling exponents.) Adapted, with permission, from *Ann. Rev. Ecol. Syst.* 14, © 1983 by Annual Reviews Inc.

Element	r-selection		K-selection
Environment	Unpredictable, early successional		Predictable, mature or climax
Population size	Variable, $<K$, recolonizing		Fairly constant, equilibrium, $\approx K$
Body size	Small		Larger
Consequences of size		*Allometry*	
Development	Rapid	$t_{growth} \propto M^{0.25}$	Slower
Reproduction	Early	$t_{sex\ maturity} \propto M^{0.29}$	Delayed
Population increase	High r_{max}	$r_{max} \propto M^{-0.26\ to\ -0.36}$	Lower resource thresholds
Mortality	ρ_n-independent, catastrophic	$\rho_n \propto M^{-0.75}$	ρ_n-dependent, directed
Life span	Short	Maximum $\propto M^{0.20}$; expected $\propto M^{0.32}$	Longer
Survivorship	Often type III[a]	Expected/max $\propto M^{0.12}$ [b]	Usually type I or II[a]
Energetics emphasis	Productivity	(Births/yr)(expected yr) $\propto (M^{-0.33})$ $(M^{0.32}) \propto M^{-0.01}$	"Efficiency"
Colonizing	Great ability	Follows from development, reproduction, and population increase	Small ability
Social behavior	Weak	A_{hr} overlap $\propto M^{0.27\ to\ 0.34}$	Frequently well developed
Competition	Less ability	Experience implied in survivorship and social behavior	Greater ability

a. Deevey, 1947.

b. This means that the proportion of maximum life span attained, on the average, increases with size, so that the larger species will tend to have a greater representation of experienced senior citizens.

size, and for *K*-selection, wherein rates are less than predicted from body size.

Other investigations into vertical allometry have related brain size to feeding strategies (Eisenberg and Wilson, 1978) and testis size to mating patterns (Harcourt et al., 1981). Doubtless there will be many additional studies in this area.

Conclusion

At best, an allometric equation can only describe the average tendency within a given class or body plan. For the study of evolutionary ecology, as for comparative physiology, this description of central tendency is an essential reference point for appreciation of the adaptations that fit the structure, physiology, reproduction, and ecology of the animal to its physical and sociobiological environment.

13 The Limits on Evolution of Body Size

A warm-blooded animal much smaller than a mouse becomes an impossibility; it could neither obtain nor yet digest the food required to maintain its constant temperature.

—D'ARCY THOMPSON, *ON GROWTH AND FORM* (1917)

What is the ultimate limit to the size of land animals? Unfortunately, we are unable to give an adequate answer, and we cannot study the question experimentally by building a bigger elephant.

—KNUT SCHMIDT-NIELSEN (1975)

WE BEGAN the study of size by looking at the known extremes in body sizes of various phylogenetic groups and life-styles, then surveyed the apparent "rules" by which characteristics of form, function, reproduction, growth, and ecology are scaled to meet the needs of various body sizes. Such a description cannot guarantee an explanation of the scaling, but the patterns described quantitatively are prerequisites to the eventual clarification and full appreciation of the constraints and consequences of size.

Thus far, we have considered the "vertical" aspects of the allometric plot; given the size, what is its predicted value on the *y*-axis? How much variation is there, and what compensations have been necessary for the deviation from the customary physical constraints? This could be termed study of the "vertical constraints" that limit, for example, the amount of muscle in a 500-g mammal to a range considerably smaller than "0% to 100%." There also are apparent constraints on the total body sizes of homeotherms, there being no adult mammals with body mass under 2 g or over 130 metric tons.

There are discontinuities which occur fairly abruptly at what appear to be the size limits for a certain type of function. In the widest range of all kinds and sizes of animals, F. W. Went (1968) has provided a fascinating account of the qualitative differences in life and the problems for various kinds of animals on either side of a discontinuity. Our lives and those of other mammals and birds are passed in surroundings dominated by gravity, a force which precludes large animals like man from walking up vertical surfaces or across free-water surfaces; these are abilities restricted to the smallest animals, those with best claws or grips, and those that do not weigh enough to break the surface tension of the water. In contrast, insects live in a world of surface tensions that allow the ant an easy vertical climb, the water strider a walk on a surface-tension film without fear of sinking, and so on. At the microscopic level, Brownian molecular motion can threaten with chaos. At the other end, the very largest mammals—the whales—depend upon buoyancy to support their giant masses.

There are also size limits for different metabolic types; for example, those that have caused insects to dominate nectar feeding up to a size approaching that of hummingbirds, and conversely a size below which the humming-birds apparently cannot evolve (Brown et al., 1978). The total range of animal sizes can be subdivided by body-temperature regime: (1) totally poikilothermic, as in the smallest insects, (2) homeothermic in activity but poikilothermic at rest, as in many of the larger insects (see Bartholomew, 1981), (3) normally homeothermic at rest, but utilizing hypothermic torpor as an energy-conserving mechanism, and (4) totally homeothermic. This has meant that there is a lower limit for homeothermy as well as an upper limit to body size for true hibernation—the latter apparently at the size of marmots, which weigh 5 kg or less, their larger bodies cooling and warming too slowly for practical use of hypothermia. There may even be a size large enough that poikilothermy is lost by virtue of great thermal inertia, provid-ing some plausible physiological, morphological, and ecological extrapola-tions back to the lives of dinosaurs before their extinction (for a review see Thomas and Olson, 1980).

The size limits, of course, depend upon what scaling rule applies. If there is an absolute upper physical limit to Y in the allometric equation $Y = aM^b$, that value will be reached at a smaller total M if b is 1.07 than if b is 0.67; if we really want to know what is going on, we must look at the appropriate relationship. Take, for example, the act of contracting a muscle during locomotion. The maximal force that can be developed is proportional to the cross-sectional area of the muscle (the number of parallel muscle fibers contracting), which scales as $M^{0.67}$. The work done by this contraction is the force times the distance of the shortening. In mammals other than the bovids, the characteristic lengths scale as $M^{-0.33}$, so the work done is proportional to $M^{0.67+0.33=1.0}$. The maximal power output or work rate is the product of work per contraction and frequency of contraction, or $P_{max} \propto$ $(M^{1.0})(M^{-0.25}) \propto M^{0.75}$.

It appears that there are physiological limits to both work and power. Power is limiting at high frequencies, whereas work is limiting at lower frequencies. There are trade-offs — one at a frequency of about 10 Hz, the upper frequency for running in mammals, and one near the lowest frequency of wingbeat in hovering insects (Weis-Fogh and Alexander, 1977; Alexander, 1982).

Limits to Largeness

It is clear that there have been evolutionary trends toward larger body size, correlated with both time (Cope's Law) and latitude (Bergmann's Rule) (Bonner, 1965, pp. 176–198; James, 1970; Stanley, 1973). The mere existence of hummingbirds and shrews, and the fossil record for therapsid skulls (Baur and Friedl, 1980), indicate that there has also been a natural selection toward smaller animals. However, these trends ultimately reach limits, and when our understanding is complete, we should be able to explain convincingly why there are not — and were not — smaller or larger extremes than those indicated in Chapter 1. For now, we can only review some of the explanations and speculations aimed at such an understanding.

The failure of natural selection to produce larger terrestrial birds and mammals has been assumed to be related to the problems of locomotion and support against the force of gravity. The skeleton of an elephant contributes 27% of total body mass, compared to 5% for a shrew (Prange et al., 1979). A blue whale weighs 15 to 20 times as much as an elephant, but because of buoyant support needs a skeleton only 15% to 18% of body mass (Smith and Pace, 1971; Anderson et al., 1979).

Back in the Oligocene and Miocene periods there was an even larger terrestrial mammal, *Baluchiterium*, a relative of the rhinoceros, that had a body mass estimated to be 20 to 30 metric tons (Schmidt-Nielsen, 1975; Economos, 1981). Extrapolation of the mammalian skeletal mass equation (4-10) "predicts" a skeletal mass at 15% of the estimated body mass (the elephant skeleton weighs twice the allometric prediction for 6,600 kg, the latter being 13% of body mass, at least for dried bones). Economos' (1981) extrapolation from an allometric equation for gravitational tolerance shows 20 metric tons to be the largest-size mammal that could tolerate terrestrial gravity. A mouse can tolerate seven times the force of gravity, a dog less than four times, and a 20-metric-ton animal would be on the 1-G line. For a wide variety of support designs (endoskeletons and exoskeletons except those of mollusks) 10% to 20% seems to be an upper limit to the tolerable support burden or proportion of body mass devoted to support (Anderson et al., 1979). Elephants seem to have been granted special exemption from this limit, unless the difference is due to a comparison of wet bones with a dry-bones regression.

With the body mass supported by two legs instead of four (one instead of two when running), birds would be limited to a maximum body mass only half that of mammals. However, it appears that another constraint enters into the picture, having evolutionarily stopped body-mass increase of the elephant bird a full order of magnitude below that of a medium-sized elephant. As mentioned in Chapter 4, this constraint could have been eggshell strength, the *Aepyornis* shell probably amounting to a tremendous 16% of egg mass yet offering little if any margin in the eggshell strength safety factor (Paganelli et al., 1974; Ar et al., 1979).

The upper limit to body mass for flying birds has been explained as the consequence of the allometry of power required to become airborne versus power available from the flight muscles (Pennycuick, 1972). The dimensions of power (P) are force times velocity. The force generated to sustain flight at a particular lift: drag ratio must be proportional to the bird's weight or mass. As we saw in Chapter 7, characteristic velocities such as the stalling velocity are proportional to $M^{1/6}$. Thus we have for power required, P_r:

$$P_r \propto (M^{1.0})(M^{0.17}) \propto M^{1.17}. \tag{13-1}$$

The forces developed by the flight muscles are proportional to the cross-sectional area of the muscles, and the power available (P_a) is the product of that contracting force and the muscle-shortening velocity, which is size independent:

$$P \propto (M^{0.67})(M^0) \propto M^{0.67}. \tag{13-2}$$

From the difference in scaling exponents, it is obvious that power available does not increase in step with power required. The smallest birds have a large margin and can hover, maintaining their weight by thrust alone for extended periods. Larger birds can develop thrust equal to weight for short periods but cannot sustain flight without forward motion, thereby generating lift to offset weight. Swans, pelicans, the kori bustard *(Ardeotis kori),* and the extinct condor *(Teratornis)* approach an empirical limit that appears to have a rational explanation, in the range of 12 kg to 20 kg (weights of 120 N to 196 N). The genes of the ratites became earthbound when the body they directed exceeded the maximum mass that could become airborne.

Limits to Smallness

The minimum size for the homeothermic bird or mammal of 2 g is not so easily explained. We know that as size decreases, the intensity (rate per gram) of metabolism increases, the pace of life is more rapid, and a number of measures have been taken to ensure adequate delivery of oxygen and blood glucose and removal of metabolic wastes (Table 5-7). The smaller mammals have greater capillary densities and higher concentrations of several enzymes. If an animal were small enough, it would be all capillaries and enzymes—which, of course, would not work. Iberall (1979) suggested that a minimum body size could be set by the requirement to have an aorta free of appreciable viscous loss. His calculated body size was 8 g, about four times the size of the smallest shrew—not bad for a theoretical calculation.

Another possibility is that the bottleneck is in the supply line at the level of the ability to feed, or to capture prey or digest the intake. With greater body-surface-to-volume ratios, small birds and mammals have less thermal inertia and a narrower, higher range of thermoneutral temperatures (Chapter 8). Thus if the ability to produce heat is impaired, they will equilibrate rapidly. Are we missing the truth by limiting our concern to adult physiology and ecology? The relative size of egg or neonate is inversely related to size; perhaps, in the long run, the constraints of reproduction and/or growth set the minimum body size.

METABOLISM AND SMALL SIZE

It has been well known that metabolic intensity is inversely related to size. Oliver Pearson (1948, 1959) pioneered the measurement of basal metabolic rates of the smallest homeotherms, shrews and hummingbirds. He found that $\dot{V}_{O_2}/g$ took a dramatic upswing from the basic $M^{-1/4}$ trend in the size range of both shrews and hummingbirds. He concluded for both classes that smaller animals would be unable to gather and consume food at rates adequate to prevent starvation.

Subsequent studies have established that shrews and hummingbirds have characteristically high metabolic rates. A 7-g rodent *(Baiomys)* and a shrew-like marsupial of 5 g to 10 g have basal metabolic rates that fall on the allometric lines for eutherian and marsupial mammals, respectively. However, shrews metabolize at over twice the Kleiber predictions for their sizes, and hummingbirds consume energy more rapidly than the general nonpasserine equation would predict, but more slowly than the passerine equation would suggest. There are several small passerines that are about three times the mass of the smallest hummingbird (Poczopko, 1971; Lindstedt, 1978; Prinzinger et al., 1981).

In view of these metabolic rates elevated above the extrapolation, it does not seem that the inability to eat fast enough to avoid starvation can explain a lower limit on body sizes for birds or mammals in general, but only for shrews and hummingbirds, if those. If they cannot eat fast enough to elevate above the Kleiber line, why do they not throttle down to the Kleiber line, where they should be able to evolve to a slightly smaller size? The desert shrew *(Notiosorex crawfordi)* does appear to have exercised such basal restraint, compared with other shrews, metabolizing only 27% above the Kleiber prediction. Adults of this smallest desert mammal, 4 g, are still twice the size of the smallest shrew, but the reduced metabolic level must help the overall heat and water balance (Lindstedt, 1978, 1980).

REGULATING THE TEMPERATURE OF A SMALL BODY MASS

A partial explanation, given by Tracy (1977) for the abrupt upswing in the metabolic rate vs. size plot, is that most of the so-called basal metabolic rates were obtained during exposure to 24° C, while the lower critical temperatures of the smallest mammals (shrews) are considerably higher, probably closer to 33° C. Tracy compares the curves for metabolism vs. mass for small mammals at temperatures of 0° C and thermoneutrality with those for shrews at 24° C to make a good case for his conclusion that "it is not possible to predict the smallest size of endothermic homeotherms without also specifying the environment in which the animal exists."

The smallest (2 g) hummingbird, *Calypte helenae,* lives on a tropical island where cold stress would be unusual, but the 2.6-g to 3.2-g *Stellula calliope* breeds in the Rocky Mountains where freezing air temperatures during incubation are not uncommon (Calder, 1971). The smallest shrew, *Suncus etruscus,* is found in the Mediterranean region and Africa. However, the smallest American mammal, the 2.3-g *Microsorex hoyi,* is distributed from Alaska and across Canada to the northeastern United States (Walker, 1975). While Tracy's conclusion that minimum size is a function of environmental temperature is undoubtedly true, that minimum size does not appear to have been reached by 2-g shrews or hummingbirds in the tropics, considering that similar animals less than a half-gram larger are obviously successful in northern habitats.

Limits to Body Size in Reproduction

In smaller species the neonatal mass is relatively larger in proportion to adult body mass, and the synthesis of this neonatal mass is therefore a significant part of the maternal energy balance. The reproduction that defines fitness cannot occur until the adult female is able to obtain energy and nutrients in excess of her own needs. Not only the proportionate neonatal mass, but the annual fecundity is inversely related to adult body mass. Furthermore, the reproduction is occurring on a faster physiological time scale in the smaller animals. The allometric characteristics of reproduction and growth suggest that the lower limit for body size may be imposed by the demands of reproduction, from either mechanical or energetic constraints.

The incubation periods of birds are proportional to $M_{egg}^{0.22}$ (Rahn and Ar, 1974; Ar and Rahn, 1978). For the smallest eggs (0.5 g) the predicted incubation period would be 10 days; the shortest incubation period clearly documented is 11 days. Mammalian gestation is proportional to $M_{adult}^{0.25}$; for the smallest mammal (2 g), this should be 13.8 days, but the shortest gestation listed by Eisenberg (1981) is 16 days for the hamster *(Mesocricetus auratus).* In both cases there appears to be a lower limit for development time that is somewhat longer than the extrapolation, which may in turn be expressed as a limitation on size for the smallest homeotherms.

Curiously, however, the smallest mammals and birds — shrews and hummingbirds — have significantly longer embryonic development times than the minimum observed in their classes. The smallest shrew, *Suncus etruscus,* has a gestation period of 27 days to 28 days and the shortest shrew pregnancies are 20 days long (*Sorex araneus* and *S. vagrans*), 25% longer than the hamster's gestation. The incubation periods of hummingbirds run from 14 days to 19 days, the shortest being 27% longer than the minimum observed in birds. Thus minimum incubation or gestation times do not come close to being a constraint upon the evolution of smaller mammals and birds. We must search for other explanations.

Another approach is to postulate a bird or mammal smaller or larger than the living range, then extrapolate to that hypothetical size to see what the consequences of such a trend might be. Brown and associates (1978) extrapolated from available data using the appropriate allometric expressions for a 1-g vs. a 2-g hummingbird. The reproductive consequences of being a 1-g hummingbird are that the life span and the opportunity for reproduction might be reduced 12%. The energy cost of egg synthesis is essentially the same proportion of the daily metabolism as for a 2-g bird, the egg surface does not scale down as rapidly as the brood patch of the female might, and the eggs warm and cool more rapidly when the female leaves the nest and returns. In times of lost feeding opportunity, the endurance on the total of calculated crop and fat energy reserves decreases by 30%, from 1.9 days to 1.3 days. These calculations do not identify an end point to evolutionary size decrease, they only give us a chance to explore the possibilities and thereby gain some quantitative idea of which might be involved.

Displacement among Body Sizes

Not so obvious or dramatic as the extremes of smallest and largest animals in a class are the observed differences in size in closely related, sympatric species. This is the phenomenon of character displacement mentioned in Chapter 1. Hutchinson, in a widely cited and influential article (1959), reported several cases, avian and mammalian, in which similar species were distinguished by having linear dimensions of trophic structures (bill or skull length) in ratios of 1.1 to 1.4 (mean 1.28).

While analysis by Løvtrop and colleagues (1974) supported this quantitative displacement, it has been widely disputed (Faaborg, 1977; Roth, 1979, 1981; Wiens and Rotenberry, 1981; and Simberloff and Boecklen, 1981).

Until the controversy is resolved, we can ask what the consequences would be if a size-displacement factor of 1.28^3 were selected. If geometric similarity is preserved among the larger and the smaller species, body volumes and masses scale as 1^3 or $1.28^3 = 2.1$. This increase in body mass would result in a 74% increase in basal metabolic rate, a 64% increase in total daily energy requirement, and a 20% increase in characteristic physiological time periods. The desynchronizing of times (or the reciprocal frequencies) might have some effect on feeding behavior. The larger species would be able to endure fasting 20% longer, and would perhaps displace from one another also in sizes of preferred food items.

14 Alternatives and Applications

Empirical science is apt to cloud the sight, and by the very knowledge of functions and processes to bereave the student of the manly contemplation of the whole. The savant becomes unpoetic.

—RALPH WALDO EMERSON, "NATURE" (1856)

But not all naturalists want to do science; many take refuge in nature's complexity as a justification to oppose any search for patterns . . . Doing science is not such a barrier to feeling or such a dehumanizing influence as is often made out. It does not take the beauty from nature.

—R. H. MACARTHUR, *GEOGRAPHICAL ECOLOGY* (1972)

I READ Emerson's warning when I was starting my graduate studies and I have treasured it ever since. It may seem on first reading that his thoughts and those of MacArthur are diametrically opposed, but I interpret them as two essential views that provide a balance—like the insulin and glucagon which regulate blood glucose concentration within upper and lower bounds, or like the populations of cold as well as heat receptors in the hypothalamus.

It is through a few empirical functions that I am able to approach contemplation of the whole. When two or more allometric equations fit together to show how a vital process is maintained across an evolutionary change in body size, I am all the more able to agree with another Emerson passage, this time from his essay "Beauty":

Beyond their sensuous delight, the forms and colors of Nature have a new charm for us in our perception, that not one ornament was added for ornament, but is a sign of some better health, or more excellent action . . . 'Tis a law of botany, that in plants, the same virtues follow the same forms. It is a rule of largest application, true in a plant, true in a loaf of bread, that in the construction of any fabric or organism, any real increase of fitness to its end, is an increase of beauty.

One by one, the allometric equations reviewed to this point may suggest a mechanical simplicity that may indeed "cloud the sight." Perhaps it is time to step back and attempt to put allometry into perspective, and to see if the ideals of Emerson and MacArthur are not, after all, compatible.

Awareness of the existence of biological scaling has been growing for over a half-century (Huxley, 1927; Kleiber, 1932), but the focus has been on the fragments: scaling of specific organs, functions, or requirements. Attempts to put the allometric animal together again have been the rare exceptions (Adolph, 1949; Stahl, 1962, 1963a,b, 1967). It is difficult or even impossible to reassemble anything when a number of pieces are missing, but fortunately many of these pieces are turning up in the current literature.

Going beyond the allometry of one characteristic to consider the scaling of everything from enzyme concentrations and cellular events to life histories and population dynamics, we may still not know "Why $M^{3/4}$?", but at least we have a different perspective. (Even without knowing which came first, the chicken or the egg, our knowledge of the facts of life improved when it was determined that chickens came from eggs and eggs came from chickens, rather than from old rags, barnacles, or hibernating frogs!)

If attention is limited to the scaling of a single feature—one's special research interest, for example—it is easy to assume that there is a gene for controlling heart size, another for brain size, and another for longevity, each subject individually to natural selection. The path to scientific explanation does not start with the most complicated and awe-inspiring assemblage, but with the reductionist selection of a relatively simple case or mechanism, in the hope of being able to generalize to a more complex situation. We need a broad appreciation too of just how much has been going on. Even with evolution of size and its many consequences and requirements, there are other interactive characteristics, such as the interdependence with shape, development time, and timing of events within that development (Alberch et al., 1979; Bonner and Horn, 1982).

Gould and Lewontin (1979) have been critical of the "adaptationist programme" or "Panglossian paradigm" which they claim has dominated evolutionary thought for several decades. According to the adaptationist programme, the animal is reduced to individual traits, each of which is acted upon by natural selection, thereby optimized for its function as if isolated from all other traits. Where suboptimality is evident, it is taken to represent a compromise or trade-off. Amid the "incomplete hierarchy of alternatives to immediate adaptation for the explanation of form, function, and behaviour" appears:

> No adaptation and no selection on the part at issue; form of the part is a correlated consequence of selection directed elsewhere . . . Here we come face to face with organisms as integrated wholes, fundamentally not decomposable into independent and separately optimized parts.

In this category Gould and Lewontin make specific mention of allometry as an example. This revives what the physiologist Adolph stated exactly three decades earlier in his 1949 classic:

> It seems likely that an organism is an integrated system by virtue of the fact that none of its properties is entirely uncorrelated, but that most are demonstrably interlinked; and not just by single chains, but by a great number of crisscrossed linkages.

While criticism of the adaptationist programme has sparked considerable controversy (see Brown, 1982), the similarity of these conclusions from essentially opposite ends of the spectrum suggests to me a sort of convergent evolution in our description and appreciation of life, one in which allometry has much to contribute.

Suppose that competition has led, as a matter of character displacement (sympatric size evolution), to a discrete dichotomy or discontinuity in body size of a type of rodent, or that an adaptive response to climate has given rise according to Bergmann's Rule to a gradation toward larger sparrow size going from milder to colder environments (allopatric size evolution). Whether the new animal was produced gradually or in a single leap, there surely have been size changes in the 18 organs and systems listed in Table 3-4, in no doubt many more than the 8 enzyme concentrations and 8 densities or concentrations listed in Table 5-7, in the strength of the skeleton (Chapter 4), in the surface areas of the integument, alveoli, glomeruli, and gut (Chapter 5), in rates of metabolism and all the supporting functions (Chapter 5), in reproduction and growth (Chapters 9 and 10), and in perception of "territory" or "home range" (Chapter 11). The number of changes is staggering, to say the least.

Is there a gene for the size or amount of each of these? Would it not be better to have some sort of genetic integration that ties each characteristic to the other? The chances for the occurrence of a single beneficial mutation are said to be slight, but the permutation of simultaneously getting all that are needed for the separate scaling adjustments suggested in the preceding paragraph is beyond my own comprehension and that of my electronic calculator.

We could explore the quantification of the genetic material itself. Munro and Gray (1969) examined the DNA concentrations and total amounts in liver and muscle tissues of adult mammals ranging in size from mouse to horse. The DNA concentration of liver showed no dependence on body size, while that of muscle (in milligrams per 100 grams) was slightly but significantly and inversely related to body size in the same general fashion that we saw in the case of oxidative enzymes (Table 5-7):

$$[DNA] = 45 \, M^{-0.08}; \qquad r^2 = 0.91, \tag{14-1}$$

from which we might predict, based on the linear proportion of muscle mass to total body mass:

$$\text{total DNA} = 202 \, M^{0.92}. \tag{14-2}$$

Total liver DNA is roughly proportional to liver mass, of course. While this is interesting as another aspect of body composition, I see no implications for genetic control of body and life-history allometry. Allometric control seems to be beyond the vision of this present survey. When body size increases or decreases, the accompanying quantitative changes are many. How they are accomplished is a mystery, the solution of which might resolve the controversy between practitioners and critics of the adaptationist programme!

$M^{1/3}$ Scaling — Still a Possibility?

This book was never intended to be anything more than a survey of apparent patterns to stimulate further examination, testing, and search for interlinkage of scaling at the morphological, physiological, and ecological levels. It is, however, motivated by the desire to see life as something more than a catalog of random facts, and this motivation leads to expectations that may bias one's perceptions. Ultimately, the patterns should suggest the physical basis for natural selection of the scaling; but first the measurement of body-mass exponents must be refined, so that we are looking at actual scaling factors, not approximations loaded with statistical "noise."

We have seen that there are three models from which the scaling of physical and morphological variables might be predicted: the geometric model, in which linear dimensions scale as $M^{1/3}$; the elastic similarity model, which predicts that linear dimensions will scale as $M^{1/4}$; and the dynamic similarity model, which can scale either way. Despite shortcomings and criticisms, empirically most of the scaling does seem to fit the $M^{1/4}$ scaling of the elastic or dynamic scaling models, and thus a bias for the expectation of $M^{1/4}$ scalings becomes almost inevitable.

Now it is time to balance this by collecting a list of the many variables that approximate $M^{1/3}$ scaling. Since the basis for the scaling models is usually physical support, support that is less sturdy the longer the lengths, let us review first some of the $M^{\sim 1/3}$ (geometric scaling) lengths:

Bone and limb lengths, body lengths (mammals) $\propto M^{1/3+}$

Aorta, vena cava, and tracheal lengths (mammals) $\propto M^{0.30, 0.33, 0.40}$

Wing lengths (birds and insects combined) $\propto M^{1/3}$

Fur thickness (mammals) $\propto M^{1/3}$.

Frequencies tend to be inversely related to characteristic lengths, and the periods or times are reciprocals of frequencies or rates. According to the elastic similarity model, most of these frequencies would scale as $M^{-1/4}$, and the times as $M^{+1/4}$. Indeed, the majority do, but we still have an impressive list of times that come closer to an $M^{1/3}$ scaling. These are shown in Table 14-1. If you had not seen the list of $M^{1/4}$ scalings first, there would seem to be

Table 14-1   Some scalings that approximate the cube root.

Function	Scaling
Gut beat duration	$\propto M^{0.31}$
Respiration cycle duration (nonpasserine birds)	$\propto M^{0.31}$
Plasma albumen half-life (mammals)	$\propto M^{0.32}$
Wingbeat period (birds)	$\propto M^{0.38}$
Cooling time, $1°C$ drop (mammals)	$\propto M^{0.35}$
Euthermic interruption of mammalian hibernation	$\propto M^{0.32}$
Postembryonic doubling of mass (birds)	$\propto M^{0.34}$
Minimum time for population doubling (mammals, derived from Caughley and Krebs, 1983)	$\propto M^{0.36}$
Standing crop turnover time (mammals, according to Banse and Mosher, 1980)	$\propto M^{0.32}$
Time to puberty (mammals)	$\propto M^{0.30}$
Life expectancy, 1 in 1,000 (mammals)	$\propto M^{0.32}$
Minimum time difference between sound arrival at each ear (mammals)	$\propto M^{0.30}$
Time to metabolize 1% of body mass as fat at estimated field metabolic rate	$\propto M^{0.33}$

a good case for a pattern that is compatible with geometric scaling—although it is more than likely that several of these exponents are not statistically distinguishable from the $M^{1/4}$ scaling. Nevertheless, it is a possibility to which we must be open as the allometric analyses are refined by the addition of further data, or by more rigorous criteria for including data. Such a geometric scaling ($M^{1/3}$ for lengths and times, $M^{2/3}$ for metabolic rates) is also compatible with some of the exponents obtained in support of Heusner's (1982a,b) insistence that the $M^{3/4}$ metabolic scaling is merely an artifact of interspecific allometry.

Allometric Splitting and Grouping

The body-mass exponent for standard metabolic rates of birds, as determined in earlier studies, was characteristically lower than the $M^{3/4}$ of mammalian metabolism (see Lasiewski and Dawson, 1967, for a review). This suggested that there was a fundamental difference between these two classes of homeothermic vertebrates. Lasiewski and Dawson reanalyzed the data and found that birds of the order Passeriiformes (which includes the garden variety of songbirds) have distinctly higher resting metabolic rates than birds of other orders. Analyzed separately, the effect of size was seen to be the same for both groups of birds, an $M^{0.72}$ statistically indistinguishable from the mammalian metabolic $M^{3/4}$, but with a characteristically higher a coefficient for the passerines. This tended to skew the regression for all birds, because the passerines predominated at the small end of the avian body-size range.

Should the splitting be carried further to obtain better resolution of the size-metabolism relationship? Zar (1968) separated the available data further for eight orders of birds and found statistically distinct exponents for three orders: 0.63 for the Anseriformes (ducks, geese, swans), 0.65 for the Falconiformes (hawks), and 0.86 for the Columbiformes (doves). Does this mean that size dependency is different in these orders than in the others, or are we examining smaller fragments of the total regression, each of which is difficult to quantify precisely? It is hard enough to come up with an infallible explanation for a "universal" $M^{3/4}$ scaling; if exponential differences at the level of taxonomic order are real, the goal seems even more hopeless.

Subdividing the available data into more restrictive groupings not only reduces the sample sizes, but it usually reduces the size range. If in the world of reluctant experimental subjects (wild species), the data are good only to within ± 10 or 20%, the effect of a 10% or 20% size difference will be lost. To obtain a precise exponential expression of the effect of size would require a considerably greater range in sizes represented by the data. For example, Kleiber (1961) calculated that "a significant difference between proportionality to the three-fourths power of body weight and proportionality to the two-thirds power of body weight (representing roughly the surface area) could not be established with groups of animals in which the heaviest animals weighed less than nine times as much as the lightest animals." Effects of sample size can be seen in Table 14-2.

Interest in allometric splitting and the theoretical meaning of scaling exponents has been stimulated recently by papers by Heusner (1982a,b), Harvey (1982), and Feldman and McMahon (1983). Heusner reexamined what Eberhardt (1969) and Wilkie (1977) had shown, using published mass and metabolism values for intraspecific allometric analysis. A principal criterion for use of the data was that "the mass range for a given species studied by the same investigator was sufficiently large to yield a significant correlation between body mass and basal metabolism." This could be done for seven species of mammals, all of which were domesticated except *"Peromyscus M."* The mass exponents for these intraspecific regressions ranged from 0.52 to 0.91. Since the "mean intragroup slope is equal to $b =$

Table 14-2 Some examples of the effects of sample size or handling on allometric exponents.

Relationship	n	Source	Generally accepted relationship	n
$\dot{E}_{sm,mammals} =$ $0.037\ m^{0.68}$	7	Table 3-3		
			$\dot{E}_{sm} = 0.018\ m^{0.76};$ $0.020\ m^{0.76}$	26; 349
$\dot{E}_{sm,mammals} \propto m^{0.67}$ (mean of intra-specific exponents)	7	Heusner, 1982a		
$\dot{E}_{sm,birds} = 3.48\ M^{0.72}$	4	Lasiewski and Calder, 1971; see Figure 3-2		
			$\dot{E}_{sm,birds} = 3.79\ M^{0.72}$	72
$\dot{E}_{sm,birds},\ 0.5$ (error in largest datum) $=$ $2.77\ M^{0.68}$	4	Lasiewski and Calder, 1971; see Figure 3-2		

0.67 ± 0.03 . . . and is significantly different from 0.75," Heusner reasoned that "the mass exponent is the same and equal to ⅔" for the seven species, consistent with the theory of biological similitude (see below). He further concluded that the ¾ exponent is an artifact caused by improper interspecific regression, which does not accurately describe the relation of basal metabolism to body mass. Harvey, however, concluded that intraspecific allometries "do not allow distinction between exponential and linear models."

The theory of biological similitude says essentially that if two animals of different sizes are similar geometrically, the body-mass exponent should be predicted by the product of the dimensions of metabolic rate — energy/time $\propto (M)(l^2)(t^{-3})$ — expected to be ⅔ according to the "surface law" (see Chapters 3 and 5). One could perhaps argue that the exponents for *"Peromyscus M."* ($M^{0.91}$) and cattle ($M^{0.52}$) are hardly those resembling surface area and that only the mean and two of these seven species appear to have surface-area-to-mass scaling. However, the intraspecific size ranges are very small, the ratios of largest to smallest ranging from 1.5 to only 4.5. Furthermore, most of these are domestic species for which natural selection has been relaxed, if not downright forgotten. Consider, for example, the assortment of genetic defects that plague many domestic breeds of dogs: disc protrusions of dachshunds (Figure 4-1), the obvious respiratory impairments of prognathous breeds, appearances of acromegaly in basset hounds, and symptoms of dwarfism, congenital baldness, and hyperpituitarism that grace the kennels of purebreds we think of as "cute." The intraspecific size ranges and the ways they scale in domesticated varieties could very well be irrelevant to the evolution of size and scaling in nature.

Sufficient information has been drawn together in this volume to convince me that neither the metabolic rate nor any other physiological, morphological, or ecological variable has been scaled in isolation from the rest of the functions that constitute a life history. While there has been some obvious flexibility within a species, adaptive or neutral in its effects, there are beyond the relatively small intraspecific body-size ranges some general trends that are amply demonstrated. Within the limits of Bergmannian size changes, specific ontogenetic changes, and artificial selection, there may have been some options for other scalings. However, when the full range of mammal or bird body sizes is taken into account — not only in basal metabolism, but also in renal physiology, reproductive timing, and population dynamics — the physical constraints appear to have imposed a scaling predominantly $t \propto M^{1/4}$, $\dot{V} \propto M^{3/4}$. This is fortunate, because very few species are known well enough for intraspecific analysis of function and life history. Each discipline has its standard or favorite subjects in which other specialists show no interest, and the only way to get the whole picture is to reassemble a generalized animal from regression of the pieces. One is unlikely to have the chance to see numbers for each detail in the life of one species.

To the body-mass exponent "statistical artifact" controversy, Feldman and McMahon (1983) have made a convincing case for $M^{3/4}$ as a real and precise scaling in interspecific summaries of basal metabolism. They utilize the same data that Heusner used to demonstrate parallel $M^{2/3}$ intraspecific scaling, but subdivide the actual value of log basal metabolic rate (y_{ij}) of an individual j of species i into the following components:

(1) The grand mean of log basal metabolism among all species in the data set (μ);

(2) A correction for the deviation of mean log body mass of the species i from the grand mean, $\bar{b}(\bar{x}_i - \bar{x})$;

(3) An adjustment for the qualitative difference in the basal metabolism of species i not due to mean log body mass of that species (μ_i);

(4) An adjustment for the deviation of individual j's body mass from the mean for its species, $\hat{b}(x_{ij} - \bar{x}_i)$; and

(5) Random variation in basal metabolism among all animals of a given species and mass (ϵ_{ij}):

$$y_{ij} = \mu + \bar{b}(\bar{x}_i - \bar{x}) + \mu_i + \hat{b}(x_{ij} - \bar{x}_i) + \epsilon_{ij}, \tag{14-3}$$

where $\bar{b}$ is the slope of the best-fit line for the species mean points (x_i, y_i), and $\hat{b}$ is the single value for parallel intraspecific slopes found by Heusner. This is seen in Figure 14-1.

From this approach, Feldman and McMahon found the interspecific slope ($\bar{b}$) to be 0.7517 ± 0.0039! The difference ($\bar{b} - \hat{b}$) is due to the fact that the intraspecies intercept, or log mass coefficient (log a in aM^b) increases as a function of size (x_i). They concluded that both exponents are real, $b = 2/3$ for intraspecific scaling but $b = 3/4$ for interspecific scaling. "The statistics say the relation is there, and we are challenged to understand it."

Let us pursue this $M^{2/3}$ vs. $M^{3/4}$ basal metabolic scaling a bit further. In effect, Heusner has separated the mass dependency of the Kleiber equation into two components, both of which are size dependent. The log transformation of the allometric equation has the form

$$\log \dot{E}_{sm} = \log a + b \log M. \tag{14-4}$$

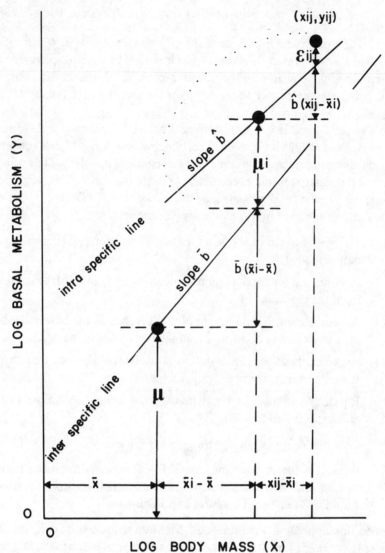

14-1 The log basal metabolic rate ($Y = y_{ij}$, x_{ij}, *upper right dot*) of an individual j of species i is a composite of statistical differences from the mean metabolic rate of mammals *(lower left dot)*. First, species i has a different body mass (x_i) from the average mammal, pushing Y up the interspecific line by an increment of mass difference times interspecific slope $\bar{b}$, to which is added the vertical allometry of species i's departure μ from the interspecific prediction, placing it on Heusner's (1982a) intraspecific line. Individual j is heavier (x_{ij}), so Y is increased by the product of intraspecific slope $\hat{b}$ and mass increase. Finally the random variation (ϵ_{ij}) of individual j's metabolic rate determination (within-population differences, imprecision of measurement) is added. (From Feldman and McMahon, 1983, *Respiration Physiology* 52:149–163; with permission.)

Heusner found different coefficients a for each of the seven species he analyzed, and these increased with animal size approximately as follows:

$$\log a = 0.52 + 0.115 \log M. \qquad (14\text{-}5)$$

If we substitute Heusner's average intraspecific regression exponent (0.67) and the regression for log a from Eq. (14-5) into Eq. (14-4), we have:

$$\log \dot{E}_{sm} = 0.52 + 0.115 \log M + 0.67 \log M. \qquad (14\text{-}6)$$

When the log M terms are combined, this reduces to

$$\log \dot{E}_{sm} = 0.52 + 0.79 \log M \qquad (14\text{-}7)$$

and we have returned to Kleiber's exponent! This suggests that the ¾ scaling exponent is a combination of surface-area consequences and some other factor that scales as $M^{0.12}$. One possibility suggested by Economos (1979) is that the ¾ exponent combined a surface-law term ($M^{0.67}$) and a gravitational-cost term, which he derived from maximum tolerance to increased gravitational force (in centrifuging mice, rats, and dogs) to be scaled as $M^{0.14}$, perhaps not statistically distinct from $M^{0.12}$. If that were true, why would it appear as a step function between species, but not intraspecifically? Perhaps because of the relaxation of natural selection in the six domesticated species that made up most of Heusner's data base. Furthermore, if size accounts for 86% of the variance in log a, as it appears to, is it not proper to add the many data for basal metabolism of all the other species of eutherian mammals to a single regression based on that dominating characteristic?

Applied Allometry

PALEOECOLOGY

Man's exceptionally long life span, the evolution of detailed language, and the manual dexterity by which we can both record our experiences and observations and use tools have permitted us to enjoy a sense of history, though our species has seen only a brief and perhaps terminal snatch of the history of life on this planet. The sole records that predate man are fossils. Reconstruction of the immense span of missing observations requires a lot of imagination, plus some general "rules." We observe life around us today, make correlations and attempt to discern the principles of interaction. Then, assuming that the principles and correlations were valid in the past, we extrapolate back through time. The growing assemblage of consistent patterns in contemporary allometry suggests that it is indeed reasonable to use allometry in our time machine. Significantly, some of the most valuable present-day ecological allometry has been derived with paleoecological applications in mind (Farlow, 1976; Beland and Russell, 1980; Western, 1980; Damuth, 1981a,b, 1982).

In the last century the race was on between Edward Cope and Othniel Marsh to describe more species; now the goal is to interpret how the extinct animals lived, in what biotic and climatic circumstances they flourished, and what caused them to disappear from the fauna. The allometries of energetics, physical support, and brain size have all been used in the speculations on whether or not the dinosaurs were homeothermic (Thomas and Olson, 1980).

Just as it is difficult to extrapolate from bones and tracks to physiology, there is a static-to-dynamic leap from bones to population and community ecology. Can relative numbers of bones in a fossil assemblage be used to reconstruct communities of the past? Such assumptions have been challenged (Western, 1980; Western and Georgiadis, MS). Only if death rates and the weathering rates of the bones left at death exactly counterbalanced each other could the fossil assemblage represent the prior community composition.

This has been tested with data from herbivorous mammals in southern Kenya. Annual birth rates are inversely related to body mass (Western, 1979; Chapter 9), and if the population is stable, as it would tend to be over geological time, the death rate should be scaled similarly. Applying the death rate to the population size, the annual carcass production can be calculated. The ratio of observed/expected carcass numbers is a measure of the durability of bones:

$$\text{carcass ratio} = 7.8 \times 10^{-4} M^{1.35}. \tag{14-8}$$

A very small fraction of the skeletons (0.02) persisted at the surface for the small Thomson's gazelle (15 kg), but the resistance to bone erosion was strongly improved with size, to a degree that the 816-kg rhino left six times as many skeletons visible at the surface as were left in an average year—six years of skeleton production still observable. The allometry of bone production and erosion needs to be used as a correction, not ignored, in estimating former community composition from the fossil assemblage that remains. Analyses similar to those by Western and Georgiadis are necessary to see how the pattern varies in different climates and different mammal communities. Damuth (1982) proposed an allometric correction to remove "surface-process bias." This assumes that the bones of the largest species weather the least and that the bones of smaller species disappear in proportion to their relative surface areas.

What is the future for "paleoallometry"? It could be a useful tool for reconstructing life histories and community structures. Population density is inversely related to body size of contemporary herbivorous mammals. If there were equations for density and home-range size of other taxa and trophic categories, would it not be possible to estimate the numbers of animals before their declines? Speed of travel and endurance on stored energy reserves have been summarized allometrically. Would it not be possible to make reasonable estimates for the rates of range extensions across the Bering land bridge, for instance?

PHARMACOLOGY

A classic story for introducing the subject of scaling in animals is the sad demise of a zoo elephant named Tusko (Schmidt-Nielsen, 1972). It seems that some psychologists wanted to see if LSD could cause Tusko to mimic the periodic rampaging of male Asiatic elephants, known as musth. The dosage was calculated as a linear scale-up from the 0.1 mg per kilogram of body mass necessary to get an effect in cats. This amounted to 297 mg for Tusko, which was sufficient to kill the poor elephant. Drug dosages should be calculated in proportion to the rates at which the drugs will be metabolized—in this case, in proportion to $M^{3/4}$ instead of the $M^{1.0}$ used in this ill-conceived experiment. Use of the wrong scaling resulted in a dose nearly four times as large as required by size alone. In addition, there is apparently a vertical allometry for LSD sensitivity, cats being less sensitive than expected for their size. Schmidt-Nielsen extrapolated from human dosage to dosages of 3 mg based on basal metabolic rates of man and elephant, or 0.4 mg based on relative brain sizes. Thus Tusko may have received 99 to 742 times as much LSD as "necessary" to test the psychologists' idea.

Because of the importance of small rodents (such as white mice and guinea pigs) for the development and testing of new drugs, the scaling of dosage and drug clearance or metabolizing times is of great practical significance. Consequently, allometric graphs, equations, and derivations are appearing frequently in the literature of pharmacokinetics (Dedrick et al., 1970; Dedrick, 1974; Boniface et al., 1976; Weiss et al., 1977; Boxenbaum, 1980, 1982).

EXPERIMENTAL DESIGN AND ANALYSIS

Our preliminary introduction to standard metabolic rates did not include details of the experimental protocol. When an animal is placed within a metabolic chamber, certain conditions must be met if the data are to have comparative value. Consider the case of a closed system of an animal within a sealed chamber. The environment of that chamber will be changed by the experimental subject as oxygen is depleted, and as carbon dioxide and water vapor build up. If the chamber were adequately ventilated by circulation of air in an open system, O_2 depletion and CO_2 and water vapor buildup could be prevented. How much ventilation is adequate? If the turnover of air is too slow, these problems remain to some extent; if it is too fast, the oxygen content may not decrease enough for the instrumentation to detect it. We therefore want to set upper and lower limits to the dilution of exhaled air with fresh air flow. Carbon dioxide stimulates ventilation in air-breathing animals. To preclude a stimulus for hyperventilation, the CO_2 content should be kept below 0.5% (atmospheric F_{CO_2} is 0.03%). And to avoid hypoxic stress, the oxygen content should be held within 1% of the normal atmospheric 20.95%. For normal heat dissipation by evaporation, the vapor pressure of the air should be kept well below the saturation point.

As an illustration of the importance of maintaining adequate flow in an open metabolic system, consider the history of the evaluation of evaporative heat loss in birds. Prior to 1966 it had been concluded that birds were incapable of evaporating enough water to eliminate more than half of their metabolic heat production, despite the fact that desert birds obviously survived exposure to air temperatures exceeding body temperature (in which case, to avoid lethal increase of body temperature, the heat loss by evaporation would have to equal or exceed the simultaneous metabolic heat production). Then Lasiewski and coworkers (1966a,b) showed that this apparently limited capacity was entirely an artifact of humidity buildup in experimental chambers, to the point that the vapor-pressure difference between bird and surrounding air became the limiting factor.

How does one determine the proper rate of airflow? From the appropriate allometric equations and the body mass of the proposed subject, the rates of oxygen uptake and of carbon dioxide and water-vapor release can be estimated. These estimates are the basis for calculating how much dilution must be provided by the ventilation of the chamber. Subsequent fine-tuning of this variable can be done empirically.

Another requirement for standard metabolic measurement is that the animal be postabsorptive, so that digestive work and stimulation are not added complications. This is accomplished by depriving the subject of food prior to the experiment. How long must fasting be imposed to make sure that the animal is truly postabsorptive? At best we can make an educated guess. Kleiber (1961) refers to "the so-called post-absorptive period (14 hours after the last meal)" for man. Beyond that, the practice is to fast an animal overnight before metabolic measurements. However, as we have seen, physiological time is size dependent, so 12 to 14 hours' fasting is physiologically much different for the shrew than for man; in fact, it would kill the smaller shrews. Many small animals enter a state of torpor when energy reserves are seriously depleted.

From Adolph's (1948) tabulation of mammalian allometry, the duration of an intestinal contraction (gut beat duration, in hours) is

$$t_{gut\ beat} = 9.3 \times 10^{-5}\ m^{0.31}. \tag{14-9}$$

Since the mass of the gut (stomach and intestines, in grams) approaches linearity,

$$m_{gut} = 0.112\ m^{0.94}. \tag{14-10}$$

We could then assume that the time between peristaltic contractions could be taken as an index of time in gut, in which case the fasting period before an experiment should be scaled as $m^{1/4}$ to $m^{1/3}$. Better than that would be an allometric equation directly relating absorption or passage times to body size; but such is lacking from the study of animal energetics, in part because of the complexity of passage times relative to variables of dietary composition and digestive design (rumination, coprophagy, or the more common single passage; see Brandt and Thacker, 1958). The derivation of such relationships was considered in Chapter 5.

Life-History Pendulums

A popular focus of the evolutionary ecologist in recent years has been the analysis of life-history strategies or life-history tactics. Such figures of speech suggest that the survival of the fittest has been won by de novo solutions to the problems of competition for each species. The allometry seems to suggest, rather, that the laws of physics may have ruled with a firmer hand, more like Calvinistic predetermination than evolutionary free will.

The term "life history" implies a time scale. The time scales of the life histories of eutherian mammals, perhaps the most completely analyzed animals, are summarized in Figure 14-2. What we see is an overwhelming similarity of slopes, such that the life history of the large mammal and that of the small mammal are scheduled in approximately the same proportions, even though the small beast must telescope its history into a short absolute time span. This is the faster pace described by A. V. Hill over three decades ago. Of the dozen relationships shown, the smallest exponent is 0.20, for

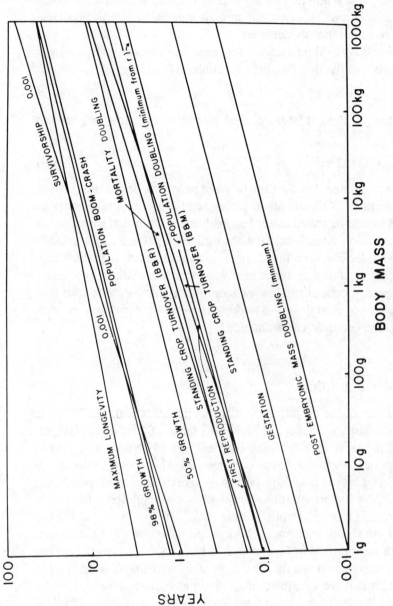

14-2 An allometric summary of the life histories of eutherian mammals, based on equations given in previous chapters. Many of these relationships need fine-tuning with more extensive data bases, but the overall pattern is clear: the quantitative details of life histories are largely a matter of size and the pace of life that the size determines. Survivorship of 0.001 of the original adult cohort and t_{sc} derived from Banse and Mosher, 1980 (B & M), have the largest scaling exponents (0.32, 0.33), compared to maximum longevity (0.20) and t_{sc} derived from Beland and Russell, 1980 (B & R; 0.29).

maximum observed life spans. The largest exponents, 0.32 to 0.36 for life expectancy, standing crop turnover, and population doubling, are consistent with one another as they should be — and with the necessary inverse function for annual natality, which scales as $M^{-0.33}$, reflecting the greater vulnerability to predation of smaller animals. Including these outliers, the mean exponent is $M^{0.26\pm0.04}$; excluding them, it is $M^{0.29\pm0.06}$. In either case physiological time scaling seems to underlie ecological scaling.

Of the dozen relationships, correlation coefficients were included in the original reports for nine. Squaring these and taking the mean, body size accounts for $75 \pm 18\%$ of the variability. The dimension of body size, like the dimension of length for a pendulum, sets the natural frequency of the cycle and the period of swing for the living pendulum. We could talk of the "strategy of swinging" on a pendulum or a playground swing, and decide whether to try to push the swing faster or slower, but we would have little effect.

It appears that the appropriate way to understand the life history of an animal is to start with the "life-history pendulum," the physiological and physical consequences of body size that account for most of the differences among animals within a class, and not with the "life-history strategy" that covers the remaining one-fourth of the variability in nature (see Figure 14-3). Perhaps it is the energetics of the food supply — how much and how big the pieces, and how concentrated the chemical energy in the foodstuffs that determines the "body-size strategy." Many details follow simply from the physical dimensions of that size, being quantitative details on a graded scale, as opposed to the "yes-no" decisions of whether to disperse or recruit, outcross or inbreed, socialize or fight, be selfish or altruistic, opt for r-selection or K-selection.

We are still far from a full understanding. We see the patterns a bit more distinctly and can perhaps ask better questions. Why are so many exponents for time allometry so close to $\frac{1}{4}$? In Chapters 4 and 7 we saw the merits and the shortcomings of elastic and dynamic similarity models, either of which could predict the $M^{1/4}$ scaling. Only in the Bovidae do the linear scalings conform to the $M^{1/4}$ scaling, yet most times seem to have been scaled as $M^{1/4}$ for mammals in general.

The progress of experimental and molecular biology in the past quarter-century has cast "descriptive biology" into an inferior or obsolete status from which substantial contributions can no longer be made, or so some would have us believe. The study of allometry is a vital and highly useful form of descriptive biology. At present, the descriptive process is outstripping our ability to explain, but it does give a better description of what we

14-3 The quantitative details of a mammal's life history may be less a matter of strategies, accounting for perhaps only 25% of the variation, and more (75% or better) a matter of the dimensions of the pendulum that swings through that life history. Each animal is allowed the same size-independent number of actions. The smaller dimensions of the mouse-sized swing result in a higher frequency of oscillation and an earlier completion of the mouse's swing through life. (Drawing by Lorene Calder.)

should *try* to explain, in a form that may suggest or expedite the explanation. I have limited my consideration for the most part to mammals and birds because more information is available, not because allometry will not work for poikilothermic vertebrate classes, insects, and other forms of life as well. In view of the ease of finding these patterns, the study of size, function, and life history should be pursued vigorously, to hasten the day when all *adaptation* will be accurately perceived as *adaptive deviation* from the basic size-dependent pattern.

Epilogue

What is man, that thou art mindful of him? and the son of man that thou visitest him? For thou hast made him a little lower than the angels, and hast crowned him with glory and honour. Thou madest him to have dominion over the works of thy hands; thou hast put all things under his feet.

—Ps. 8:4–6

Is SUCH A QUESTION only for the psalmist or the theologian to ask, or might the allometrician also inquire, "What is man?" The normal average body mass of our species is 70 kg, from which we can predict many characteristics. The observed values do not always agree. Our basal metabolism rate should be, and in fact is, 83 watts. Even in top physical condition, a runner can only burn at a rate about 20 times this amount, yet the per capita energy consumption rate of man is, in the United States, over 100 times the basal rate, and we wonder how to get more! Substituting 70 kg in Eq. (11-5) and multiplying by the earth's 133×10^6 km^2 of ice-free land, the human population should be 0.5 billion—but we have already attained nine times that number, and stability is nowhere in sight.

Allometrically, we should live about 27 years, but our life expectancy is nearly three times this; still, through gerontological research we seek a fountain of youth to reach beyond this. Even as we try valiantly to extend our life, we do not see the inconsistency of investing considerably more of our resources and technology in life-destroying weapons than in life-sustaining understanding. Allometrically, a brain of about 230 g would be adequate to control a 70-kg mammal, but we are endowed with a brain about six times larger. We have performed some incredible feats with this brain, it is true, but so far we have not figured out how to live in harmony with nature and with one another. In many senses we live to excess — but, given what we have, we fall short of our potential. Can we not do better with our abundance of brains, years, and energy?

Such philosophical asides may seem out of place in a basic biological book, but they are not without precedent. *Bioenergetics and Growth* (1945) is a classic of metabolic allometry, into which Samuel Brody tucked a section on "social homeostasis," which shows that he was equal in sensitivity to the peace movement of the 1980s — and almost four decades ahead of it:

> Who can predict what the third world war will cost and whether human society will react to these changes so as to bring itself to normal? . . . The greatest immediate need seems to be the development of methods for selecting leaders who will utilize the tools of science, which grow cumulatively ever more powerful, for peaceful construction rather than for warlike destruction. (pp. 256–257)

Max Kleiber ended *The Fire of Life* (1961) with a chapter entitled "Must Man Starve?":

> There is more profit in the manufacture of atom bombs and the construction of foreign bases for missiles than in the feeding of hungry people. That is why people starve today . . . Next to the atom bombs, the most ominous force in the world today is uncontrolled fertility . . . It is fitting that this book should end with an assertion which is beyond the realm not only of bioenergetics but of science in general . . . Now man has learned to use a power much more awful than fire, and he is able to destroy that creative eddy in the energy stream. To prevent the disastrous misuse of his power, man should follow a little light which appears weak indeed even in comparison with the small fire of life. This little light, which helps him distinguish between good and evil, is called wisdom. By following that light, men will not only extend and prolong the fire of human life, but will cause it to burn brighter. (pp. 331, 342)

Since those words were written, we have emphasized the increase of our power to destroy. But we are endowed also with power to appreciate; the four hundred post-1961 papers and books dealing with functional and life-historical consequences that I have cited bear witness to the patterns in the "creative eddy in the energy stream" and our ability to perceive them.

In a well-designed jigsaw puzzle each piece has a unique orientation in a unique location, so that it can be assembled in only one way. The present state of allometry is one of imperfect pieces. While this book may have joined more of the pieces than have been assembled in one place before, many undoubtedly will prove to have been inserted incorrectly. Some of the errors will appear when pieces have been cut more accurately and more gaps have been filled. Then adjoining fields will take different shapes.

Even if all of the pieces had been available, perfectly machined and assembled without error, it would have been only a beginning. The real challenge, one for which my background in descriptive biology has its limitations, is to explain how and why life is configured. To ponder the "how" and "why" questions of life is uniquely human and is what sets us apart from other organisms. Until our species can answer, we have no right to destroy parts of the puzzle—by poisoning, by fire, by bulldozer, by missile. If we must destroy, we must have a reason; and before that reason is established, the reasons for the predestruction beauty must be known. We must reshape our priorities to put nature and learning ahead of the race toward destruction.

Symbols, Dimensions, and Units

THIS APPENDIX is an attempt to sort out a consistent system of nomenclature that will hold across a wide range of biological disciplines from physiology to ecology, the scope of this book. While the list of symbols is not without some redundancy and conflicts, it is consistent with the case I made in Calder, 1982a. M is always mass (M in kg or m in g), t is always time, V is always volume, and E is energy (energy stored as work is W; heat energy in heat-exchange studies is H). Many of the symbols below are also initials, making them easy to remember and recognizable, regardless of context. Furthermore, they can be modified with subscripts and other supplementary symbols to clarify their meaning. For instance, V_{O_2} stands for volume of oxygen, T_b for body temperature, H_{sm} for standard or basal heat production. A bar over a symbol or subscript denotes the mean value (thus $C_{\bar{V}_{O_2}}$ is mean venous oxygen concentration). A dot over the symbol indicates a rate or first time derivative, so that $\dot{V}_{O_2}$ stands for rate of oxygen consumption, $\dot{E}_m$ for metabolic rate, and $\dot{H}_m$ for metabolic heat production rate.

The physical dimensions of the following entities are given in terms of the fundamentals m = mass; l = a linear dimension; t = time; and T = temperature. The variable n signifies a pure number or a dimensionless count or ratio. Units are given as appropriate.

A = area $\propto l^2$

a $\begin{cases} = \text{allometric coefficient} \\ = \text{acceleration} \propto lt^{-2} \end{cases}$

B = birth rate = nt^{-1}; as subscript, stands for body

b = allometric exponent or scaling factor

C or [] = concentration $\propto nl^{-3}$

c $\begin{cases} = \text{specific heat} \propto l^2 t^{-2} T^{-1} \\ = \text{wave velocity} \propto lt^{-1} \end{cases}$

D = diffusion coefficient $\propto l^2 t^{-1}$

d = diameter $\propto l$

E $\begin{cases} = \text{energy} \propto ml^2 t^{-2} \\ = \text{elastic (Young's) modulus,} \\ \quad \text{stress/strain} \end{cases}$

e $\begin{cases} = \text{efficiency (nondimensional)} \\ = \text{base of natural logarithms} = \\ \quad 2.718 \end{cases}$

F $\begin{cases} = \text{force} \propto mlt^{-2} \\ = \text{fraction} \propto n \end{cases}$

f = frequency $\propto t^{-1}$

G = conductance $\begin{cases} \propto m^{-1} l^4 t \\ \quad \text{for fluids} \\ \propto ml^2 t^{-3} T^{-1} \\ \quad \text{for heat} \end{cases}$

g = gravitational acceleration $\propto lt^{-2}$

H = heat energy $\propto ml^2 t^{-2}$

h = heat transfer coefficient or "thermal conductance" $\propto ml^2 t^{-3} T^{-1}$

I $\begin{cases} = \text{intensity} \propto mt^{-2} \\ = \text{inertial moment} \propto ml^2 t^{-1} \end{cases}$

i =

J = flux or fluid flow $\propto l^3 t^{-1}$

j = $\Big\}$ subscripts for individuals in a series

K $\begin{cases} = \text{numerical constant} \\ \quad \text{(dimensions vary)} \\ = \text{carrying capacity} \propto nl^{-2} \end{cases}$

k = a constant, assignable

L = length, distance $\propto l$

l = length $\propto l$

M = mass, in kg, $\propto m$

m = mass, in g

N = number, as avogadro's number

n $\begin{cases} = \text{number by count or} \\ \quad \text{number of moles} \\ = \text{number of data in} \\ \quad \text{statistical sample} \end{cases}$

O = confused with zero

o = subscript for original or starting value

P = power $\propto ml^2 t^{-3}$

p = pressure $\propto ml^{-1} t^{-2}$

Q = quantity of energy or mass

q = mortality or disappearance rate $\propto nt^{-1}$

R = resistance to flow or pressure difference required per unit flow $\propto ml^{-4}t^{-1}$

r {
= radius $\propto l$
= intrinsic (Malthusian) rate of increase $\propto nt^{-1}$
= correlation coefficient (dimensionless)
}

S {
= survivorship $\propto nt^{-1}$
= entropy $\propto ml^2t^{-2}T^{-1}$
}

s {
= speed $\propto lt^{-1}$
= standard error (subscripts a for coefficient, b for exponent)
}

T = temperature = T

t = time = t

U = velocity $\propto lt^{-1}$

u = velocity $\propto lt^{-1}$

V = volume $\propto l^3$

v = velocity $\propto lt^{-1}$

W = work $\propto ml^2t^{-2}$

w = fitness (relative and dimensionless)

X = independent variable, dimensions assigned

Y = dependent variable, dimensions assigned

x = }
y = } a thickness (also τ) $\propto l$; algebraic substitutes

Z {
= a variable in a third dimension, orthogonal to X- and Y-axes
= impedance (electrical, but may be applied by analogy as resistance to flow in an oscillating system)
}

z {
= a distance, usually vertical, as elevation or altitude $\propto l$
= electrical charge or valence
}

$\$$ = cost, usually in energy or in amount of some limiting resource

ρ = density $\propto ml^{-3}$

α = annual incremental or actuarial increase in mortality rate, as in the Gompertz equation

ρ_n = numerical density $\propto nl^{-2}$ or nl^{-3}

π = circumference/diameter = 3.1416

θ = tension $\propto mt^{-2}$

σ = stress $\propto ml^{-1}t^{-2}$

μ = kinematic viscosity $\propto l^2t^{-1}$

ϵ = strain = $\Delta l/l_o$ (dimensionless)

ν = dynamic viscosity $\propto ml^{-1}t^{-1}$

κ = compliance = $\Delta V/\Delta p \propto m^{-1}l^4t^{-2}$

τ = thickness $\propto l$

Interpretation and Prediction from Allometric Exponents

THE ELECTRONIC pocket calculator, possessed by virtually every analytical biologist, permits instantaneous solution of allometric predictions for a given body size. However, the display is limited, so it is still difficult to get a sense of the meaning of an exponent, or of a difference in exponents from an allometric cancellation, across a size range. For that reason I have provided in Appendix Table 1 the scaling consequences of commonly encountered allometric exponents relative to the 1-g intercept, arbitrarily given a value of one. (The figures in the other columns are factors for the increase or decrease of Y as a function of size.) Thus the basal metabolism ($\propto M^{0.734}$) of a 100-kg mammal would be predicted as 4,677/5.42, or 863 times that of a 10-g mouse, for example, and the heart rate ($\propto M^{-0.25}$) would be expected to be 0.06/0.56, or 10.7% of the mouse's heartbeat.

When the exponent is a residual mass exponent (r.m.e.) from an allometric cancellation, as in text equations (3-25) or (11-60), for example, one must consider that the r.m.e. may not be significant; there may be a size-independent, dimensional or nondimensional constant of design. Significance or its absence is established on the basis of whether or not the 95% confidence interval of one exponent includes the other. The confidence interval (c.i.) is calculated as the produce of the t-value at the 0.05 level for the number of degrees of freedom in the sample and the standard error of the mean for that exponent, ts_b. The value of t may be obtained from the Student's t-table in any statistics text. Unfortunately, the s_b and 95% c.i. are missing from most of the older literature, and are not automatically calculated in pocket or desk-top calculator programs. From now on, it should be the solemn responsibility of the authors of technical papers to provide these data for comparison purposes, just as body-mass values should always be provided.

The significance of apparent differences in coefficients (a) or intercept values of allometric equations and plots is tested in the same way, using the appropriate t-value multiplied by s_a, the standard error of the coefficient (often designated s_{yx}), to determine the 95% confidence limits of the position of the regression line on the Y-axis.

These procedures extend to specific as well as to general comparisons; that is, they are applicable in comparing actual observations with general size-related trends. Chapter 12, for example, is concerned with vertical allometry, the interesting adaptive or phylogenetic departures from the allometric trend within a class. If one measures a particular variable for a species and the datum does not fall on the line, is the departure significant or is it just more of the variability that went into the previous allometric summary? To answer this, one would calculate the 95% confidence intervals for both the exponent and the coefficient, determining thereby a range of predictions. If these include the observation, or if the 95% confidence limit of the observations includes the allometric prediction, one cannot argue that the apparent vertical allometry represents a special adaptative strategy.

Appendix Table 1 The scaling consequences of exponents, including residual mass exponents. Starting with an arbitrary one as a variable in a 1-g animal, subsequent columns show the relative increase (or decrease, in the case of negative exponents) in Y for larger animals.

Exponent of M				Mass of animal			
	1 g	10 g	100 g	1 kg	10 kg	100 kg	1,000 kg
1.30	1	19.95	398.1	7,943	1.58×10^5	3.16×10^6	6.31×10^7
1.09	1	12.30	151.4	1,862	22,909	2.81×10^5	3.49×10^6
1.02	1	10.47	109.7	1,148	12,023	12.6×10^5	1.32×10^6
1.00	1	10.00	100.0	1,000	10,000	100,000	1.00×10^6
0.99	1	9.77	95.5	933	9,120	89,125	8.71×10^5
0.98	1	9.55	91.2	871	8,318	79,433	7.59×10^5
0.95	1	8.91	79.4	708	6,310	56,234	5.01×10^5
0.90	1	7.94	63.10	501	3,981	31,623	2.51×10^5
0.85	1	7.08	50.12	355	2,512	17,783	1.26×10^5
0.80	1	6.31	39.81	251	1,585	10,000	6.31×10^4
0.75	1	5.62	31.62	178	1,000	5,623	3.16×10^4
0.734	1	5.42	29.38	159	863.0	4,677	2.54×10^4
0.723	1	5.28	27.93	148	779.8	4,121	2.18×10^4
0.667	1	4.65	21.58	100	465.6	2,163	1.00×10^4
0.50	1	3.16	10.00	31.6	100.0	316.2	1,000

0.33	1	2.14	4.57	9.77	20.89	44.67	95.50
0.25	1	1.78	3.16	5.62	10.00	17.78	31.62
0.10	1	1.28	1.58	2.00	2.51	3.16	3.98
0.09	1	1.23	1.51	1.86	2.29	2.82	3.47
0.08	1	1.20	1.45	1.74	2.09	2.51	3.02
0.07	1	1.17	1.38	1.62	1.91	2.24	2.63
0.06	1	1.15	1.32	1.51	1.74	2.00	2.29
0.05	1	1.12	1.26	1.41	1.58	1.78	2.00
0.04	1	1.10	1.20	1.32	1.45	1.58	1.74
0.03	1	1.07	1.15	1.23	1.32	1.41	1.51
0.02	1	1.05	1.10	1.15	1.20	1.26	1.32
0.01	1	1.02	1.05	1.07	1.10	1.12	1.15
0.00	1	1.00	1.00	1.00	1.00	1.00	1.00
−0.01	1	0.98	0.95	0.93	0.91	0.89	0.87
−0.02	1	0.98	0.91	0.87	0.83	0.79	0.76
−0.05	1	0.89	0.79	0.71	0.63	0.56	0.50
−0.10	1	0.79	0.63	0.50	0.40	0.32	0.25
−0.25	1	0.56	0.32	0.18	0.10	0.06	0.03

* Stahl (1967) did not consider an r.m.e. in this range to be significantly different from M^0 for respiratory variables.

Study Problems ("Allometricks")

ALTHOUGH ALLOMETRY is basically quite simple, it is not without a few "tricks," whereby simple mistakes can send one astray. It is useful to review some typical errors, as a test of our grasp of the principles.

1. From the data in the table below, here is one conclusion drawn: "Since the tree shrews produce less heat than insectivores and more than the primates, the placement of the tree shrews in an intermediate order, the Menotyphia, by some paleontologists is supported by a physiological parameter—that is, metabolic rate."

Species	Mass	Metabolic rate
Shrew	3.5 g	66.7 cal/(kg · hr)
Tree shrew	12.5 g	10.0 cal/(kg · hr)
Chimpanzee	48.0 kg	3.5 cal/(kg · hr)
Man	70.0 kg	1.2 cal/(kg · hr)

Provide an alternative conclusion via allometric analysis and a comparison with an appropriate equation from the text.

2. One response to an allometric analysis of hominid evolution stated: "Metabolic rates in mammals increase by the 0.75 power of body mass. If tooth surfaces increased by the 0.75 power of body mass, they might be considered metabolically scaled." The authors of the paper in question countered with: "The metabolic (physiological) explanation may well apply, but other ecological rationales are equally compelling." Does the metabolic explanation apply?

3. Regarding data for metabolic rates plotted on logarithmic coordinates, a reporter for *Science* quoted a primatologist as stating: "When you do this for basal metabolic rate, you get a slope of ¾. In other words, larger animals are more energy efficient than small ones." Does the second sentence follow from the first?

4. This portrayal of the log-log plot of basal metabolic rates was captioned: "The slope of the rising curve, 0.73, means that an increase in body weight of 100% is associated with an increase in basal metabolism of about 73%." The original expression for the rising curve for whole-animal metabolism was kcal/day = $70.5 \text{ kg}^{0.734}$. One cal = 4.184 J. How many errors can you find in the math and its interpretation? If you can find three, you understand allometry.

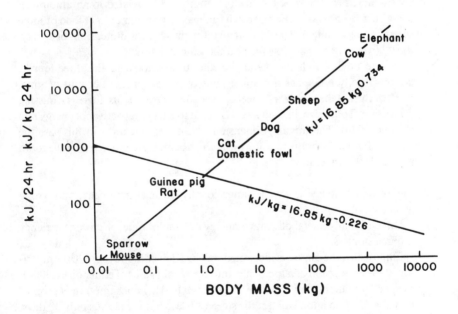

5. "Interspecific increases in brain and body size appear to occur at the same rate . . . when the amount of available energy is taken into account." This was concluded when brain weight was regressed on "body weight adjusted for

BMR," the BMR used being "amount of O_2 consumed per body weight." The adjusted regression was

$$\log \text{brain weight} = -2.11 + 1.026 \log [(\text{body weight}) (\text{BMR})]$$

What was actually being correlated?

MY ANSWERS

1. An allometric analysis of these data shows the following:

   (a) A simple allometric exponent of $M^{-0.29}$ in cal/(kg · hr), similar to the mass-specific form of the Kleiber equation (3-13a):

   $$\dot{E}_{sm}/M^{1.0} = 67.6 \, M^{0.76}/M^{1.0} = 67.6 \, M^{-0.24};$$

   (b) An apparent ambiguity-error of "cal" instead of "kcal" for the energy unit;

   (c) Metabolic rates approximately twice the basal level after correction to kcal;

   (d) Differences among the animals that were not phylogenetic, but allometric!

2. Attempts to explain empirical power functions must incorporate dimensional consistency. For example, the size of a tooth can be related to an amount of energy if a given cross-section of tooth can mash a given amount of food of some specific energy content. The derived quantity thus has the dimensions of energy, ml^2t^{-2} (where m = mass, l = linear dimension, t = time).

   On the other hand, the metabolic rate, proportional to the three-fourths power of body mass, has the dimensions of power (energy per unit of time, ml^2t^{-3}). Even if there were a perfect coincidence of exponents, there is a missing dimension. Tooth size, like gut size, is related to energy capacity, not energy rate (Calder, 1974). The missing dimension, t^{-1}, is a frequency—in this case the chewing frequency or the rate at which lumps of food are processed by the tooth area under consideration.

3. The second sentence is wrong dimensionally, wrong empirically, and wrong allometrically.

   Dimensionally, metabolic rate is energy per unit of time, or power; efficiency is dimensionless.

   Empirically, the matter is illustrated well by Kleiber's (1961, p. 320) comparison of the conversion of a ton of hay into the weight gains of either one 1,300-lb steer or 300 rabbits that together weigh 1,300 lb. Hay consumption is proportional to $M^{3/4}$, so a ton will last the steer 120 days but the rabbits only 30 days. However, the aggregate weight gains are the same—240 lb of steer or rabbits—and the heat losses are the same—20,000 kcal/day times 120 days for the steer, and 80,000 kcal/day times 30 days for the rabbits. The energy efficiency is the same (as it would be in S.I. units today).

Allometrically, physiological time has been overlooked. The smaller rabbits live at a faster pace, growing and turning over more rapidly than a large steer. Physiological time passes in approximate proportion to $M^{1/4}$. Thus the lifetime energy costs per kilogram of animal are the following size-independent product:

$$(\dot{E})(M^{-1})(t) \propto (M^{3/4})(M^{-1})(M^{1/4}) \propto M^0.$$

4. The slope (body-mass exponent) is not a percentage increase. An increase in body mass of 100% is a doubling of mass; $2^{0.734} = 1.66$. Thus doubling the size results in a 66% increase. The error is small for doubling of mass, but across the shrew-to-elephant mass range of Table 3-3, the error would be 105%!

    There are other errors. The metabolic rate of a 1-kg animal is the same as its metabolic rate per kilogram of mass (the intersection of the two slopes). From the vertical axis, this appears to be about 316 kJ/day, not 16.85 as the lines were labeled. Now, 70.5 kcal $\times$ 4.184 kJ/kcal = 295 kJ/day, whereas $70.5 \div 4.184 = 16.85$. The lines seem to be plotted correctly, but the S.I. conversion was apparently made erroneously by dividing instead of multiplying. Finally, if the whole-animal scaling is $M^{0.734}$, the scaling per kilogram would be $M^{0.734}/M^{1.0}$, solved by subtracting exponents. Alas, $0.734 - 1.0$ is -0.266, not -0.226.

5. By canceling out the two body-weight terms,

    (body weight)(amount of O_2 consumed)/body weight

    can be reduced to "amount of O_2 consumed" (actually rate of consumption, amount/time), making this a correlation of brain mass and metabolic rate. The original paper's introductory statements claimed that "the newer and larger sets of points may have disproportionately increased the numbers of small mammals with relatively small brains and this alone could produce a steeper slope (the 0.75 of Martin, 1981 vs. past analyses with 0.67 exponents)." Two issues later, the editors posted a special notice that the journal was taking "drastic steps to reduce its backlog of accepted reports . . . by a sharp temporary reduction in the rate of acceptance of manuscripts." Lock the barn, the horse got away!

References

Adolph, E. F. 1949. Quantitative relations in the physiological constitutions of mammals. *Science* 109:579–585.

Agostoni, E., Thimm, F., and Fenn, W. O. 1959. Comparative features of the mechanics of breathing. *J. Appl. Physiol.* 14:679–683.

Alberch, P., Gould, S. J., Oster, G. F., and Wake, D. B. 1979. Size and shape in ontogeny and phylogeny. *Paleobiol.* 5:296–317.

Alerstam, T., and Högstedt, G. 1983. Regulation of reproductive success towards $e^{-1}(=37\%)$ in animals with parental care. *Oikos* 40:140–145.

Alexander, R. McN. 1968. *Animal Mechanics.* Seattle: University of Washington Press. 346 pp.

—— 1976. Mechanics of bipedal locomotion. In *Perspectives in Experimental Biology,* vol. 1, ed. P. S. Davies. New York: Pergamon Press, pp. 493–504.

—— 1977. Allometry of the limbs of antelopes (Bovidae). *J. Zool.* (London) 183:125–146.

—— 1982. Size, shape, and structure for running and flying. In *A Companion to Animal Physiology,* eds. C. R. Taylor, K. Johansen, and L. Bolis. New York: Cambridge University Press, pp. 309–324.

Alexander, R. McN., Jayes, A. S., Maloiy, G. M. O., and Wathuta, E. M. 1979. Allometry of the limb bones of mammals from shrews (Sorex) to elephant (Loxodonta). *J. Zool.* (London) 189:305–314.

Allison, A. C. 1960. Turnovers of erythrocytes and plasma proteins in mammals. *Nature* 188:37–40.

Altman, P. L., and Dittmer, D. S., eds. 1972. *Biology Data Book.* Bethesda: Fed. Amer. Soc. Exper. Biol. 3 vols.

Amadon, D. 1947. An estimated weight of the largest known bird. *Condor* 49:159–164.

American Society for Testing and Materials. 1980. Standard for metric practice. In *Annual Book of ASTM Standards.* Philadelphia. Pt. 41, pp. 504–545.

Andersen, H. T. 1966. Physiological adaptations in diving vertebrates. *Physiol. Rev.* 46:212–243.

Anderson, J. F., Rahn, H., and Prange, H. D. 1979. Scaling of supportive tissue mass. *Q. Rev. Biol.* 54:139–148.

Ar, A., and Rahn, H. 1978. Interdependence of gas conductance, incubation length, and weight of the avian egg. In *Respiratory Function in Birds, Adult and Embryonic,* ed. J. Piiper. New York: Springer-Verlag, pp. 227–236.

Ar, A., and Yom-Tov, Y. 1978. The evolution of parental care in birds. *Evolution* 32:655–669.

Ar, A., Rahn, H., and Paganelli, C. V. 1979. The avian egg: mass and strength. *Condor* 81:331–337.

Armitage, K. B. 1981. Sociality as a life-history tactic of ground squirrels. *Oecologia* (Berlin) 48:36–49.

Aschoff, J. 1981a. Der Tagesgang der Körpertemperatur von Vögeln als Funktion des Körpergewichtes. *J. Ornithol.* (Berlin) 122:129–151.

—— 1981b. Thermal conductance in mammals and birds: its dependence on body size and circadian phase. *Comp. Biochem. Physiol.* 69A:611–619.

—— 1982. The circadian rhythm of body temperature as a function of body size. In *A Companion to Animal Physiology,* eds. C. R. Taylor, K. Johansen, and L. Bolis. New York: Cambridge University Press, pp. 173–188.

Aschoff, J., and Pohl, H. 1970a. Rhythmic variations in energy metabolism. *Fed. Proc.* 29:1541–52.

—— 1970b. Der Ruheumsatz von Vögeln als Funktion der Tageszeit und der Körpergröss. *J. Ornithol.* (Berlin) 111:28–47.

Austin, O. L., Jr. 1961. *Birds of the World.* New York: Golden Press, p. 316.

Bakken, G. S. 1976. An improved method for determining thermal conductance and equilibrium body temperature with cooling curve experiments. *J. Thermal Biol.* 1:169–175.

Banko, W. E. 1960. *The Trumpeter Swan.* North American Fauna, 63. Washington, D.C.: Government Printing Office. 214 pp.

Banse, K., and Mosher, S. 1980. Adult body mass and annual production/biomass relationships of field populations. *Ecol. Monogr.* 50:355–379.

Barrett, G. W. 1969. Bioenergetics of a captive least shrew, *Cryptotis parva. J. Mammal.* 50:629–630.

Bartholomew, G. A. 1981. A matter of size: an examination of endothermy in insects and terrestrial vertebrates. In *Insect Thermoregulation,* ed. B. Heinrich. New York: John Wiley, pp. 45–78.

—— 1982. Energy metabolism. In *Animal Physiology: Principles and Adaptations,* ed. M. S. Gordon. New York: Macmillan, pp. 57–110.

Bartholomew, G. A., and Cade, T. J. 1963. The water economy of land birds. *Auk* 80:504–539.

Barton, A. D. and Laird, A. K. 1969. Analysis of allometric and non-allometric differential growth. *Growth* 33:1–16.

Baudinette, R. V. 1978. Scaling of heart rate during locomotion in mammals. *J. Comp. Physiol.* 127B:337–342.

Baumann, F. H., and Baumann, R. 1977. A comparative study of the respiratory properties of bird blood. *Respir. Physiol.* 31:333–343.

Baur, M. E., and Friedl, R. R. 1980. Application of size-metabolism allometry to therapsids and dinosaurs. In *A Cold Look at the Warm-Blooded Dinosaurs,* eds. R. D. K. Thomas and E. C. Olson. AAAS Selected Symp. 28. Boulder: Westview Press, pp. 253–286.

Bech, C., and Johansen, K. 1980. Ventilation and gas exchange in the mute swan, *Cygnus olor. Respir. Physiol.* 39:285–295.

Bech, C., Johansen, K., and Maloiy, G. M. O. 1979. Ventilation and expired gas composition in the flamingo, *Phoenicopterus ruber,* during normal respiration and panting. *Physiol. Zool.* 52:313–328.

Beland, P., and Russell, D. A. 1980. Dinosaur metabolism and predator/prey ratios in the fossil record. In *A Cold Look at the Warm-Blooded Dinosaurs,* eds. R. D. K. Thomas and E. C. Olson. AAAS Selected Symp. 28. Boulder: Westview Press, pp. 85–102.

Benedict, F. G. 1938. *Vital Energetics: A Study in Comparative Basal Metabolism.* Washington, D.C.: Carnegie Inst. Wash., publ. 503.

Bennett, A. F. 1982. The energetics of reptilian activity. In *Biology of the Reptilia,* eds. C. Gans and F. H. Pough. New York: Academic Press, vol. 13, pp. 155–199.

Bennett, A. F., and Dawson, W. R. 1976. Metabolism. In *Biology of the Reptilia,* eds. C. Gans and F. H. Pough. New York: Academic Press, vol. 5, pp. 127–223.

Bennett, F. M., and Tenney, S. M. 1982. Comparative mechanics of mammalian respiratory system. *Respir. Physiol.* 49:131–140.

Berger, M., and Hart, J. S. 1974. Physiology and energetics of flight. In *Avian Biology,* eds. D. S. Farner and J. R. King. New York: Academic Press, pp. 415–477.

Bergmann, H. H. 1976. Konstitutionsbedingte Merkmale in Gesangen und Rufen europaischer Grasmucken (Gattung Sylvia). *A. Tierpsychol.* 42:315–329.

Bernard, C. 1865. *An Introduction to the Study of Experimental Medicine.* New York: Dover (1957 repub. of 1927 English trans.), pp. 129–130.

Bertalanffy, L. von 1957. Quantitative laws in metabolism and growth. *Q. Rev. Biol.* 32:217–231.

Biewener, A. A. 1983. Scaling relative mechanical advantage: implications for muscle function in different sized animals. *Fed. Proc.* 42:469.

Birkebak, R. C. 1966. Heat transfer in biological systems. *Int. Rev. Gen. Exper. Zool.* 2:269–344.

Blake, B. H. 1977. The effects of kidney structure and the annual cycle on water requirements in golden-mantled ground squirrels and chipmunks. *Comp. Biochem. Physiol.* 58A:413–419.

Bland, D. K., and Holland, R. A. B. 1977. Oxygen affinity and 2, 3-diphosphoglycerate in blood of Australian marsupials of differing body size. *Respir. Physiol.* 31:279–290.

Blueweiss, L., Fox, H., Kudzma, V., Nakashima, D., Peters, R., and Sams, S. 1978. Relationships between body size and some life history parameters. *Oecologia* (Berlin) 37:257–272.

Blum, J. J. 1977. On the geometry of four-dimensions and the relationship between metabolism and body mass. *J. Theor. Biol.* 64:599–601.

Bond, C. F., and Gilbert, P. W. 1958. Comparative study of blood volume in representative aquatic and nonaquatic birds. *Amer. J. Physiol.* 194:519–521.

Bone, Q. 1978. Locomotor muscle. In *Fish Physiology,* eds. W. S. Hoar and D. J. Randall. New York: Academic Press, vol. 7, pp. 361–362.

Boniface, J., Picone, D., Schebalin, M., Zfass, A. M., and Makhlouf, G. M. 1976. Clearance rate, half-life, and secretory potency of human gastrin-17-I in different species. *Gastroenterology* 71:291–294.

Bonner, J. T. 1965. *Size and Cycle.* Princeton: Princeton University Press. 219 pp.

Bonner, J. T., and Horn, H. S. 1982. Selection for size, shape, and development timing. In *Evolution and Development,* ed. J. T. Bonner. New York: Springer-Verlag, pp. 259–276.

Botkin, D. B., and Miller, R. S. 1974. Mortality rates and survival of birds. *Amer. Nat.* 108:181–192.

Bourlière, F. 1964. *The Natural History of Mammals,* 3rd ed. New York: Knopf. 364 pp.

——— 1975. Mammals, small and large: the ecological implications of size. In *Small Mammals: Their Productivity and Population Dynamics,* eds. F. B. Golley, K. Petrusewicz, and L. Ryszkowski. Cambridge: Cambridge University Press, pp. 1–8.

Boxenbaum, H. 1980. Interspecies variation in liver weight, hepatic blood flow, and antipyrine intrinsic clearance: extrapolation of data to benzodiazepines and phenytoin. *J. Pharmacokin. Biopharm.* 8:165–176.

——— 1982. Interspecies scaling, allometry, physiological time, and the ground plan of pharmacokinetics. *J. Pharmacokin. Biopharm.* 10:201–227.

Brackenbury, J. H. 1977. Physiological energetics of cock crow. *Nature* (London) 270:433–435.

——— 1980. Respiration and production of sounds by birds. *Biol. Rev.* 55:363–378.

Bradley, S. R., and Deavers, D. R. 1980. A re-examination of the relationship between thermal conductance and body weight in mammals. *Comp. Biochem. Physiol.* 65A:465–476.

Bramble, D. M., and Carrier, D. R. 1983. Running and breathing in mammals. *Science* 219:251–256.

Brandborg, S. M. 1955. *Life History and Management of the Mountain Goat.* Idaho Wildlife Bull., 2. Boise: Idaho Dept. of Fish and Game.

Brandt, C. S., and Thacker, E. J. 1958. A concept of rate of food passage through the gastro-intestinal tract. *J. Anim. Sci.* 17:218–223.

Braun, E. J., and Dantzler, W. H. 1972. Function of mammalian-type and reptilian-type nephrons in kidney of desert quail. *Amer. J. Physiol.* 222:617–629.

——— 1974a. Effects of ADH on single-nephron glomerular filtration rates in the avian kidney. *Amer. J. Physiol.* 226:1–8.

——— 1974b. Effect of a water load on single-nephron glomerular filtration rate (SNGFR) in desert quail. *Physiologist* 17:186.

——— 1975. Effects of water load on renal glomerular and tubular function in desert quail. *Amer. J. Physiol.* 229:222–228.

Brockway, J. M., and Gessman, J. A. 1977. The energy cost of locomotion on the level and on gradients for the red deer *(Cervus elaphus). Q. J. Exper. Physiol.* 62:333–339.

Brody, J., Comfort, J. E., and Matthews, J. S. 1928. Growth and development with special reference to domestic animals. XI. Further investigations on surface area with special reference to its significance in energy metabolism. *University of Missouri College of Agriculture, Agricultural Experiment Station Research Bull.* 115:3–60.

Brody, S. 1945. *Bioenergetics and Growth* (rpt. 1968). New York: Hafner. 1,023 pp.

Brown, J. H., and Bowers, M. A. MS. Community organization in hummingbirds: relationships between morphology and ecology.

Brown, J. H., Calder, W. A. III, and Kodrick-Brown, A. 1978. Correlates and consequences of body size in nectar-feeding birds. *Amer. Zool.* 18:687–700.

Brown, J. L. 1982. The adaptationist program. *Science* 217:884–886.

Buckner, C. H. 1964. Metabolism, food capacity, and feeding behavior in four species of shrews. *Can. J. Zool.* 42:259–279.

Burt, W. H., and Grossenheider, R. P. 1964. *A Field Guide to the Mammals.* Boston: Houghton Mifflin. 284 pp.

Butler, P. J., and Woakes, A. J. 1980. Heat rate, respiratory frequency and wing beat frequency of free-flying barnacle geese *Branta lencopis. J. Exper. Biol.* 82:213–226.

Cabana, G., Frewin, A., Peters, R. H., and Randall, L. 1982. The effect of sexual size dimorphism on variations in reproductive effort of birds and mammals. *Amer. Nat.* 120:17–25.

Cahalane, V. H. 1961. *Mammals of North America.* New York: Macmillan. 682 pp.

Calder, W. A. 1968. Respiratory and heart rates of birds at rest. *Condor* 70:358–365.

—— 1969. Temperature relations and underwater endurance of the smallest homeothermic diver, the water shrew. *Comp. Biochem. Physiol.* 30:1075–82.

—— 1970. Respiration during song in the canary *(Serinus canaria) Comp. Biochem. Physiol.* 32:251–258.

—— 1971. Temperature relationships and nesting of the calliope hummingbird. *Condor* 73:314–321.

Calder, W. A. III. 1974. The consequences of body size for avian energetics. In *Avian Energetics,* ed. R. A. Paynter, Jr. Cambridge, Mass.: Nuttall Ornithol. Club, publ. 15, pp. 86–157.

—— 1975. Day length and the hummingbirds' use of time. *Auk* 92:81–97.

—— 1976a. Aging in vertebrates: allometric considerations of spleen size and lifespan. *Fed. Proc.* 35:96–97.

—— 1976b. Energetics of small body size and high latitude: the rufous hummingbird in coastal Alaska. *Int. J. Biometeor.* 20:23–35.

—— 1978. The kiwi. *Sci. Amer.* 239:132–142.

—— 1979. The kiwi and egg design: evolution as a package deal. *Bioscience* 29:461–467.

—— 1981. Scaling of physiological processes in homeothermic animals. *Ann. Rev. Physiol.* 43:301–322.

—— 1982a. A proposal for the standardization of units and symbols in ecology. *Bull. Ecol. Soc. Amer.* 63:7–10.

—— 1982b. The pace of growth: an allometric approach to comparative embryonic and post-embryonic growth. *J. Zool.* (London) 198:215–225.

—— 1982c. A tradeoff between space and time: dimensional constants in mammalian ecology. *J. Theor. Biol.* 98:393–400.

—— 1982d. The relationship of the Gompertz constant and maximum potential lifespan to body mass. *Exper. Geront.* 17:383–385.

—— 1983a. An allometric approach to population cycles of mammals. *J. Theor. Biol.* 100:275–282.

—— 1983b. Ecological scaling: mammals and birds. *Ann. Rev. Ecol. and Sys.* 14:213–230.

—— 1983c. Commentary on "Ecological energetics: what are the questions?" In *Perspectives in Ornithology,* eds. G. A. Clark, Jr., and A. Brush. Cambridge: Cambridge University Press, pp. 158–164.

—— 1983d. Body size, mortality, and longevity. *J. Theor. Biol.* 102:135–144.

Calder, W. A. III, and Booser, J. 1973. Hypothermia of broad-tailed hummingbirds during incubation in nature with ecological correlations. *Science* 180:751–753.

Calder, W. A. III, and Braun, E. J. 1983. Scaling osmotic regulation and body size in mammals and birds. *Amer. J. Physiol.* 244:R601–R606.

Calder, W. A. III, and Dawson, T. J. 1978. Resting metabolic rates of ratite birds: the kiwis and the emu. *Comp. Biochem. Physiol.* 60A:379–481.

Calder, W. A. III, and King, J. R. 1972. Body weight and the energetics of temperature regulation: a re-examination. *J. Exper. Biol.* 56:775–780.

—— 1974. Thermal and caloric relations of birds. In *Avian Biology,* eds. D. S. Farner and J. R. King. New York: Academic Press, pp. 259–413.

Calder, W. A. III, Hiebert, S. M., Waser, N. M., Inouye, D. W., and Miller, S. J. 1983. Site-fidelity, longevity, and population dynamics of broad-tailed hummingbirds: a ten-year study. *Oecologia* (Berlin) 56:359–364.

Calow, P. 1978. *Life Cycles.* Berlin: Springer-Verlag.

Campbell, G. S. 1977. *An Introduction to Environmental Biophysics.* Berlin: Springer-Verlag.

Case, T. J. 1978. On the evolution and adaptive significance of post-natal growth rates in the terrestrial vertebrates. *Q. Rev. Biol.* 53:243–282.

Casey, T. M. 1981. Insect flight energetics. In *Locomotion and Energetics in Arthropods,* eds. C. V. Herreid II, and C. R. Fourtner. New York: Plenum Press, pp. 419–452.

Caughley, G. 1970. Eruption of ungulate populations, with emphasis on Himalayan Thar in New Zealand. *Ecology* 51:53–71.

—— 1977. *Analysis of Vertebrate Populations.* New York: John Wiley. 234 pp.

Caughley, G., and Krebs, C. J. 1983. Are big mammals simply little mammals writ large? *Oecologia* (Berlin) 59:7–17.

Cheke, R. A. 1971. Temperature rhythms in African montane sunbirds. *Ibis* 113:500–506.

Cheverud, J. M. 1982. Relationships among ontogenetic, static, and evolutionary allometry. *Amer. J. Phys. Anthro.* 59:139–149.

Chew, R. M. 1965. Water metabolism of mammals. In *Physiological Mammalogy,* eds. W. V. Mayer and R. G. Van Gelder. New York: Academic Press, vol. 2, pp. 43–178.

Christoph, H. C. 1975. *Diseases of Dogs.* New York: Pergamon Press. 496 pp.

Clark, G. A., Jr. 1979. Body weights of birds: a review. *Condor* 81:193–202.

Clements, J. A., Nellenbogen, J., and Trahan, H. J. 1970. Pulmonary surfactant and evolution of the lungs. *Science* 169:603–604.

Cock, A. G. 1966. Genetical aspects of metrical growth and form in animals. *Q. Rev. Biol.* 41:131–190.

Cohen, R. R. 1967. Total circulating erythrocyte and plasma volumes of ducks measured simultaneously with Cr^{51}. *Poultry Sci.* 46:1539–44.

Cole, L. C. 1954. The population consequences of life history phenomena. *Q. Rev. Biol.* 29:103–137.

Corrisin, S. 1982. A more explicit estimate for the "implications of athlete's brady-cardia on lifespan." *J. Theor. Biol.* 96:683–688.

Cracraft, J. 1974. Phylogeny and evolution of the ratite birds. *Ibis* 116:494–521.

Crampton, E. W., and Harris, L. E. 1969. *Applied Animal Nutrition,* 2nd ed. San Francisco: W. H. Freeman, p. 154.

Crawford, E. C., Jr., and Kampe, G. 1971. Resonant panting in pigeons. *Comp. Biochem. Physiol.* 40A:549–552.

Crawford, E. C., Jr., and Lasiewski, R. C. 1968. Oxygen consumption and respiratory evaporation of the emu and rhea. *Condor* 70:333–339.

Crawford, E. C., Jr., and Schmidt-Nielsen, K. 1967. Temperature regulation and evaporative cooling in the ostrich. *Amer. J. Physiol.* 212:347–353.

Crosfill, M. L., and Widdicombe, J. G. 1961. Physical characteristics of the chest and lungs and the work of breathing in different mammalian species. *J. Physiol.* 158:1–14.

Crowcroft, P. 1954. The daily cycle of activity in British shrews. *Proc. Zool. Soc., London (J. Zool.)* 123:715–728.

Currey, J. D. 1968. The effect of protection on the impact strength of rabbits' bones. *Acta Anat.* 71:87–93.

Cutler, R. G. 1978. Evolutionary biology of senescence. In *The Biology of Aging,* eds. J. A. Behnke, C. E. Finch, and G. B. Moment. New York: Plenum Press, pp. 311–360.

Damuth, J. 1981a. Population density and body size in mammals. *Nature* 290:699–700.

——— 1981b. Home range, home range overlap, and species energy use among herbivorous mammals. *Biol. J. Linn. Soc.* 15:185–193.

——— 1982. Analysis of the preservation of community structure in assemblages of fossil mammals. *Paleobiology* 8:434–446.

Dantzler, W. H. 1970. Kidney function in desert vertebrates. *Mem. Soc. Endocrin.* 18:157–190.

Davis, C. N., Davis, L. E., and Powers, T. E. 1975. Comparative body compositions of the dog and goat. *Amer. J. Vet. Res.* 36:309–311.

Davis, D. D. 1962. Allometric relationship in lions vs. domestic cats. *Evolution* 16:505–514.

Dawson, T. J. 1973. "Primitive" mammals. In *Compartive Physiology of Thermo-regulation,* ed. G. C. Whittow. New York: Academic Press, vol. 3, pp. 1–46.

——— 1983. *Monotremes and Marsupials: The Other Mammals.* London: Edward Arnold. 90 pp.

Dawson, T. J., and Dawson, W. R. 1982. Metabolic scope and conductance in response to cold of some Dasyurid marsupials and Australian rodents. *Comp. Biochem. Physiol.* 71A:59–64.

Dawson, T. J., and Hulbert, A. J. 1970. Standard metabolism, body temperature, and surface areas of Australian marsupials. *Amer. J. Physiol.* 218: 1233–38.

Dawson, T. J., and Needham, A. D. 1981. Cardiovascular characteristics of two resting marsupials: an insight into the cardio-respiratory allometry of marsupials. *J. Comp. Physiol.* B145:95–100.

Dawson, T. J., and Schmidt-Nielsen, K. 1966. Effect of thermal conductance on water economy in the antelope jack rabbit, *Lepius alleni. J. Cell Physiol.* 67:463–471.

Dawson, T. J., Grant, T. R., and Fanning, D. 1979. Standard metabolism of monotremes and the evolution of homeothermy. *Aust. J. Zool.* 27:511–515.

Dawson, W. R., and Carey, C. 1976. Seasonal acclimatization to temperature in cardueline finches. I. Insulative and metabolic adjustments. *J. Comp. Physiol.* 112:317–338.

Dedrick, R. L. 1974. Animal scale-up. In *Pharmacology and Pharmacokinetics,* eds. T. Teorell, R. L. Dedrick, and P. G. Condliffe. New York: Plenum Press, pp. 117–145.

Dedrick, R. L., Bischoff, K. B., and Zaharko, D. S. 1977. Interspecies correlation of plasma concentration history of methotrexate. *Cancer Chemother. Rep.* 54:95–101.

Deevey, E. S., Jr. 1947. Life tables for the natural populations of animals. *Q. Rev. Biol.* 22:283–314.

Degen, A. A., Pinshow, B., and Alkon, P. U. 1982. Water flux in chukar partridges *(Alextoris chukar)* and a comparison with other birds. *Physiol. Zool.* 55:64–71.

Dejours, P. 1981. *Principles of Comparative Respiratory Physiology.* New York: Elsevier/North-Holland. 265 pp.

Delcomyn, F. 1980. Neural basis of rhythmic behavior in animals. *Science* 210:492–498.

Denny, M. J. S., and Dawson, T. J. 1975. Comparative metabolism of tritiated water by macropodid marsupials. *Amer. J. Physiol.* 228:1794–99.

Dooling, R. J. 1980. Behavior and psychophysics of hearing in birds. In *Comparative Studies of Hearing in Vertebrates,* eds. A. N. Popper and R. R. Fay. New York: Springer-Verlag, pp. 261–285.

Drabkin, D. L. 1950. The distribution of the chromoproteins, hemoglobin, myoglobin, and cytochrome c, in the tissues of different species, and the relationship of the total content of each chromoprotein to body mass. *J. Biol. Chem.* 182:317–333.

Drent, R. H. 1975. Incubation. In *Avian Biology,* eds. D. S. Farner and J. R. King. New York: Academic Press, vol. 5, pp. 333–420.

Drent, R. H., and Daan, S. 1980. The prudent parent: energetic adjustments in avian breeding. *Ardea* 68:225–252.

Drent, R. H., and Stonehouse, B. 1971. Thermoregulatory responses of the Peruvian penguin, *Spheniscus humboldti. Comp. Biochem. Physiol.* 40A:689–710.

Dubach, M. 1981. Quantitative analysis of the respiratory system of the house sparrow, budgerigar and violet-eared hummingbird. *Respir. Physiol.* 46:43–60.

Eberhardt, L. L. 1969. Similarity, allometry, and food chains. *J. Theor. Biol.* 24:43–55.

Eckert, R., and Randall, D. 1983. *Animal Physiology.* San Francisco: W. H. Freeman. 830 pp.

Economos, A. C. 1979. Gravity, metabolic rate, and body size of mammals. *Physiologist* 22:S71–S72.

—— 1980. Taxonomic differences in the mammalian life span–body weight relationship and the problem of brain weight. *Gerontology* 26:90–98.

—— 1981a. The largest land mammal. *J. Theor. Biol.* 89:211–215.

—— 1981b. Beyond rate of living. *Gerontology* 27:258–265.

—— 1982. On the origin of biological similarity. *J. Theor. Biol.* 94:25–60.

—— 1983. Elastic and/or geometric similarity in mammalian design? *J. Theor. Biol.* 103:167–172.

Edwards, N. A. 1975. Scaling of renal functions in mammals. *Comp. Biochem. Physiol.* 52A:63–66.

Eisenberg, J. F. 1981. *The Mammalian Radiations.* Chicago: University of Chicago Press. 610 pp.

Eisenberg, J. F., and Wilson, D. E. 1978. Relative brain size and feeding strategies in the Chiroptera. *Evolution* 32:740–751.

Elias, H., and Schwartz, D. 1969. Surface areas of the cerebral cortex of mammals determined by stereological methods. *Science* 166:111–113.

Emerson, R. W. 1856. "Nature," in *English Traits;* 1860. "Beauty," in *Conduct of Life.* In *Collected Works of Ralph Waldo Emerson.* Masterworks Library. New York: Greystone (undated). 478 pp.

Emmett, B., and Hochachka, P. W. 1981. Scaling of oxidative and glycolytic enzymes in mammals. *Respir. Physiol.* 45:261–272.

Epstein, M. 1979. Effects of aging on the kidney. *Fed. Proc.* 38:168–172.

Epting, R. J. 1980. Functional dependence of the power for hovering on wing disc loading in hummingbirds. *Physiol. Zool.* 53:347–357.

Faaborg, J. 1977. Metabolic rates, resources, and the occurrence of nonpasserines in terrestrial avian communities. *Amer. Nat.* 111:903–916.

Farlow, J. O. 1976. A consideration of the trophic dynamics of a late Cretaceous large-dinosaur community (Oldman Formation). *Ecology* 57:841–857.

Farner, D. S. 1955. Bird banding in the study of population dynamics. In *Recent Studies in Avian Biology,* ed. A. Wolfson. Urbana: University of Illinois Press, pp. 297–499.

Fedak, M. A., and Seehermann, H. J. 1979. Reappraisal of energetics of locomotion shows identical cost in bipeds and quadrupeds including ostrich and horse. *Nature* 282:713–716.

Fedak, M. A., Heglund, N. C., and Taylor, C. R. 1982. Energetics and mechanics of terrestrial locomotion: II. Kinetic energy changes of the limbs and body as a function of speed and body size in birds and mammals. *J. Exper. Biol.* 79:23–40.

Feldman, M. A., and McMahon, T. A. 1983. The ¾ mass exponent for energy metabolism is not a statistical artifact. *Respir. Physiol.* 52:149–163.

Felicetti, S. A., Wolff, R. K., and Muggenbuz, B. A. 1981. Comparison of tracheal mucous transport in rats, guinea pigs, rabbits, and dogs. *J. Appl. Physiol.* 51:1612–17.

Fenchel, T. 1974. Intrinsic rate of natural increase: the relationship with body size. *Oecologia* (Berlin) 14:317–326.

Fischer, R. 1968. On the steady state nature of evolution, learning, perception, hallucination and dreaming. In *Quantitative Biology of Metabolism*, ed. A. Locker. New York: Springer-Verlag, pp. 245–256.

Fisher, J. T., and Mortola, J. P. 1981. Statics of the respiratory system and growth: an experimental and allometric approach. *Amer. J. Physiol.* 241:R336–R341.

Fleharty, E. D., Krause, M. E., and Stinnett, D. P. 1973. Body composition, energy content and lipid cycles of four species of rodents. *J. Mammal.* 54:426–438.

Fleming, T. H. 1975. The role of small mammals in tropical ecosystems. In *Small Mammals: Their Productivity and Population Dynamics,* eds. F. B. Golley, K. Petrusewicz, and L. Ryszkowski. Cambridge: Cambridge University Press, pp. 269, 286–287, 292–298.

Forbes, W. H., and Dill, D. B. 1940. Resignation or equanimity? *Science* 92:605–606.

Frazer, J. F. D. 1977. Growth of young vertebrates in the egg or uterus. *J. Zool.* (London) 183:189–201.

Frazer, J. F. D., and Huggett, A. St. G. 1974. Species variations in the foetal growth rates of eutherian mammals. *J. Zool.* (London) 174:481–509.

French, A. R. MS "a." Body size and the duration of arousal episodes during mammalian hibernation.

—— MS "b." The patterns of mammalian hibernation: trade-offs between time and energy.

French, N. R., Stoddart, D. M., and Bobek, B. 1975. Patterns of demography in small mammal populations. In *Small Mammals: Their Productivity and Population Dynamics,* eds. F. B. Golley, K. Petrusewicz, and L. Ryszkowski. Cambridge: Cambridge University Press, pp. 73, 76–93, 100–101.

Fried, G. H., and Tipton, S. R. 1953. Comparison of respiratory enzyme levels in tissues of mammals of different sizes. *Proc. Soc. Exp. Biol. Med.* 82:531–532.

Galileo, G. 1637. Proposition VIII in *Dialogues Concerning the Two New Sciences,* trans. H. Crew and A. de Salvio (1933). New York: Macmillan.

Ganong, W. F. 1981. *Review of Medical Physiology.* Los Altos: Lange, p. 628.

Gans, C. 1979. Momentarily excessive construction as the basis for protoadaptation. *Evolution* 33:227–233.

Garland, T. 1983a. Scaling maximum running speed and body mass in terrestrial mammals. *J. Zool.* (London) 199:157–170.

———— 1983b. Scaling the ecological cost of transport to body mass in terrestrial mammals. *Amer. Nat.* 121:571–587.

Gauthreaux, S. A., Jr. 1980. *Animal Migration, Orientation, and Navigation.* New York: Academic Press. 387 pp.

Geelhaar, A., and Weibel, E. R. 1971. Morphometric estimation of pulmonary diffusion capacity. III. The effect of increased oxygen consumption in Japanese waltzing mice. *Respir. Physiol.* 11:354–366.

Gehr, P., and Erni, H. 1980. Morphometric estimation of pulmonary diffusion capacity in two horse lungs. *Respir. Physiol.* 41:199–210.

Gehr, P., Sehovic, S., Burri, P. H., Claassen, H., and Weibel, E. R. 1980. The lung of shrews: morphometric estimation of diffusion capacity. *Respir. Physiol.* 40:33–47.

Gehr, P., Mwangi, D. K., Ammann, A., Maloiy, G. M. O., Taylor, C. R., and Weibel, E. R. 1981a. Design of the mammalian respiratory system. V. Scaling morphometric pulmonary diffusing capacity to body mass: wild and domestic mammals. *Respir. Physiol.* 44:61–86.

Gehr, P., Siegwart, B., and Weibel, E. R. 1981b. Allometric analysis of the morphometric pulmonary diffusing capacity in dogs. *J. Morph.* 168:5–15.

Gold, A. 1973. Energy expenditure in animal locomotion. *Science* 181:275–276.

Gould, E., Negus, N. C., and Novick, A. 1964. Evidence for echolocation in shrews. *J. Exper. Zool.* 156:19–38.

Gould, S. J. 1963. "Irish elk"—positive allometry of antlers. *Nature* 244:375.

———— 1966. Allometry and size in ontogeny and phylogeny. *Biol. Rev.* 41:587–640.

———— 1971. Geometric similarity in allometric growth: a contribution to the problem of scaling. *Amer. Nat.* 105:113–136.

———— 1974. The origin and function of "bizarre" structures: antler size and skull size in the "Irish elk," *Megaloceros giganteus. Evolution* 28:191–220.

———— 1975. On the scaling of tooth size in mammals. *Amer. Zool.* 15:351–362.

———— 1977. *Ontogeny and Phylogeny.* Cambridge, Mass.: Harvard University Press. 501 pp.

———— 1982. Change in developmental timing as a mechanism of macroevolution. In *Evolution and Development,* ed. J. T. Bonner. New York: Springer-Verlag, pp. 333–346.

Gould, S. J., and Lewontin, R. C. 1979. The spandrels of San Marco and the Panglossian paradigm: a critique of the adaptationist programme. *Proc. Royal Soc. Lond.* (B) 205:581–598.

Grant, P. R. 1983. The relative size of Darwin's Finch eggs. *Auk* 100:228–229.

Grant, T. R., and Dawson, T. J. 1978. Temperature regulation in the platypus, *Ornithorhychus anatinus:* production and loss of metabolic heat in air and water. *Physiol. Zool.* 51:315–332.

Gray, I. E. 1954. Comparative study of the gill area of marine fishes. *Biol. Bull.* 107:219–225.

Greegor, D. H. 1975. Renal capabilities of an Argentina desert armadillo. *J. Mammal.* 56:626–632.

Greenewalt, C. H. 1962. Dimensional relationships for flying animals. *Smithsonian Misc. Coll.* 144(2):1–16.

—— 1975a. The flight of birds. *Trans. Amer. Phil. Soc.* 65:1–67.

—— 1975b. Could Pterosaurs fly? *Science* 188:676.

Grodzinski, W., and Wunder, B. A. 1975. Ecological energetics of small mammals. In *Small Mammals: Their Productivity and Population Dynamics,* eds. F. B. Golley, K. Petrusewicz, and L. Ryszkowski. Cambridge: Cambridge University Press, pp. 173–204.

Günther, B. 1972. Allometric ratios, invariant numbers and the theory of biological similarities. *Pflügers Archiv* 331:283–293.

—— 1975. On theories of biological similarity. *Fortschritte der experimentellen und theoretischen Biophysic,* vol. 19. Leipzig: Georg Thieme. 111 pp.

Günther, B., and Guerra, E. 1957. Theory of biological similarity applied to some data of comparative physiology. *Acta Physiol. Latinoamer.* 7:95–103.

Günther, B., and Martinoya, C. 1968. Operational time and theory of biological similarities. *J. Theor. Biol.* 20:107–111.

Hafez, E. S. E., Asdell, S. A., and Blandau, R. J. 1972. Propagation: mammals. In *Biology Data Book,* eds. P. L. Altman and D. S. Dittmer. Bethesda: Fed. Amer. Soc. Exper. Biol., vol. 1, pp. 138–139.

Hainsworth, F. R. 1981. *Animal Physiology: Adaptations in Function.* Reading, Mass.: Addison-Wesley, p. 170.

—— MS. On mammal trails on mountain slopes: the physics of moving up inclined planes.

Hainsworth, F. R., and Wolf, L. L. 1972. Crop volume, nectar concentration and hummingbird energetics. *Comp. Biochem. Physiol.* 42A:359–366.

—— 1978. The economics of temperature regulation and torpor in nonmammalian organisms. In *Strategies in Cold: Natural Torpidity and Thermogenesis,* eds. L. C. H. Wang and J. W. Hudson. New York: Academic Press, pp. 67–86.

Haldane, J. B. S. 1928. On being the right size. In *A Treasury of Science,* eds. H. Shapley, S. Rapport, and H. Wright. New York: Harper, repr. 1958, pp. 321–325.

Hamilton, T. H. 1961. The adaptive significances of intraspecific trends of variation in wing length and body size among bird species. *Evolution* 15:180–195.

Hanwell, A., and Peaker, M. 1977. Physiological effects of lactation on the mother. In *Comparative Aspects of Lactation,* ed. M. Peaker. *Symp. Zool. Soc. London* 41:297–312.

Harcourt, A. H., Harvey, P. H., Larson, S. G., and Short, R. V. 1981. Testis weight, body weight and breeding system in primates. *Nature* 293:55–56.

Harestad, A. S., and Bunnell, F. L. 1979. Home range and body weight—a re-evaluation. *Ecology* 60:389–402.

Harris, C. C. 1973. Lilliput revisited, or how fed-up was Gulliver? *Chem. Tech.* 3:600–602.

Harrison, D. E. 1978. Is limited cell proliferation the clock that times aging? In *The Biology of Aging,* eds. J. A. Behnke, C. E. Finch, and G. B. Moment. New York: Plenum Press, pp. 33–55.

Hart, J. S. 1971. Rodents. In *Comparative Physiology of Thermoregulation,* ed. G. C. Whittow. New York: Academic Press, vol. 2, pp. 1–149.

Hart, J. S., and Berger, M. 1972. Energetics, water economy, and temperature regulation during flight. *Proc. XV. Int. Ornithol. Congr.* pp. 189–199.

Hart, R. W., Sacher, G. A., and Moskins, T. L. 1979. DNA repair in a short- and long-lived rodent species. *J. Geront.* 34:808–817.

Hartman, F. A. 1961. Locomotor mechanisms of birds. *Smithsonian Misc. Coll.* 143:1–91.

Harvey, P. H. 1982. On rethinking allometry. *J. Theor. Biol.* 95:37–41.

Harvey, P. H., and Clutton-Brock, T. H. 1981. Primate home-range size and metabolic needs. *Behav. Ecol. Sociobiol.* 8:151–155.

Harvey, P. H., and Mace, G. M. 1982. Comparisons between taxa and adaptive trends: problems of methodology. In *Current Problems in Sociobiology,* ed. King's College Sociobiology Group. Cambridge: Cambridge University Press, pp. 343–361.

Haukioja, E., and Hakala, T. 1979. On the relationship between avian clutch size and life span. *Ornis Fennica* 56:45–55.

Heffner, H., and Masterton, B. 1980. Hearing in glires: domestic rabbit, cotton rat, feral house mouse, and kangaroo rat. *J. Acoust. Soc. Amer.* 68:1584–99.

Heffner, R., and Heffner, H. 1980. Hearing in the elephant *(Elephas maximus). Science* 208:518–520.

Heglund, N. C., Cavagna, G. A., and Taylor, C. R. 1982a. Energetics and mechanics of terrestrial locomotion. III. Energy changes of the center of mass as a function of speed and body size in birds and mammals. *J. Exper. Biol.* 79:41–56.

Heglund, N. C., Fedak, M. A., Taylor, C. R., and Cavagna, G. A. 1982b. Energetics and mechanics of terrestrial locomotion. IV. Total mechanical energy changes as a function of speed and body size in birds and mammals. *J. Exper. Biol.* 79:57–66.

Heglund, N. C., Taylor, C. R., and McMahon, T. A. 1974. Scaling stride frequency and gait to animal size: mice to horses. *Science* 186:1112–13.

Heinrich, B., and Bartholomew, G. A. 1971. An analysis of pre-flight warm-up in the sphinx moth, *Manduca sexta. J. Exper. Biol.* 55:223–239.

Heldmaier, G. 1971. Zitterfreie Wärmebildung und Korpergrösse bei Säugetieren. *Z. vergl. Physiol.* 73:222–248.

Hemmingsen, A. M. 1960. Energy metabolism as related to body size and respiratory surfaces, and its evolution. *Repts. Steno Mem. Hosp. Nord. Insulinlab.* 9(pt. 2):7–110.

Hennemann, W. W. 1983. Relationship among body mass, metabolic rate and the intrinsic rate of natural increase in mammals. *Oecologia* (Berlin) 56:104–108.

Heptonstall, W. B. 1970. Quantitative assessment of the flight of *Archaeopteryx. Nature* 228:185–186.

Herreid, C. F., II, and Kessell, B. 1967. Thermal conductance in birds and mammals. *Comp. Biochem. Physiol.* 21:405–414.

Heusner, A. A. 1982a. Energy metabolism and body size. I. Is the 0.75 mass exponent of Kleiber's equation a statistical artifact? *Respir. Physiol.* 48:1–12.

——— 1982b. Energy metabolism and body size. II. Dimensional analysis and energetic non-similarity. *Respir. Physiol.* 48:13–25.

Heuwinkel, H. 1978. Der Gesang des Teichrohrsängers *(Acrocephalus scirpaceus)* unter besonderer Berücksichtigung der Schalldruckpegel-("Lautstärke"-)Verhältnisse. *J. Ornithol.* (Berlin) 119:450–461.

Hill, A. V. 1950. The dimensions of animals and their muscular dynamics. *Sci. Prog.* 38:209–230.

Hinds, D. S., and Calder, W. A. 1971. Tracheal dead space in the respiration of birds. *Evolution* 25:429–440.

——— 1973. Temperature regulation of the Pyrrhuloxia and the Arizona cardinal. *Physiol. Zool.* 46:55–71.

Holt, A. B., Cheek, D. B., Mellits, E. D., and Hill, D. E. 1975. Brain size and the relation of the primate to the nonprimate. In *Fetal and Postnatal Growth; Hormones and Nutrition,* ed. D. B. Cheek. New York: John Wiley, pp. 23–44.

Holt, J. P., and Rhode, E. A. 1976. Similarity of renal glomerular hemodynamics in mammals. *Amer. Heart J.* 92:465–472.

Holt, J. P., Rhode, E. A., and Kines, H. 1968. Ventricular volumes and body weight in mammals. *Amer. J. Physiol.* 215:704–715.

Holt, J. P., Rhode, E. A., Holt, W. W., and Kines, H. 1981. Geometric similarity of aorta, venae cavae, and certain of their branches in mammals. *Amer. J. Physiol.* 241:R100–R104.

Holt, W. W., Rhode, E. A., and Holt, J. P., Sr. 1978. Geometric similarity in the vascular system. *Fed. Proc.* 37:823.

Hoppeler, H., Mathieu, O., Weibel, E. R., Krauer, R., Lindstedt, S. L., and Taylor, C. R. 1981. Design of the mammalian respiratory system VIII. Capillaries in skeletal muscles. *Respir. Physiol.* 44:129–150.

Hoyt, D. F., and Rahn, H. 1980. Respiration of avian embryos—a comparative analysis. *Respir. Physiol.* 39:255–264.

Hoyt, D. F., and Taylor, C. R. 1981. Gait and the energetics of locomotion in horses. *Nature* 292:239–240.

Huggett, A. St. G., and Widdas, W. F. 1951. The relationship between mammalian foetal weight and conception age. *J. Physiol.* 114:306–317.

Hughes, G. M. 1966. The dimensions of fish gills in relation to their function. *J. Exper. Biol.* 45:177–195.

———— 1983. Allometry of gill dimensions in some British and American decapod crustacea. *J. Zool.* (London) 200:83–97.

Hughes, G. M., and Morgan, M. 1973. The structure of fish gills in relation to their respiratory function. *Biol. Rev.* 48:419–475.

Hughes, M. R. 1970. Relative kidney size in nonpasserine birds with functional salt glands. *Condor* 72:164–168.

Hugh-Jones, P., Barter, C. E., Hime, J. M., and Rusbridge, M. M. 1978. Dead space and tidal volume of the giraffe compared with some other mammals. *Respir. Physiol.* 35:53–58.

Humphreys, W. F. 1979. Production and respiration in animal populations. *J. Anim. Ecol.* 48:427–453.

———— 1981. Towards a simple index based on live-weight and biomass to predict assimilation in animal populations. *J. Anim. Ecol.* 50:543–561.

Hutchinson, G. E. 1959. Homage to Santa Rosalia or why are there so many kinds of animals. *Amer. Nat.* 93:145–159.

Hutchinson, G. E., and MacArthur, R. H. 1959. A theoretical ecological model of size distributions among species of animals. *Amer. Nat.* 93:117–125.

Huxley, J. S. 1927. On the relation between egg-weight and body-weight in birds. *J. Linn. Soc. Zool.* 36:457–466 (5 pl.).

———— 1972. *Problems of Relative Growth* (rpt. of 1932 ed.). New York: Dover. 312 pp.

Huxley, J. S., and Teissier, G. 1936. Terminology of relative growth. *Nature* 137:780–781.

Iberall, A. S. 1979. Some comparative scale factors for mammals: comments on Milnor's paper concerning a feature of cardiovascular design. *Amer. J. Physiol.* 237:R7-R9.

Irving, L. 1939. Respiration in diving mammals. *Physiol. Rev.* 19:112–134.

Itô, Y. 1980. *Comparative Ecology.* Cambridge: Cambridge University Press. 436 pp.

James, F. C. 1970. Geographic size variation in birds and its relationship to climate. *Ecology* 51:365–390.

Jarman, P. J. 1974. The social organization of antelope in relation to their ecology.

Behaviour 48:215–267.

Jerison, H. J. 1961. Quantitative analysis of evolution of the brain in mammals. *Science* 133:1012–1014.

—— 1968. Brain evolution and archaeopteryx. *Nature* 219:1381–82.

Johnson, O. W. 1968. Some morphological features of avian kidneys. *Auk* 85:216–228.

Johnston, D. W. 1968. Body characteristics of palm warblers following an overwater flight. *Auk* 85:13–18.

Johnston, D. W., and McFarlane, R. W. 1967. Migration and bioenergetics of flight in the Pacific golden plover. *Condor* 69:156–168.

Jones, M. L. 1982. Longevity of captive mammals. *Zool. Garten:* in press.

Kanwisher, J. 1977. Temperature regulation. In *Introduction to Comparative Physiology*, ed. L. Goldstein. New York: Holt, Rinehart & Winston, pp. 477–514.

Kanwisher, J., and Sundnes, G. 1966. Thermal regulation in cetaceans. In *Whales, Dolphins and Porpoises*, ed. K. S. Norris. Berkeley: University of California Press, pp. 397–409.

Kay, R. F. 1975. Allometry and early hominids. *Science* 189:61–63.

Keith, L. B. 1963. Hypotheses of cycle causes. In *Wildlife's Ten-Year Cycle*. Madison: University of Wisconsin Press, pp. 100–120.

Kendeigh, S. C. 1969. Tolerance of cold and Bergmann's rule. *Auk* 86:13–25.

—— 1970. Energy requirements for existence in relation to size of birds. *Condor* 72:60–65.

Kendeigh, S. C., Dol'nik, V. R., and Gavrilov, V. M. 1977. Avian energetics. In *Granivorous Birds in Ecosystems*, eds. J. Pinowski and S. C. Kendeigh. Internat. Biol. Programme, 12. Cambridge: Cambridge University Press, pp. 127–204.

King, J. R. 1974. Energetics of reproduction in birds. In *Breeding Biology of Birds*, ed. D. S. Farner. Washington, D.C.: Natl. Acad. Sci., pp. 78–107.

—— 1980. Ornithological theory: whence and whither? *Auk* 97:415–418.

Kinnear, J. E., and Brown, G. D. 1967. Physiology: minimum heart rates of marsupials. *Nature* 215:1501.

Kirkwood, J. K. 1983. A limit to metabolisable energy intake in mammals and birds. *Comp. Biochem. Physiol.* 75A:1–3.

Kleiber, M. 1932. Body and size and metabolism. *Hilgardia* 6:315–353.

—— 1950. Physiological meaning of regression equations. *J. Appl. Physiol.* 2:417–423.

—— 1961. *The Fire of Life.* New York: John Wiley. 453 pp.

—— 1975. Metabolic turnover rate: a physiological meaning of the metabolic rate per unit body weight. *J. Theor. Biol.* 53:199–204.

Klein, D. R. 1968. The introduction, increase, and crash of reindeer on St. Matthew Island. *J. Wildlife Mgt.* 32:350–367.

Knudsen, E. I. 1980. Sound localization in birds. In *Comparative Studies of Hearing in Vertebrates,* eds. A. N. Popper and R. R. Fay. New York: Springer-Verlag, pp. 289–322.

Koeppl, J. W., and Hoffman, R. S. 1981. Comparative postnatal growth of four ground squirrel species. *J. Mammal.* 62:44–57.

Kooyman, G. L. 1966. Maximum diving capacities of the Weddell seal, *Leptonychotes weddelli. Science* 151:1553–54.

Kramer, R., and Drexler, G. 1981. Representative breast size of reference female. *Health Physics* 40:913.

Krebs, C. J., and Myers, J. H. 1974. Population cycles in small mammals. In *Advances in Ecological Research,* ed. A. Macfayden. New York: Academic Press, vol. 9, pp. 267–399.

Kunkel, H. O., and Campbell, J. E., Jr. 1952. Tissue cytochrome oxidase activity and body weight. *J. Biol. Chem.* 198:229–236.

Kunkel, H. O., Spalding, J. F., de Franciseis, G., and Futrell, M. F. 1956. Cytochrome oxidase activity and body weight in rats and in three species of large animals. *Amer. J. Physiol.* 186:203–206.

Lack, D. 1968. *Ecological Adaptations for Breeding in Birds.* London: Methuen, pp. 284–295, 306–309.

Laird, A. K. 1965. Dynamics of relative growth. *Growth* 29:249–263.

——— 1966a. Dynamics of embryonic growth. *Growth* 30:263–275.

——— 1966b. Postnatal growth of birds and mammals. *Growth* 30:349–363.

Laird, A. K., Barton, A. D., and Tyler, S. A. 1968. Growth and time: an interpretation of allometry. *Growth* 32:347–354.

Lake, P. E. 1975. Gamete production and the fertile period with particular reference to domesticated birds. In *Avian Physiology,* ed. M. Peaker. *Symp. Zool. Soc. London,* 35:225–244.

Lande, R. 1979. Quantitative genetic analysis of multivariate evolution, applied to brain: body size allometry. *Evolution* 33:402–416.

Larimer, J. J., and Schmidt-Nielsen, K. 1960. A comparison of blood carbonic anhydrase of various mammals. *Comp. Biochem. Physiol.* 1:19–23.

Lasiewski, R. C., and Calder, W. A. 1971. A preliminary allometric analysis of respiratory variables in resting birds. *Respir. Physiol.* 11:152–166.

Lasiewski, R. C., and Dawson, W. R. 1967. A re-examination of the relation between standard metabolic rate and body weight in birds. *Condor* 69:13–23.

——— 1969. Calculation and miscalculation of equations relating avian standard metabolism to body weight. *Condor* 71:335–336.

Lasiewski, R. C., and Lasiewski, R. J. 1967. Physiological responses of the blue-throated and rivoli's hummingbirds. *Auk* 84:34–48.

Lasiewski, R. C., and Seymour, R. S. 1972. Thermoregulatory responses to heat stress in four species of birds weighing approximately 40 grams. *Physiol. Zool.* 45:106–118.

Lasiewski, R. C., Acosta, A. L., and Bernstein, M. H. 1966a. Evaporative water loss in birds. I. Characteristics of the open flow method of determination, and their relation to estimates of thermoregulatory ability. *Comp. Biochem. Physiol.* 19:445–457.

—— 1966b. Evaporative water loss in birds. II. A modified method for determination by direct weighing. *Comp. Biochem. Physiol.* 19:459–470.

Lasiewski, R. C., Weathers, W. W., and Bernstein, M. H. 1967. Physiological responses of the giant hummingbird, *Patagonia gigas. Comp. Biochem. Physiol.* 23:797–813.

Latimer, H. B. 1925. The relative postnatal growth of the systems and organs of the chicken. *Anat. Rec.* 31:233–253.

Lavigne, D. M. 1982. Similarity in energy budgets of animal populations. *J. Anim. Ecol.* 51:195–206.

Ledger, H. P., Sachs, R., and Smith, N. S. 1967. Wildlife and food production with special reference to the semi-arid areas of tropics and sub-tropics. *World Rev. Anim. Production* 3:13–37.

Lee, M. O., and Fox, E. L. 1933. Surface area in a monkey, *Macacus rhesus. Amer. J. Physiol.* 106:91–94.

Leighton, A. T., Jr., Siegel, P. B., and Seigel, H. S. 1966. Body weight and surface area of chickens *(Gallus domesticus). Growth* 30:229–238.

Leitch, I., Mytten, F. E., and Billewicz, W. Z. 1959. The maternal and neonatal weights of some mammalians. *Proc. Zool. Soc.,* London *(J. Zool.)* 133:11–28.

Lin, H. 1982. Fundamentals of zoological scaling. *Am. J. Physiol.* 50:72–81.

Lindsey, C. C. 1966. Body sizes of poikilotherm vertebrates at different latitudes. *Evolution* 20:456–465.

Lindstedt, S. L. 1980a. Energetics and water economy of the smallest desert mammal. *Physiol. Zool.* 53:82–97.

—— 1980b. Regulated hypothermia in the desert shrew. *J. Comp. Physiol.* 137:173–176.

—— 1980c. The smallest insectivores: coping with scarcities of energy and water. In *Comparative Physiology: Primitive Mammals,* eds. K. Schmidt-Nielsen, L. Bolis, and C. R. Taylor. Cambridge: Cambridge University Press, pp. 163–169.

—— MS. Scaling of respiratory function in mammals.

Lindstedt, S. L., and Calder, W. A. 1976. Body size and longevity in birds. *Condor* 78:91–94.

—— 1981. Body size, physiological time, and longevity of homeothermic animals. *Q. Rev. Biol.* 56:1–16.

Lindstedt, S. L., and Jones, J. H. 1980. Desert shrews. *Nat. Hist.* 89(1):46–53.

Lowery, G. H., Jr., and Newman, R. J. 1955. Direct studies of nocturnal bird migration. In *Recent Studies in Avian Biology,* ed. A. Wolfson. Urbana: University of Illinois Press, pp. 238–263.

Lorenz, K. Z. 1953. *King Solomon's Ring.* London: Reprint Society, pp. 111–131.

Løvtrop, S., Rahemtulla, F., and Höglund, N.-G. 1974. Fisher's axiom and the body size of animals. *Zool. Scripta* 3:53–58.

Lutz, P. L., Longmuir, I. S., and Schmidt-Nielsen, K. 1974. Oxygen affinity of bird blood. *Respir. Physiol.* 20:325–330.

MacArthur, R. H. 1972. *Geographical Ecology.* New York: Harper & Row.

Mace, G. M., and Harvey, P. H. 1983. Energetic constraints on home-range size. *Amer. Nat.* 121:120–132.

MacMillen, R. E., and Nelson, J. E. 1969. Bioenergetics and body size in dasyurid marsupials. *Amer. J. Physiol.* 217:1246–51.

Mallouk, R. S. 1975. Longevity in vertebrates is proportional to relative brain weight. *Fed. Proc.* 34:2102–03.

———— 1976. Author's reply. *Fed. Proc.* 35:97–98.

Maloiy, G. M. O., Alexander, R. M., Njau, R., and Jayes, A. S. 1979. Allometry of the legs of running birds. *J. Zool.* (London) 187:161–167.

Martin, R. D. 1981. Relative brain size and basal metabolic rate in terrestrial vertebrates. *Nature* 293:57–60.

Martin, R. R., and Haines, H. 1970. Application of LaPlace's law to mammalian hearts. *Comp. Biochem. Physiol.* 34:959–962.

Mathieu, O., Krauer, R., Hoppeler, H., Gehr, P., Lindstedt, S. L., McNeill, R., Taylor, C. R., and Weibel, E. R. 1981. Design of the mammalian respiratory system. VII. Scaling mitochondrial volume in skeletal muscle to body mass. *Respir. Physiol.* 44:113–128.

Matthews, C. A., Swett, W. W., and McDowell, R. E. 1975. External form and internal anatomy of Holsteins and Jerseys. *J. Dairy Sci.* 58:1453–75.

May, R. M. 1978. The dynamics and diversity of insect faunas. In *Diversity of Insect Faunas,* eds. L. A. Mound and N. Waloff. *Symp. Royal Entomol. Soc. London* 9:188–204.

———— 1979. Production and respiration in animal communities. *Nature* 282:443–444.

May, R. M., and Rubenstein, D. I. MS. Reproductive strategies. In *Reproductive Fitness: Reproduction in Mammals,* vol. 4, eds. C. R. Austin and R. V. Short. Cambridge: Cambridge University Press, forthcoming.

McGlashan, M. L. 1970. Manual of symbols and terminology for physiochemical quantities and units. *Int. Un. Pure Appl. Chem.* 21(1):3–44.

McMahon, T. A. 1973. Size and shape in biology. *Science* 179:1201–4.

———— 1975a. Allometry and biomechanics: limb bones in adult ungulates. *Amer. Nat.* 109:547–563.

———— 1975b. Using body size to understand the structural design of animals: quadrupedal locomotion. *J. Appl. Physiol.* 39:619–627.

———— 1980. Scaling physiological time. *Lectures on Mathematics in the Life Sciences* (Amer. Math. Soc.) 12:131–163.

McNab, B. K. 1963. Bioenergetics and the determination of home range size. *Amer. Nat.* 97:133–140.

———— 1971. On the ecological significance of Bergmann's rule. *Ecology* 52:845–854.

McNeill, S., and Lawton, J. H. 1970. Annual production and respiration in animal populations. *Nature* 225:472–474.

Meadows, S. D., and Hakonson, T. E. 1982. Contribution of tissues to body mass in elk. *J. Wildl. Management* 46:838–841.

Mech, D. 1966. *The Wolves of Isle Royale.* Fauna of the National Parks of the United States. Fauna Series, 7. Washington, D.C.: Government Printing Office. 210 pp.

Mechtly, E. A. 1973. The international system of units. *Physical Constants and Conversion Factors,* 2nd ed. NASA SP-7012. Washington, D.C.: National Aeronautics and Space Administration.

Metzler, D. E. 1977. *Biochemistry.* New York: Academic Press, p. 528.

Meyer, E., and Neuumann, E. G. 1972. *Physical and Applied Acoustics.* New York: Academic Press. 467 pp.

Millar, J. S. 1977. Adaptive features of mammalian reproduction. *Evolution* 31:370–386.

Millar, J. S., and Zammuto, R. M. 1983. Life histories of mammals: an analysis of life tables. *Ecology* 64:631–635.

Milnor, W. R. 1979. Aortic wavelength as a determinant of the relation between heart rate and body size in mammals. *Amer. J. Physiol.* 237:R3–R6.

Mitchell, R. 1974. Scaling in ecology. *Science* 184:1131.

Mohr, C. O. 1940. Comparative populations of game, fur, and other mammals. *Amer. Midl. Nat.* 24:581–584.

Moog, F. 1948. Gulliver was a bad biologist. *Sci. Amer.* 179(5):52–55.

Moore-Ede, M. C., Sulzman, F. M., and Fuller, C. A. 1982. *The Clocks That Time Us.* Cambridge, Mass.: Harvard University Press. 448 pp.

Morrison, P. 1960. Some interrelations between weight and hibernation function. *Bull. Mus. Comp. Zool.* 124:75–91.

Morrison, P. R., and Tietz, W. 1957. Cooling and thermal conductivity in three small Alaskan mammals. *J. Mammal.* 38:79–86.

Morton, S. R., Hinds, D. S., and MacMillen, R. E. 1980. Cheek pouch capacity in heteromyid rodents. *Oecologia* (Berlin) 46:143–146.

Munro, H. N. 1969. Evolution of protein metabolism in mammals. In *Mammalian Protein Metabolism,* ed. H. N. Munro. New York: Academic Press, vol. 3, pp. 133–182.

Munro, H. N., and Gray, J. A. M. 1969. The nucleic acid content of skeletal muscle and liver in mammals of different body size. *Comp. Biochem. Physiol.* 28:897–905.

Murray, B. G., Jr. 1979. *Population Dynamics: Alternative Models.* New York: Academic Press. 212 pp.

Murrish, D. E. 1970. Responses to diving in the dipper, *Cinclus mexicanus. Comp. Biochem. Physiol.* 34:853–858.

Muybridge, E. 1902. *Animals in Motion.* London: Chapman & Hall, pp. 183, 199, 201, 203.

Nagy, K. A. 1982. Energy requirements of free-living iguanid lizards. In *Iguanas of the World: Their Behavior, Ecology, and Conservation,* eds. G. M. Burghardt and A. S. Rand. Park Ridge, N.J.: Noyes Publications, pp. 49–59.

Needham, A. D., and Dawson, T. J. 1982. Heart size and body mass in marsupials. *Bull. Aust. Mammal. Soc.* 7:42.

Needham, J. 1931. *Chemical Embryology,* vol. 1. Cambridge: Cambridge University Press.

Nirmalan, G. P., and Robinson, G. A. 1972. Effect of age, sex, and egg laying on the total erythrocyte volume ($^{51}CrO_4$ label) and the plasma volume (^{125}I-Serum Albumin label) of Japanese quail. *Can. J. Physiol. Pharmacol.* 50:6–10.

Noordergraaf, A., Li, J. K.-J., and Campbell, K. B. 1979. Mammalian hemodynamics: a new similarity principle. *J. Theor. Biol.* 79:485–489.

Oliver, J. 1968. *Nephrons and Kidneys.* New York: Harper & Row.

O'Rourke, M. F. 1981. Comments on a paper by W. R. Miller. *Amer. J. Physiol.* 240:R393–R395.

Pace, N., and Smith, A. H. 1981. Gravity and metabolic scale effects in mammals. *Physiologist* 24:S37–S40.

Pace, N., Rahlmann, D. F., and Smith, A. H. 1979. Scale effects in the musculoskeletal system, viscera and skin of small terrestrial mammals. *Physiologist* 22:S51–S52.

Paganelli, C. V., Olszowska, A., and Ar, A. 1974. The avian egg: surface area, volume and density. *Condor* 76:319–325.

Paladino, F. V., and King, J. R. 1979. Energetic cost of terrestrial locomotion: biped and quadruped runners compared. *Rev. Can. Biol.* 38:321–323.

Palomeque, J., Palacios, L., and Planas, J. 1980. Comparative respiratory functions of blood in some passeriform birds. *Comp. Biochem. Physiol.* 66A:619–624.

Parer, J. T., and Metcalfe, J. 1967. Oxygen transport by blood in relation to body size. *Nature* 215:653–654.

Pearson, O. P. 1947. The rate of metabolism of some small mammals. *Ecology* 28:127–145.

——— 1948. Metabolism of small mammals, with remarks on the lower limit of mammalian size. *Science* 108:44.

——— 1950. The metabolism of hummingbirds. *Condor* 52:145–152.

Pedley, T. J., ed. 1977. *Scale Effects in Animal Locomotion.* New York: Academic Press. 545 pp.

Pennycuick, C. J. 1969. The mechanics of bird migration. *Ibis* 11:525–556.

——— 1972. *Animal Flight.* London: Edward Arnold. 68 pp.

——— 1975. Mechanics of flight. In *Avian Biology,* eds. D. S. Farner and J. R. King. New York: Academic Press, vol. 5, pp. 5–17.

——— 1979. Energy costs of locomotion and the concept of "foraging radius." In

Serengeti: Dynamics of an Ecosystem, eds. A. R. E. Sinclair and M. Norton-Griffiths. Chicago: University of Chicago Press, pp. 164–184.

Peterson, J. A., Benson, J. A., Ngai, M., Morin, J., and Ow, C. 1982. Scaling in tensile "skeletons": structures with scale-independent length dimensions. *Science* 217:1267–69.

Peterson, R. O., Page, R. E., and Dodge, K. M. MS. Wolves, moose and the allometry of population cycles.

Pettingill, O. S. 1970. *Ornithology in Laboratory and Field,* 4th ed. Minneapolis: Burgess. 524 pp.

Pianka, E. R. 1983. *Evolutionary Ecology.* New York: Harper & Row. 416 pp.

Pilbeam, D., and Gould, S. J. 1974. Size and scaling in human evolution. *Science* 186:892–901.

Pinshow, B., Degen, A. A., and Alkon, P. U. 1983. Water intake and existence energy, and responses to water deprivation in the sand partridge *(Ammoperdix heyi)* and the chukar *(Alectoris chukar):* two phasianids of the Negev Desert. *Physiol. Zool.:* in press.

Pitts, G. C., and Bullard, T. R. 1968. Some interspecific aspects of body composition in mammals. In *Body Composition in Animals and Man,* ed. Nat. Res. Coun. Washington, D.C.: Nat. Acad. of Sci., pp. 45–79.

Platt, T., and Silvert, W. 1981. Ecology, physiology, allometry, and dimensionality. *J. Theor. Biol.* 93:855–860.

Poczopko, P. 1971. Metabolic rates in adult homeotherm. *Acta Theriologica* 16:1–21.

Porter, W. P., and Gates, D. M. 1969. Thermodynamic equilibria of animals with environment. *Ecol. Mono.* 39:227–244.

Porter, W. P., Parkhurst, D., and McClure, P. A. MS. Critical radius of endotherms.

Prange, H. D. 1977. The scaling and mechanics of arthropod exoskeletons. In *Scale Effects in Animal Locomotion,* ed. T. J. Pedley. New York: Academic Press, pp. 169–181.

Prange, H. D., and Christman, S. P. 1976. The allometrics of rattlesnake skeletons. *Copeia* 3:542–545.

Prange, H. D., Anderson, J. F., and Rahn, H. 1979. Scaling of skeletal mass to body mass in birds and mammals. *Amer. Nat.* 113:103–122.

Prinzinger, R., and Hänssler, I. 1980. Metabolism-weight relationship in some small nonpasserine birds. *Experientia* 36:1299–1300.

Prinzinger, R., Krüger, K., and Schuchman, K. L. 1981. Metabolism-weight relationship in 17 hummingbird species at different temperatures during day and night. *Experientia* 37:1307–8.

Prosser, C. L. 1973. *Comparative Animal Physiology,* ed. C. L. Prosser. Philadelphia: W. B. Saunders. 966 pp.

Prothero, J. W. 1979. Maximal oxygen consumption in various animals and plants. *Comp. Biochem. Physiol.* 64A:463–466.

———— 1980. Scaling of blood parameters in mammals. *Comp. Biochem. Physiol.* 67A:649–657.

———— 1982. Organ scaling in mammals: the liver. *Comp. Biochem. Physiol.* 71A:567–577.

Quiring, D. P. 1950. *Functional Anatomy of the Vertebrates.* New York: McGraw-Hill, pp. 528–572.

Rahn, H. 1982. Comparison of embryonic development in birds and mammals: birth weight, time and cost. In *A Companion to Animal Physiology,* eds. C. R. Taylor, K. Johansen, and L. Bolis. New York: Cambridge University Press, pp. 124–137.

Rahn, H., and Ar, A. 1974. The avian egg: incubation time and water loss. *Condor* 76:147–152.

———— 1980. Gas exchange of the avian egg: time, structure, and function. *Amer. Zool.* 20:477–484.

Rahn, H., and Paganelli, C. V. 1981. *Gas Exchange in Avian Eggs: Publications in Gas Exchange, Physical Properties, and Dimensions of Bird Eggs.* Buffalo: Dept. of Physiology, SUNY. 358 pp.

Rahn, H., Paganelli, C. V., and Ar, A. 1974. The avian egg: air-cell, gas tension, metabolism, and incubation time. *Respir. Physiol.* 22:297–309.

———— 1975. Relation of avian egg weight to body weight. *Auk* 92:750–765.

Rashevsky, N. 1960. *Mathematical Biophysics: Physico-Mathematical Foundations of Biology.* New York: Dover, vol. 2, pp. 262–305.

Rayner, J. M. V. 1979. A new approach to animal flight mechanics. *J. Exper. Biol.* 80:17–54.

———— 1981. Flight adaptations in vertebrates. In *Vertebrate Locomotion,* ed. M. H. Day. Symp. Zool. Soc. London, 48:137–172.

Regoeczi, E., and Hatton, M. W. C. 1980. Transferring catabolism in mammalian species of different body size. *Amer. J. Physiol.* 238:R306–R310.

Regoeczi, E., and Taylor, P. 1978. The net weight of the rat liver. *Growth* 42:451–456.

Reichman, O. J., and Aitchison, S. 1981. Mammal trails on mountain slopes: optimal paths in relation to slope angle and body weight. *Amer. Nat.* 117:416–420.

Reid, B. 1971a. The weight of the kiwi and its egg. *Notornis* 18:245–249.

———— 1971b. Composition of a kiwi egg. *Notornis* 18:250–252.

———— 1977. The energy value of the yolk reserve in a North Island brown kiwi chick *(Apteryx australis mantelli). Notornis* 24:194–195.

Reid, K. H. 1971. Periodical cicada: mechanism of sound production. *Science* 172:949–951.

Reynolds, W. W. 1977. Skeleton weight allometry in aquatic and terrestrial vertebrates. *Hydrobiologia* 56:35–37.

Richards, D. G. 1981. Estimation of distance of singing conspecifics by the Carolina wren. *Auk* 98:127–133.

Richmond, C. R., Langham, W. H., and Trujillo. T. T. 1962. Comparative metabolism of tritiated water by mammals. *J. Cell. Comp. Physiol.* 59:45–53.

Ricklefs, R. E. 1973. Fecundity, mortality, and avian demography. In *Breeding Biology of Birds,* ed. D. S. Farner. Washington, D.C.: Natl. Acad. Sci., pp. 336–347.

———— 1974. Energetics of reproduction in birds. In *Avian Energetics,* ed. R. A. Paynter, Jr. Cambridge, Mass.: Nuttall Ornith. Club, publ. 15, pp. 152–297.

———— 1979. Adaptation, constraint, and compromise in avian postnatal development. *Biol. Rev.* 54:269–290.

Riggs, D. S. 1963. *The Mathematical Approach to Physiological Problems.* Baltimore: Williams and Wilkins, pp. 2–40.

Rosen, P., Woodhead, A. D., and Thompson, K. H. 1981. The relationship between the Gompertz constant and maximum potential lifespan: its relevance to theories of aging. *Exper. Geront.* 16:131–135.

Rosenzweig, M. L. 1968. The strategy of body size in mammalian carnivores. *Amer. Midl. Nat.* 80:299–315.

Ross, D. M. 1981. Illusion and reality in comparative physiology. *Can. J. Zool.* 59:2151–58.

Roth, V. L. 1979. Can quantum leaps in body size be recognized among mammalian species? *Paleobiol.* 5:318–336.

———— 1981. Constancy in the size ratios of sympatric species. *Amer. Nat.* 118:394–404.

Russell, E. M. 1982. Patterns of parental care and parental investment in marsupials. *Biol. Rev.* 57:423–486.

Rytand, D. A. 1938. The number and size of mammalian glomeruli as related to kidney and to body weight, the methods for their enumeration and measurement. *Amer. J. Anat.* 62:507–520.

Sacher, G. A. 1959. Relation of lifespan to brain weight and body weight in mammals. *Ciba Found. Coll.* 5:115–141.

———— 1976. Evaluation of the entropy and information terms governing mammalian longevity. *Interdiscipl. Topics Geront.* 9:69–82.

———— 1978. Evolution of longevity and survival characteristics in mammals. In *The Genetics of Aging,* ed. E. L. Schneider. New York: Plenum Press, pp. 151–168.

Sacher, G. A., and Hart, R. W. 1978. Longevity, aging, and comparative cellular and molecular biology of the house mouse, *Mus musculus* and the white-footed mouse, *Peromyscus maniculatus.* In *Genetic Effects on Aging,* eds. D. Bergsma and D. E. Harrison. New York: A. R. Liss, pp. 71–96.

Sacher, G. A., and Staffeldt, E. F. 1974. Relation of gestation time to brain weight for placental mammals: implications for the theory of vertebrate growth. *Amer. Nat.* 108:593–615.

Sachs, R. 1967. Live weights and body measurements of Serengeti game animals. *E. Afr. Wildl. J.* 5:24–36.

Saito, N. 1980. Structure and function of the avian ear. In *Comparative Studies of*

Hearing in Vertebrates, eds. A. N. Popper and R. R. Fay. New York: Springer-Verlag, pp. 241–249.

Sawbridge, D. F., and Bell, M. A. M. 1969. Pacific shores. *Science* 164:1089.

Sayre, R. 1982. Creatures. *Audubon* 84(5):28–30.

Scheffer, V. B. 1951. The rise and fall of a reindeer herd. *Sci. Monthly* 73:356–362.

Schmidt-Nielsen, K. 1964. *Desert Animals: Physiological Problems of Heat and Water.* London: Clarendon Press. 277 pp.

——— 1972. *How Animals Work.* London: Cambridge University Press. 114 pp.

——— 1975. Scaling in biology: the consequences of size. *J. Exper. Zool.* 194:287–308.

——— 1977. Problems of scaling: locomotion and physiological correlates. In *Scale Effects in Animal Locomotion,* ed. T. J. Pedley. New York: Academic Press, pp. 1-21.

——— 1979. *Animal Physiology: Adaptions and Environment,* 2nd ed. London: Cambridge University Press. 560 pp.

Schmidt-Nielsen, K., and Larimer, J. L. 1958. Oxygen dissociation curves of mammalian blood in relation to body size. *Amer. J. Physiol.* 195:425–428.

Schmidt-Nielsen, K., and Pennycuick, P. 1961. Capillary density in mammals in relation to body size and oxygen consumption. *Amer. J. Physiol.* 200:746–750.

Schmidt-Nielsen, K., Dawson, R. J., and Crawford, E. C., Jr. 1966. Temperature regulation in the echidna *Tachyglossus aculeatus. J. Cell. Physiol.* 67:63–72.

Schnell, G. G., and Hellack, J. J. 1979. Bird flight speeds in nature: optimized or a compromise? *Amer. Nat.* 113:53–66.

Schoener, T. W. 1968. Sizes of feeding territories among birds. *Ecology* 49:123–131.

Scholander, P. F. 1940. Experimental investigations on the respiratory function in diving mammals and birds. Hvålradets *Skrifter Norske Videnskaps-Acad., Oslo* 22:1–131.

Schroeder, L. A. 1981. Consumer growth efficiencies: their limits and relationships to ecological energetics. *J. Theor. Biol.* 93:805–828.

Schwartzkopff, J. 1968. Structure and function of the ear and of the auditory brain areas in birds. In *Hearing Mechanisms in Vertebrates,* eds. A. V. S. DeReuck and J. Knight. Boston: Little, Brown, pp. 41–49.

Scott, D. M., and Ankney, C. D. 1983. Do Darwin's finches lay small eggs? *Auk* 100:226–227.

Silyn-Roberts, H. 1983. The pore geometry and structure of the egg shell of the North Island brown kiwi, *Apteryx australis mantelli. J. Micros.* 130:23–36.

Simberloff, D., and Boecklen, W. 1981. Santa Rosalia reconsidered: size ratios and competition. *Evolution* 35:1206–28.

Simon, L. M., and Eugene, D. R. 1971. Relationship of cytochrome oxidase activity to vertebrate total and organ oxygen consumption. *Int. J. Biochem.* 2:569–573.

Simpson, G. G. 1961. *Horses.* Natural History Library. Garden City, N.Y.: Doubleday. 323 pp.

Skadhauge, E. 1974. Renal concentrating ability in selected west Australian birds. *J. Exper. Biol.* 61:269–276.

—— 1975. Renal and cloacal transport of salt and water. *Symp. Zool. Soc. Lond.* 35:97–106.

—— 1981. *Osmoregulation in Birds.* New York: Springer-Verlag.

Smith, A. T. 1978. Comparative demography of pikas *(Ochotona):* effect of spatial and temporal age-specific mortality. *Ecology* 59:133–139.

Smith, H. W. 1959. *From Fish to Philosopher.* Summit, N.J.: CIBA. 304 pp.

Smith, R. E. 1956. Quantitative relations between liver mitochondria metabolism and total body weight in mammals. *Ann. N.Y. Acad. Sci.* 62:403–422.

Smith, R. J. 1980. Rethinking allometry. *J. Theor. Biol.* 87:97–111.

Snow, D. W. 1958. Climate and geographical variation in birds. *New Biol.* 25:64–84.

Sorenson, M. W. 1962. Some aspects of water shrew behavior. *Amer. Midl. Nat.* 68:445–462.

Spaan, G., and Klussmann, F. W. 1970. Die frequency des Kültezittens bei tierarten verschiedesen Grösse. *Pflug. Archiv* 320:318–333.

Spear, L. 1980. Band loss from the western gull on southeast Farallon Island. *J. Field Ornith.* 51:319–328.

Spells, K. E. 1968. Some physical considerations relevant to the dimensions of lung alveoli. *Nature* 219:64–66.

Sperber, I. 1944. Studies on the mammalian kidney. *Zoologiska Bidrag Från Uppsala* 22:249–432.

Stahl, W. R. 1962. Similarity and dimensional methods in biology. *Science* 137:205–212.

—— 1963a. Similarity analysis of physiological systems. *Perspect. Biol. Med.* 6:291–321.

—— 1963b. The analysis of biological similarity. In *Advances in Biological and Medical Physics,* eds. J. H. Lawrence and J. W. Gofman. New York: Academic Press, vol. 9, pp. 355–464.

—— 1965. Organ weights in primates and other mammals. *Science* 150:1039–42.

—— 1967. Scaling of respiratory variables in mammals. *J. Appl. Physiol.* 22:453–460.

Stahl, W. R., and Gummerson, J. Y. 1967. Systematic allometry in five species of adult primates. *Growth* 31:21–34.

Stanley, S. M. 1973. An explanation for Cope's rule. *Evolution* 27:1–26.

Stearns, S. C. 1976. Life history tactics: a review of the ideas. *Q. Rev. Biol.* 51:3–47.

—— 1983a. The influence of size and phylogeny on patterns of covariation among life-history traits in the mammals. *Oikos:* in press.

—— 1983b. The effects of size and phylogeny on patterns of covariation in the life-history traits of reptiles. *Amer. Nat.:* in press.

Stitt, J. T., Hardy, J. D., and Nadel, E. R. 1971. Surface area of the squirrel monkey in relation to body weight. *J. Appl. Physiol.* 31:140–141.

Storer, R. W. 1971. Adaptive radiation of birds. In *Avian Biology,* eds. D. S. Farner and J. R. King. New York: Academic Press, vol. 1, pp. 150–188.

Strehler, B. L. 1959. Origin and comparison of the effects of time and high-energy radiations on living systems. *Q. Rev. Biol.* 34:117–142.

Streit, B. 1982. Water turnover rates and half-times in animals studied by use of labelled and non-labelled water. *Comp. Biochem. Physiol.* 72A:445–454.

Studier, E. H. 1970. Evaporative water loss in bats. *Comp. Biochem. Physiol.* 35:935–943.

Sweet, S. S. 1980. Allometric inference in morphology. *Amer. Zool.* 20:643–652.

Swift, J. 1726. *Travels into Several Remote Nations of the World by Lemuel Gulliver* (rpt. 1946). New York: Ronald Press, p. 32.

Symbols Committee of the Royal Society. 1971. *Quantities, Units, and Symbols.* London: Royal Society. 48 pp.

Syrovy, I., and Gutmann, E. 1973. Myosin from fast and slow skeletal and cardiac muscles of mammals of different size. *Physiol. Bohemoslov.* 24:325–334.

Tanner, J. M. 1949. Fallacy of per-weight and per-surface area standards and their relation to spurious correlation. *J. Appl. Physiol.* 2:1–15.

Taylor, C. R. 1977a. Why big animals? *Cornell Vet.* 67:155–175.

——— 1977b. Exercise and environmental heat loads: different mechanisms for solving different problems? In *International Review of Physiology, Environmental Physiology* II, ed. D. Robertshaw. Baltimore: University Park Press, vol. 15, pp. 119–146.

——— 1977c. The energy of terrestrial locomotion and body size in vertebrates. In *Scale Effects in Animal Locomotion,* ed. T. J. Pedley. New York: Academic Press, pp. 127–151.

Taylor, C. R., and Weibel, E. R. 1981. Design of the mammalian respiratory system. I. Problem and strategy. *Respir. Physiol.* 44:1–10.

Taylor, C. R., Schmidt-Nielsen, K., and Raab, J. L. 1970. Scaling of energetic cost of running to body size in mammals. *Amer. J. Physiol.* 219:1104–7.

Taylor, C. R., Caldwell, S. L., and Rowntree, V. J. 1972. Running up and down hills: some consequences of size. *Science* 178:1096–97.

Taylor, C. R., Seeherman, H. S., Maloiy, G. M. O., Heglund, N. C., and Kamau, J. M. Z. 1978. Scaling maximum aerobic capacity ($\dot{V}_{O_2}$max) to body size in mammals. *Fed. Proc.* 37:473.

Taylor, C. R., Maloiy, G. M. O., Weibel, E. R., Langman, V. A., Kamau, J. M. Z., Seeherman, H. J., and Heglund, N. C. 1981. Design of the mammalian respiratory system. III. Scaling maximum aerobic capacity to body mass: wild and domestic mammals *Respir. Physiol.* 44:25–37.

Taylor, C. R., Heglund, N. C., and Maloiy, G. M. O. 1982. Energetics and mechanics of terrestrial locomotion. I. Metabolic energy consumption as a function of speed and body size in birds and mammals. *J. Exper. Biol.* 97:1–21.

Taylor, St.C. S. 1965. A relation between mature weight and time taken to mature in mammals. *Anim. Prod.* 7:203–220.

——— 1968. Time taken to mature in relation to mature weight for sexes, strains and species of domesticated mammals and birds. *Anim. Prod.* 10:157–169.

Tenney, S. M., and Bartlett, D. 1967. Comparative quantitative morphology of the mammalian lung: trachea. *Respir. Physiol.* 3:130–135.

Tenney, S. M., and Morrison, D. H. 1967. Tissue gas tension in small wild mammals. *Respir. Physiol.* 3:160–165.

Tenney, S. M., and Remmers, J. E. 1963. Comparative quantitative morphology of the mammalian lung: diffusing area. *Nature* 197:54–56.

Thomas, R. D. K., and Olson, E. C. 1980. *A Cold Look at the Warm-Blooded Dinosaurs.* AAAS Selected Symp. 28. Boulder: Westview Press. 514 pp.

Thompson, D'A. W. 1917. *On Growth and Form* (rpt. 1942). Cambridge: Cambridge University Press, pp. 22–77.

——— 1961. *On Growth and Form,* abr., ed. J. T. Bonner. Cambridge: Cambridge University Press. 346 pp.

Thoreau, H. D. 1854. Walden. In *The Portable Thoreau,* ed. C. Bode (rpt. 1947). New York: Viking Press.

Timiras, P. S. 1978. Biological perspectives on aging. *Amer. Sci.* 66:605–613.

Tolmasoff, J. M., Ono, T., and Cutler, R. G. 1980. Superoxide dismutase: correlation with life-span and specific metabolic rate in primate species. *Proc. Natl. Acad. Sci. USA* 77:2777–81.

Tracy, C. R. 1977. Minimum size of mammalian homeotherms: role of the thermal environment. *Science* 198:1034–35.

Tucker, V. A. 1970. Energetic cost of locomotion in animals. *Comp. Biochem. Physiol.* 34:841–846.

——— 1973. Bird metabolism during flight: evaluation of a theory. *J. Exper. Biol.* 49:527–555.

——— 1974. Energetics of natural avian flight. In *Avian Energetics,* ed. R. A. Paynter, Jr. Cambridge, Mass.: Nuttall Ornithol. Club, publ. 15, pp. 298–328.

——— 1975a. Flight energetics. *Symp. Zool. Soc. London* 35:49–63.

——— 1975b. The energetic cost of moving about. *Amer. Sci.* 63:413–419.

Tucker, V. A., and Schmidt-Koenig, K. 1971. Flight speeds of birds in relation to energetics and wind directions. *Auk* 88:97–107.

Turček, F. J. 1966. On plumage quantity in birds. *Ekol. Pol. Ser. A.* 14:617–634.

Turner, F. B., Jennrich, R. I., and Weintraub, J. D. 1969. Home range and body size of lizards. *Ecology* 50:1076–81.

Turner, M. E., Jr. 1978. Allometry and multivariate growth. *Growth* 42:434–450.

Umminger, B. L. 1975. Body size and whole blood sugar concentration in mammals. *Comp. Biochem. Physiol.* 52A:455–458.

Vácha, J., and Znojil, V. 1981. The allometric dependence of the life span of erythrocytes on body weight in mammals. *Comp. Biochem. Physiol.* 69A:357–362.

Van Tyne, J., and Berger, A. J. 1976. *Fundamentals of Ornithology,* 2nd ed. New York: John Wiley. 808 pp.

Van Valen, L. 1973. Body size and numbers of plants and animals. *Evolution* 27:27–35.

Vleck, C. M., Hoyt, D. F., and Vleck, D. 1979. Metabolism of avian embryos: patterns in altricial and precocial birds. *Physiol. Zool.* 52:363–377.

Vleck, C. M., Vleck, D., and Hoyt, D. F. 1980. Patterns of metabolism and growth in avian embryos. *Amer. Zool.* 20:405–416.

Vogel, P. 1980. Metabolic levels and biological strategies in shrews. In *Comparative Physiology: Primitive Mammals,* eds. K. Schmidt-Nielsen, L. Bolis, and C. R. Taylor. Cambridge: Cambridge University Press, pp. 170–180.

Vogel, S. 1981. *Life in Moving Fluids: The Physical Biology of Flow.* Boston: Willard Grant Press. 352 pp.

Von Schelling, H. 1954. Mathematical deductions from empirical relations between metabolism, surface area, and weight. *Ann. N.Y. Acad. Sci.* 56:1143–64.

Walker, A., and Hughes, M. A. 1978. Total body water volume and turnover rate in fresh water and sea water adapted glaucous-winged gulls, *Larus glaucescens. Comp. Biochem. Physiol.* 61A:233–237.

Walker, E. P. 1975. *Mammals of the World,* 3rd ed. Baltimore: Johns Hopkins University Press.

Walsberg, G. E. MS. Avian ecological energetics. In *Avian Biology,* eds. D. S. Farner and J. R. King. New York: Academic Press, vol. 7, forthcoming.

——— 1983. Ecological energetics: what are the questions? In *Perspectives in Ornithology,* eds. G. A. Clark, Jr. and A. H. Brush. New York: Cambridge University Press, pp. 135–158.

Walsberg, G. E., and King, J. R. 1978. The relationship of the external surface area of birds to skin surface area and body mass. *J. Exper. Biol.* 76:185–189.

Wangensteen, O. D., and Rahn, H. 1970/71. Respiratory exchange by the avian embryo. *Respir. Physiol.* 11:31–45.

Wangensteen, O. D., Wilson, D., and Rahn, H. 1970/71. Diffusion of gases across the shell of the hen's egg. *Respir. Physiol.* 11:16–30.

Warham, J. 1977. Wing loadings, wing shapes, and flight capabilities of Procellariiformes. *New Zealand J. Zool.* 4:73–83.

Weast, R. C., ed. 1975. *Handbook of Chemistry and Physics.* Cleveland: CRC Press.

Weathers, W. W. 1981. Physiological thermoregulation in heat-stressed birds: consequences of body size. *Physiol. Zool.* 54:345–361.

Webb, A. I., and Weaver, B. M. Q. 1979. Body composition of the horse. *Equine Vet. J.* 11:39–47.

Webster, D. B. 1962. A function of the enlarged middle-ear cavities of the kangaroo rat, *Dipodomys. Physiol. Zool.* 35:248–255.

Webster, D. B., and Webster, M. 1975. Auditory systems of Heteromyidae: functional morphology and evolution of the middle ear. *J. Morph.* 146:343–376.

Weibel, E. R. 1970/71. Morphometric estimation of pulmonary diffusion capacity. *Respir. Physiol.* 11:54–75.

Weibel, E. R., and Taylor, C. R., eds. 1981. Design of the mammalian respiratory system. *Respir. Physiol.* 44:1–164.

Weibel, E. R., Gehr, P., Cruz-Orive, L. M., Muller, A. E., Mwangi, D. K., and Haussener, V. 1981a. Design of the mammalian respiratory system. IV. Morphometric estimation of pulmonary diffusing capacity: critical evaluation of a new sampling method. *Respir. Physiol.* 44:39–59.

Weibel, E. R., Taylor, C. R., Gehr, P., Hoppeler, H., Mathieu, O., and Maloiy, G. M. O. 1981b. Design of the mammalian respiratory system. IX. Functional and structural limits for oxygen flow. *Respir. Physiol.* 44:151–164.

Weis-Fogh, T., and Alexander, R. McN. 1977. The sustained power output from striated muscle. In *Scale Effects in Animal Locomotion,* ed. T. J. Pedley. New York: Academic Press, pp. 511–525.

Weiss, M., Sziegoleit, W., and Förster, W. 1977. Dependence of pharmacokinetic parameters on the body weight. *Int. J. Clin. Pharmacol.* 15:572–575.

Welty, J. C. 1955. Birds as flying machines. *Sci. Amer.* 192:88–96.

—— 1982. *The Life of Birds,* 3rd ed. Philadelphia: W. B. Saunders. 546 pp.

Went, F. W. 1968. The size of man. *Amer. Sci.* 56:400–413.

Western, D. 1979. Size, life history, and ecology in mammals. *Afr. J. Ecol.* 17:185–204.

—— 1980. Linking the ecology of past and present mammal communities. In *Fossils in the Making: Vertebrate Taphonomy and Paleoecology,* eds. A. K. Behrensmeyer and A. P. Hill. Chicago: University of Chicago Press, pp. 41–54.

—— 1983. Production, reproduction, and size in mammals. *Oecologia* (Berlin) 59:269–271.

Western, D., and Georgiadis, N. MS. Bone production processes in a mammalian community and the implications for paleoecology.

Western, D., and Ssemakula, J. 1982. Life history patterns in birds and mammals and their evolutionary interpretation. *Oecologia* (Berlin) 54:281–290.

White, J. F., and Gould, S. J. 1965. Interpretation of the coefficient in the allometric equation. *Amer. Nat.* 99:5–18.

White, L., Haines, H., and Adams, T. 1968. Cardiac output related to body weight in small mammals. *Comp. Biochem. Physiol.* 27:559–565.

Whitehead, P. E., and McEwan, E. H. 1982. Contribution of tissues to body mass in elk. *J. Wildl. Mgt.* 46:838–841.

Wiens, J. A., and Rotenberry, J. T. 1981. Morphological size ratios and competition in ecological communities. *Amer. Nat.* 117:592–599.

Wilkie, D. R. 1977. Metabolism and body size. In *Scale Effects in Animal Locomotion,* ed. T. J. Pedley. New York: Academic Press, pp. 23–36.

Willoughby, E. J., and Peaker, M. 1979. Birds. In *Comparative Physiology of Osmoregulation in Animals,* ed. G. M. O. Maloiy. New York: Academic Press, vol. 2, pp. 1–55.

Wilson, E. O. 1975. *Sociobiology.* Cambridge, Mass.: Harvard University Press. 697 pp.

Winsett, R. E., ed. 1939. *Radio and Revival Special: Fine Hymns and Evangelistic Songs.* Dayton, Tenn.: R. E. Winsett, p. 58.

Withers, P. C., Casey, T. M. and Casey, K. K. 1979. Allometry of respiratory and haematological parameters of arctic mammals. *Comp. Biochem. Physiol.* 64A:343–350.

Wolf, L. L., and Hainsworth, F. R. 1971. Time and energy budgets of territorial hummingbirds. *Ecology* 52:980–988.

Worth, C. B. 1940. Egg volumes and incubation period. *Auk* 57:44–60.

Wunder, B. A., and Morrison, P. R. 1974. Red squirrel metabolism during incline running. *Comp. Biochem. Physiol.* 48A:153–161.

Yablokov, A. V. 1966. *Variability of Mammals.* Moscow: Nauka.

Yalden, D. W. 1971. Flying ability of *Archaeopteryx. Nature* 231:127.

Yates, F. E. 1979. Comparative physiology: compared to what? *Amer. J. Physiol.* 237:R1–R2.

——— 1982. Outline of a physical theory of physiological systems. *Can. J. Physiol. Pharmacol.* 60:217–248.

Young, J. Z., and Hobbs, M. J. 1975. *The Life of Mammals,* 2nd ed. London: Oxford University Press. 528 pp.

Zar, J. H. 1968a. Calculation and miscalculation of the allometric equation as a model in biological data. *Bioscience* 18:1118–20.

——— 1968b. Standard metabolism comparisons between orders of birds. *Condor* 70:278.

——— 1969. The use of the allometric model for avian standard metabolism–body weight relationships. *Comp. Biochem. Physiol.* 29:227–234.

ADDENDUM

The following papers came to my attention too late to be incorporated in the text, but are appended here because of their relevance to the topics discussed.

Alexander, R. McN. 1983. Allometry of the leg bones of moas (Dinornithes) and other birds. *J. Zool.* (London) 200:215–231.

——— 1983. On the massive legs of a moa (Pachyornis elephantopus, Dinornithes). *J. Zool.* (London) 201:363–376.

Alexander, R. McN., and Jayes, A. S. 1983. A dynamic similarity hypothesis for the gaits of quadrupedal mammals. *J. Zool.* (London) 201:135–152.

Biewener, A. A. 1983. Allometry of quadrupedal locomotion: the scaling of duty factor, bone curvature, and limb orientation to body size. *J. Exp. Biol.* 105:147–171.

Campbell, K. E., and Tonni, E. P. 1983. Size and locomotion in Teratorns (Aves: Teratornithidae). *Auk* 100:390–403.

Casey, T. M. 1983. Moth energetics. *Physiol. Zool.* 56:160–173.

Clarke, A. 1983. Life in cold water: the physiological ecology of polar marine ectotherms. *Oceanogr. Marine Biol. Ann. Rev.* 21:341–453.

Cutler, R. G. 1983. Reply. *Gerontology* 29:113–120.

Daan, S., and Aschoff, J. 1982. Circadian contributions to survival. In *Vertebrate Circadian Systems,* eds. S. Daan and J. Aschoff. New York: Springer-Verlag, pp. 305–321.

Gittleman, J. L., and Harvey, P. H. 1982. Carnivore home-range size, metabolic needs and ecology. *Behav. Ecol. Sociobiol.* 10:57–63.

Hofman, M. A. 1983. Evolution of brain size in neonatal and adult placental mammals: a theoretical approach. *J. Theor. Biol.* 105:317–332.

——— 1983. Energy metabolism, brain size and longevity in mammals. *Q. Rev. Biol.* 58:495–512.

Jørgensen, C. B. 1983. Ecological physiology: background and perspectives. *Comp. Biochem. Physiol.* 75A:5–7.

Leith, D. E. 1983. Mammalian tracheal dimensions: scaling and physiology. *J. Appl. Physiol. Respir. Environ. Exercise Physiol.* 55:196–200.

Price, M. V. 1983. Ecological consequences of body size: a model for patch choice in desert rodents. *Oecologia* (Berlin) 59:384–392.

Prothero, J. 1984. Organ scaling in mammals: the kidneys. *Comp. Biochem. Physiol.* 77A:133–138.

——— 1984. Scaling of standard energy metabolism in mammals. I. Neglect of circadian rhythms. *J. Theor. Biol.* 106:1–7.

Robinson, W. R., Peters, R. H., and Zimmermann, J. 1983. The effects of body size and temperature on metabolic rates of organisms. *Can. J. Zool.* 61:281–288.

Rutberg, A. T. 1983. The evolution of monogamy in primates. *J. Theor. Biol.* 104:93–112.

Seim, E., and Soether, B. F. 1983. On rethinking allometry: which regression model to use? *J. Theor. Biol.* 104:161–168.

Sullivan, J. L. 1982. Superoxide dismutase, longevity, and specific metabolic rate. *Gerontology* 28:242–244.

Weiser, W. 1984. Low production "efficiency" of homeotherm populations: a misunderstanding. *Oecologia* (Berlin) 61:53–54.

Index